The Coming Crisis

Nuclear Proliferation, U.S. Interests, and
World Order

The BCSIA Studies in International Security book series is edited at the Belfer Center for Science and International Affairs at Harvard University's John F. Kennedy School of Government and published by The MIT Press. The series publishes books on contemporary issues in international security policy, as well as their conceptual and historical foundations. Topics of particular interest to the series include the spread of weapons of mass destruction, internal conflict, the international effects of democracy and democratization, and U.S. defense policy.

A complete list of BCSIA Studies appears at the back of this volume.

The Coming Crisis

Nuclear Proliferation, U.S. Interests, and World Order

Editor
Victor A. Utgoff

BCSIA Studies in International Security

with sponsorship of the Institute
for Defense Analyses

MIT Press
Cambridge, Massachusetts
London, England

Library of Congress Cataloging-in-Publication Data

The coming crisis : nuclear proliferation, U.S. interests, and world
order / edited by Victor A. Utgoff.
      p.   cm.—(BCSIA studies in international security)
      Includes bibliographical references and index.
          ISBN 0-262-21015-0 (hc. : alk. paper)—
          ISBN 0-262-71005-6 (pbk. : alk. paper)
      1. Weapons of mass destruction. 2. Nuclear nonproliferation.
      3. National security—United States. 4. United States—Military
      policy. 5. International relations. I. Utgoff, Victor A. II. Series.
U793.C65   2000
327.1'747—dc21   98-31554

Contents

Foreword

The accelerating worldwide advance of technology and knowledge creates rapidly expanding opportunities for humankind—opportunities that have been of great benefit to the developed world, and especially to the United States. But the advance of technology also provides opportunities for increasing numbers of states to build weapons that in minutes to hours can cause levels of destruction that once took years of warfare with massive forces. Further, more states actually are building such weapons, or positioning themselves to be able to build them quickly. The evidence of this half-century trend is clear and compelling. The nuclear weapons tests carried out by India and Pakistan in May 1998 and Iraq's continued resistance to UN efforts to eliminate its program for weapons of mass destruction are recent reminders.

Some hope that the proliferation of weapons of mass destruction will ultimately lead potential aggressors to conclude that war has become too dangerous. But centuries of history, including the past five decades, lead most observers of the international scene to be deeply skeptical that a more proliferated world would be more peaceful. It seems more likely that highly destructive wars would increase as the number of actors armed with these weapons rises. Thus, efforts to limit or roll back proliferation remain a national priority.

There is reason for some optimism about the outcome of such efforts. Looking back, international nonproliferation efforts, coupled with the self-restraint exercised by many nations, have been surprisingly effective. Predictions made decades ago of the number of states that would have weapons of mass destruction by 2000 have proven pessimistic. While the large majority of the world's states are now capable of building weapons of mass destruction, only a minority appear to have done so, or to be purposely moving toward such weapons.

Many factors are involved in explaining this divergence between capabilities to build such weapons and the choice to do so. Among the most important is the belief that the major states will continue to play their post–World War II role of keeping sovereign states from conquering or destroying one another. But proliferation raises the risk involved in intervention, and the end of the global contest for power with the former Soviet Union causes some to believe that the outcomes of regional wars are less important to the United States. This combination could undermine confidence in the capability and the will of the United States to continue to play the key stabilizing role the world has come to expect of it.

I believe the United States will continue in its stabilizing role for at least three reasons. First, U.S. political leaders, whatever their political philosophy, have always found it difficult to keep the nation on the sidelines in the face of massive violence or destabilizing developments. Second, the United States will seldom, if ever, find it in its national interest to be deterred from standing up to aggression. Third, I believe that the United States remains willing to accept risks—even large risks— for an important cause. And the prevention, suppression, and defeat of aggression backed by the use or threat of weapons of mass destruction will continue to be seen as important to the peace and stability that serve U.S. national interests. In addition to the more immediate costs of failing to deal with such aggression, history tells us that such failures are likely to lead to a far more proliferated, dangerous, and less cooperative world.

Thus, the United States is and should be committed to deterring and, if need be, to defending against aggression, especially when backed by weapons of mass destruction. The military forces needed to do this will include nuclear forces that provide a credible threat of devastating retaliation for the use of weapons of mass destruction. At the same time, minimizing the risks posed to the forces and citizens of the United States, its allies, and other nations when confronting such aggression will require other substantial preparations—preparations that increase the power of deterrence by further reducing an aggressor's confidence that the gain from such attacks is worth their potentially very high costs. These preparations will include better ways to prevent attacks with weapons of mass destruction, to interdict such attacks when launched, and to reduce the effects that such attacks can have on their targets. They are also likely to include substantial changes in how the United States organizes, deploys, and operates its forces when faced by such threats.

The preparations that the United States must make are neither cheap nor easy. The effort required may come at the cost of other things with more obvious appeal. Thus, making the needed preparations will require a keen understanding of the larger dangers posed by continued proliferation of weapons of mass destruction, and what must be done to reduce them.

This collection of essays contributes to that crucial understanding. It presents a variety of perspectives on important policy and strategy problems posed by the continued proliferation of weapons of mass destruction. These problems and their solutions are substantially different from those posed during the Cold War, when the United States and the Soviet Union confronted each other with massive arsenals of nuclear weapons.

I hope the reader will reflect on the insightful essays presented here. And beyond that, I hope that other policymakers and experts will be encouraged to contribute further to the understanding of this broad and important topic. Better and more up-to-date analysis and understanding of the challenges posed by the proliferation of weapons of mass destruction are vital to finding paths to a safer and better world.

General Larry D. Welch, USAF (ret.)
April 1999
Alexandria, Virginia

Part I
Pressures for Nuclear Proliferation
and Crises

Chapter 1

The Specter of Nuclear, Biological, and Chemical Weapons Proliferation

Victor A. Utgoff

In the past decade, the United States and other responsible nations have become increasingly concerned that growing numbers of states and even sub-state organizations will obtain nuclear, biological, or chemical (NBC) weapons capable of causing massive destruction. These types of weapons are spreading.[1] India and Pakistan have both recently carried out multiple nuclear weapons tests. The agreement under which North Korea suspended its nuclear weapons program appears to be unraveling.[2] And a number of antagonistic states, such as Iraq, North Korea, Iran, and Libya are trying to obtain NBC weapons.[3]

The public's awareness of the harm these weapons could cause is being heightened. For example, retired military officers who once commanded nuclear arsenals have highlighted the dangers of maintaining these forces.[4] The chemical attacks by Japanese terrorists that caused

---

1. For a review of the problem of NBC proliferation as seen by the United States Department of Defense, see *Proliferation: Threat and Response* (Washington, D.C.: United States Department of Defense, November 1997). Another useful view of the proliferation problems is provided by Randall Forsberg, William Driscoll, Gregory Webb, and Jonathan Dean, *Nonproliferation Primer: Preventing the Spread of Nuclear, Chemical, and Biological Weapons* (MIT Press, 1995).

2. See Brad Roberts, "The Future of Nuclear Weapons in Asia" (Institute for Defense Analyses, forthcoming), for a comprehensive review of the potential for nuclear proliferation in Asia in the aftermath of the May 1998 nuclear tests by Indian and Pakistan.

3. Beyond the five declared nuclear powers, "at least 25 countries already have or may be developing nuclear, biological, or chemical weapons, or their missile delivery systems." *Report on Activities and Programs for Countering Proliferation and NBC Terrorism, Counterproliferation Program Review Committee* (Washington, D.C.: United States Department of Defense, May 1998), p. 3-1.

4. "Retired Nuclear Warrior Sounds Alarm on Weapons: Ex-SAC Commander Calls Policy 'Irrational'," *Washington Post,* December 4, 1996, p. 1.

nearly 20 deaths and 6,000 injuries captured public attention worldwide.[5] And the horrors of biological weapons have been publicized in a variety of recent literary works and television programs.[6]

The U.S. government is sufficiently concerned about the potential for use of biological weapons on the battlefield to take action. During the Gulf War, it vaccinated as many troops as possible against anthrax. And the Department of Defense has begun a program that will ultimately provide vaccinations against anthrax to all active-duty military service members and reservists.[7]

Most importantly, policymakers and the public sense that the proliferation of NBC weapons may lead the nation to a most difficult dilemma: If important U.S. overseas interests are challenged by states newly armed with such weapons, the United States must choose between running the sharply increased risks of defending its interests, or compromising those interests, together with its reputation for military preeminence and a willingness to protect allies and friends.

These concerns have led to new initiatives aimed at slowing or reversing the proliferation of NBC weapons. In recent years, the U.S. government has brokered agreements that have led three newly independent states to give up the nuclear arsenals they had inherited from the Soviet Union. Multiyear legislation sponsored by Senators Sam Nunn, Richard Lugar, and Pete Domenici has provided funds to reduce Soviet and now Russian nuclear forces and to minimize the prospects that the materials and expertise necessary for creating nuclear weapons will leak out of the former Soviet Union.

In addition, the United States continues to support a sputtering and still incomplete United Nations (UN) program to root out Iraq's NBC programs. The United States was also instrumental in winning the indefinite extension of the Nuclear Non-Proliferation Treaty (NPT) in 1995. And, together with the other declared nuclear powers, the United States has suspended its nuclear testing program with the expectation that all

---

5. *The Continuing Threat from Weapons of Mass Destruction* (Nonproliferation Center, Office of the Director of U.S. Central Intelligence, March 1996), p. 5.

6. For example, see Richard Preston, *The Cobra Event* (New York: Random House, 1997); and John F. Case, *The First Horseman* (Fawcett, 1998); in addition, television series such as *Seven Days* and *The X-Files* have dealt with the concept of biological warfare. Finally, numerous nonfictional documentaries and reports have been filed by the news media in relation to domestic anthrax scares, the Iraqi biological weapons program, and revelations that Russia continues to work on biological weapons.

7. "Total Force Anthrax Vaccinations To Begin," DefenseLINK Release No. 430-98, August 14, 1998 (http://www.defenselink.mil/news/Aug1998/b08141998_bt430-98.html).

nuclear and nuclear-capable states will eventually join the Comprehensive Test Ban Treaty.

In recent years, the United States has also taken new steps to counter the capabilities of proliferators to effectively threaten or actually use NBC weapons. These steps include a multiyear Counter-Proliferation Initiative (CPI) to develop new technologies that can allow these weapons to be attacked and destroyed before they can be used, or intercepted after they have been launched but before they reach their targets.[8] The CPI is also improving the protection of U.S. forces against chemical and biological agents that do arrive in their vicinities. In addition, U.S. military planners are developing operational concepts and plans for employing forces so that they can perform their missions with minimal risks of defeat or of suffering historically unprecedented losses from NBC attacks. Finally, some initial steps have been taken toward cooperative counterproliferation efforts with key allies.[9]

Impressive as these various nonproliferation and counterproliferation actions may be, they are only a start toward the goal of denying proliferators the potential destructive and coercive power of NBC weapons. Among the larger efforts that lie ahead, three efforts stand out. First, the creation of an effective defense against the kinds of NBC capabilities that proliferators might aspire to—especially considering the many different forms that these weapons and their means of delivery might take—is a task with substantial technical difficulties and costs.

Second, for political as well as practical reasons, the United States cannot bear all the burdens of countering NBC weapons. Other states that can be threatened by these weapons, or that are relatively capable of contributing to efforts to counter them, must be convinced to participate and to take the necessary actions, including cooperative efforts to protect against NBC attacks. In addition, partners will need to be visibly involved if they are to share adequately the responsibility for military actions that might be required against an NBC-armed regional challenger. Such involvement requires cooperative efforts to prepare other states' forces to fight effectively alongside those of the United States. It also means involving prospective partners in the key decisions regarding military objectives and the possible retaliatory use of nuclear weapons, should that prove necessary.

---

8. See the five *Reports to the Congress on Activities and Programs for Countering Proliferation (and NBC Terrorism [1998]), Counterproliferation Program Review Committee* (Washington, D.C.: United States Department of Defense, May 1994, 1995, 1996, 1997, and 1998).

9. *Proliferation: Threat and Response,* pp. 62–63.

Third, the United States and other cooperating governments must develop a better public awareness of the need to prevent and counter the proliferation of NBC weapons. In particular, publics must be prepared to face the possibility of challenges to important interests by NBC-armed regional aggressors and to support the necessary political and military preparations. Waiting until such a challenge materializes to clarify the potential stakes and risks and the pros and cons of alternative courses of action increases the chances of political confusion and devastating mistakes, and the chances that such challenges would arise in the first place.

Clearly, the overall political and technical effort required to halt, roll back, and counter the continued proliferation of NBC weapons is very substantial. Will the United States prove willing over the long haul to bear the costs and other burdens involved?

The answer is far from clear. Rather than defend against the Soviet Union's nuclear capabilities in any significant way, for decades the United States accepted a mutual nuclear deterrence relationship. Its willingness to compromise its policy of punishing Pakistan and India for pursuing nuclear weapons and to overlook Israel's nuclear weapons program demonstrates that nonproliferation is not always the highest priority for the United States. In addition, while the frightening specters of NBC attacks on U.S. forces or cities are disturbing, they are hard for the U.S. public to take too seriously—the public tolerated such fears for the decades of the Cold War. Moreover, it is even easier to discount the possibility of such attacks by renegade states that have not been seen as major powers in the past, and whose military capabilities are so modest compared to those of the United States and its allies.

But the possibility of such attacks cannot be discounted—and the preparations that the United States makes to meet such challenges will strongly affect the outcome of such an attack. Rather than wait until an NBC-armed state challenges an important regional interest, rather than wait until the discomforts of accommodating to a world in which NBC proliferation gives otherwise minor powers influence disproportionate to their populations, productivity, or moral considerations, we must find the motivation now to face the problem of proliferation more seriously. A deeper and broader appreciation of the eventual implications of continued proliferation of NBC weapons will allow the United States and its allies to trade the risks and discomforts of dangerous confrontations and twisted world orders for the burdens of preparation and avoidance.

The goal of this book is to help develop such an appreciation. It is an attempt to anticipate some of the ways in which continued proliferation of NBC weapons is likely to pose challenges to the United States and other supporters of a gracefully evolving liberal world order. It is also a hard look at the kinds of painful dilemmas and actions that will likely be forced on the responsible world community if strong measures to counter proliferation are not taken. In this book, six academics join with several analysts at the Institute for Defense Analyses to explore some of the implications of continued and uncountered proliferation of NBC weapons. I invited the authors to address any of the following list of questions, or any alternative question my list suggested:

- What changes might continuing proliferation of NBC weapons be expected to have in the long run on the nature of international relations? How would such changes affect the interests of the United States and the larger global community? What could be done to mitigate these effects if proliferation cannot be halted?
- What political-military problems are involved in creating and maintaining international coalitions for intervening against an NBC-armed regional challenger?
- How must a war against a regional challenger that threatens or employs NBC weapons end?
- What can be learned about a nation's biases toward the acquisition and use of NBC weapons from studying its "strategic personality"?

Every prospective explorer approached accepted this invitation with alacrity. Their explorations, presented in the following eight chapters, provide many arguments and insights that are contrary to the conventional wisdom in this area.

While all the chapters were drafted independently, some have influenced others. This resulted from a two-day meeting of the authors at the Institute for Defense Analyses to present their drafts to a small group of experts.

The first part of the book looks at some of the different motivations states see for acquiring nuclear weapons, and how proliferation is creating the potential for dangerous crises in which nuclear weapons might get used. The second part explores other potential consequences of continued nuclear proliferation, and in particular, how a crisis in which a nuclear-armed aggressor challenges the United States might evolve. In order of presentation, then, the main arguments of the chapters are as follows.

Pressures for Nuclear Proliferation and Crises

In Chapter 2, "Rethinking the Causes of Nuclear Proliferation: Three Models in Search of a Bomb," Scott D. Sagan notes the scant attention paid to the question of why states build nuclear weapons. Sagan argues that this lack of attention follows from a near consensus that nuclear weapons are only built to meet security threats that cannot be met by other means (the security model). Sagan challenges this assumption, presenting evidence that nuclear weapons also serve other less obvious, parochial purposes. Nuclear weapons are important objects in bureaucratic struggles and internal debates (the domestic politics model), and they can serve as important symbols of a state's modernity and identity (the norms model).

Sagan points out that the most appropriate nonproliferation policies for a state depend upon the model that best explains why it might seek nuclear weapons, and that some of the policies called for by different models can be contradictory. For example, while the security model calls for extending nuclear deterrence assurances to states facing threats that might otherwise lead them to build their own weapons, the norms model argues against giving nuclear weapons the importance that such a role would suggest. Similarly, the perceived value of nuclear weapons is raised if the United States, with its great conventional military power, feels it must deter chemical and biological attacks with the threat of nuclear retaliation. Finally, Sagan argues that the United States is going to have to choose either to "wean" its allies away from extended nuclear deterrent guarantees or accept the equally difficult task of maintaining a norm against nuclear proliferation that it does not honor itself.

In Chapter 3, "Universal Deterrence or Conceptual Collapse? Liberal Pessimism and Utopian Realism," Richard K. Betts argues that while the "utopian realists," who see the spread of nuclear weapons as leading to universal mutual deterrence and military restraint, may correctly predict the effect of continued proliferation in nearly all cases, the "liberal pessimists," who view the spread of nuclear weapons with alarm, are probably also right in assuming that increased numbers of nuclear-armed states means an increased likelihood that nuclear weapons eventually will be used.

Betts further argues that the ramifications of a breakdown in the taboo on nuclear use are too unpredictable for anyone to want to run this experiment. He describes a variety of ways in which the taboo could break down. For example, while most states would want nuclear weapons for strictly defensive reasons, a few might become emboldened to try aggression and end up in a dangerous confrontation with a nuclear-armed superpower accustomed to intervening in areas of vital interest. Betts points out that the United States and the Soviet Union took approximately fifteen years to work out ways to avoid dangerous confrontations and had some very tense moments along the way. He also notes that the logic of deterrence theory may not be obvious to individuals in countries for whom these questions are new, and that madness and irrationality do sometimes occur in the behavior of political leaders.

Finally, Betts argues that while the shock of the next use of nuclear weapons could lead either to faster proliferation or to far stronger efforts to roll it back, it seems improbable that the willingness of the United States and others to rely on nuclear deterrence would remain unshaken.

In Chapter 4, "The National Myth and Strategic Personality of Iran: A Counterproliferation Perspective," Caroline F. Ziemke argues that every nation has a strategic personality that defines how it is disposed to behave toward other nations. Ziemke states that this personality can be discerned by studying a nation's public myth, the stories and themes it uses to illuminate for itself its social and ethical norms and its collective identity. Thus, an understanding of a proliferator's national myth may provide important insights into why it might want nuclear weapons and the purposes to which it might put such weapons.

Ziemke's reading of Iran's national myth indicates that Iran is supremely confident of the superiority of its culture. It sees its troubled history since the glory of the Persian Empire solely as the result of invasions and evil influences from the outside world. Consistent with this, the United States, with its corrupting material culture and its decades of meddling in Iranian affairs, is seen as the embodiment of foreign evil, the "Great Satan."

Ziemke employs Iran's national myth to interpret its foreign policy and intentions for nuclear weapons. She argues that Iran wants most of all to win the respect that its superior culture deserves. It also wants hegemonic influence over the Persian Gulf region, which requires that the United States leave, and it wants to be safe from potential enemies, particularly Iraq. However, it is not interested in actually conquering its neighbors. Thus, Ziemke sees Iran wanting nuclear weapons to inspire respect and fear, and as insurance against invasion, but not as backing for conventional aggression.

Ziemke also argues that Iran is very unlikely to risk the first use of nuclear weapons against the United States or its allies unless it were about to be overwhelmed. Iran "knows" the "Great Satan" is perfectly willing to annihilate it in response to any first use of nuclear weapons against the United States. Its national myth also points to a ruling elite that will not risk the survival of the Iranian faithful, to whom it sees itself accountable.

Ziemke points out the contradictory natures of U.S. and Iranian foreign policies toward each other. Iran wants the United States out of the Persian Gulf, but threatens its neighbors in ways that increase their

interest in U.S. protection. The United States wants Iran to stop supporting terrorism and to halt its nuclear program, but reinforces Iranian
paranoia with the dual containment policy, and gives Iran the psychological victories it craves with every protest against Iranian actions and
every successful penetration of the U.S. arms embargo. Rationalizing
these contradictory policies will take a great deal of time and effort on
both sides, but seems worthwhile given the substantial interests the two
sides actually have in common. Clearly, a good understanding of each
other's interests and values will become even more important should Iran
create significant capabilities to threaten and use NBC weapons.

These three chapters constituting Part I of the book lead me to three
related conclusions. Sagan's chapter suggests that effective policies for
stopping nuclear proliferation are going to be even more difficult to find
and implement than the nonproliferation community has supposed.
Betts's chapter then tells us that there are many ways for dangerous crises
to emerge that would threaten the use of nuclear weapons both as
proliferation continues and even in a fully proliferated world. Third,
Ziemke's chapter suggests that nuclear challenges may be less likely to
emerge from some members of the current rogues' gallery, in this case
Iran, than is commonly supposed. Ziemke's chapter is also interesting for
its use of the concept of strategic personality to discern a state's likely
behavior. I employ this concept extensively in the concluding chapter.

Collectively, these three chapters suggest that a complete halt to
nuclear proliferation may be even more difficult to achieve than has been
commonly supposed. They also highlight the importance of tailoring
nonproliferation policies to the specific countries and regions in question.

Potential Evolution and Consequences of a Nuclear Crisis with the United States

In Chapter 5, "Nuclear Proliferation and Alliance Relations," Stephen
Peter Rosen argues that with the end of the Cold War, nuclear proliferation will now weaken alliances rather than strengthen them. He reasons
that the risks of becoming involved in a crisis with a nuclear-armed
regional aggressor have always seemed high, but that the incentives to
become involved are now much weaker. In particular, staying on the
sidelines no longer threatens some disadvantage in a long-term geopo-
litical competition for the highest stakes, nor that the unattended crisis
might somehow catalyze a global nuclear war.

Rosen explores the risks of intervention against a nuclear-armed
regional power by reasoning through a scenario in which Iraq had half
a dozen nuclear weapons when it invaded Kuwait in 1990. Rosen sees

the aggressor as able to employ nuclear weapons to prevent or collapse support from allies that the United States would need, or to destroy the ports necessary for a timely intervention before the United States could reach them. He sees few opportunities for the United States to threaten or use its nuclear weapons with both comparable military effect and destruction. Rosen also sees the United States as greatly concerned about the large numbers of forces it could lose to a nuclear strike against a crowded port.

Rosen argues that his assessment applies generally to interventions against nuclear-armed regional states, and believes that it is likely to be widely understood and appreciated both by regional allies and opponents. Thus, regional nuclear proliferation will devalue and erode U.S. alliances. This would happen particularly quickly should a crisis reveal that the United States is unwilling to intervene against a nuclear-armed regional aggressor.

In Chapter 6, "U.S. Security Policy in a Nuclear-Armed World, or What If Iraq Had Had Nuclear Weapons?" Barry R. Posen also assesses a counterfactual 1990–91 scenario in which Iraq had half a dozen survivable and deliverable nuclear weapons. Posen sees this scenario (and thus its prospective future analogue) as a defining moment for the future of nuclear weapons, U.S. credibility as a reliable protector, and the nature of world order.

Posen argues that if the United States had accepted Kuwait's conquest by a nuclear-armed Iraq, nuclear weapons would have been redefined as effective offensive weapons: they would have deterred a far stronger United States from rolling back a conquest made with conventional forces. Nuclear weapons would no longer be strictly defenders of the status quo, as they have been since World War II. Posen argues that this changed role for nuclear weapons would lead to a "hellishly competitive world" as aggressors rush to get nuclear weapons and potential victims, including shaken long-term U.S. allies, scramble to beat them to it. Incentives would rise for wars to prevent neighbors from getting nuclear weapons, or to capitalize on military advantages before they are offset. The world might ultimately settle into equilibrium of many mutual deterrence relationships, but the long transition would be "very exciting." Posen sees this world as forcing the United States into an uncomfortable isolation or into adopting very difficult and burdensome policies to counter the effects of having balked in the first place.

Thus, Posen argues in favor of intervening. To help minimize the prospects that Iraq would use its presumed nuclear weapons, he suggests making "ferocious threats" of nuclear retaliation, and clearly limiting U.S. military goals well short of threatening Iraq's total defeat. Posen admits that this reemphasis of nuclear deterrence would do some damage to U.S. nonproliferation policy, but argues that failure to intervene and to minimize the prospects of nuclear use by the opponent would do even more. Finally, Posen suggests a campaign to clarify to all states involved in the crisis the unacceptable long-term consequences for the United States of failing to roll back Iraq's gains.

Posen's analysis of the official strategy debates during the actual 1990–91 Persian Gulf War suggests that had Iraq actually been nuclear-armed, his recommendations would have gotten a sympathetic hearing. Finally, Posen argues the United States will eventually face a defining moment of this kind and should think it through in advance.

In Chapter 7, "Containing Rogues and Renegades: Coalition Strategies and Counterproliferation," Stephen M. Walt argues that history does not support the pessimism of many strategic planners about prospects for containing rogue regimes, or their warnings that the spread of nuclear weapons would have corrosive effects on important U.S. commitments. Walt says that states have usually been willing to ally with a strong power to balance threats posed by neighbors having powerful offensive forces and obvious aggressive intent.

Walt points out that such willingness to confront even powers armed with nuclear weapons has been well demonstrated by history. He also points out that great power allies, and particularly the United States, will be available, given their strong interest in preventing nuclear weapons from being used for coercion or as a shield for conventional aggression.

Walt argues that defensive coalitions for containing aggressive states are far easier to form and maintain than offensive coalitions. Defensive coalitions should prove willing to force an aggressor to relinquish any conquests, though not to try to overthrow the rogue government. To facilitate the formation of coalitions for containment, Walt recommends aggressive collection of intelligence that would reveal the intentions and capabilities of rogue states. He recommends arrangements to share the costs and risks in a reasonably equitable manner. He further recommends assuring potential aggressors that the United States will not seek their overthrow if they remain peaceful, but will oust them if they attack their neighbors with nuclear weapons. Walt also highlights the need to maintain the strong defensive capabilities required to assure threatened states that an alliance with the United States can protect them.

In Chapter 8, "The Response to Renegade Use of Weapons of Mass Destruction," George H. Quester argues that if a renegade state were to use weapons of mass destruction against the United States and its allies, the U.S. response is more likely to be guided by the norms of the U.S. law enforcement system and the historical precedents of World War II than by the Cold War theory that unconditional surrender can never be sought from a state possessing such weapons.

Quester points out that domestic criminals are jailed for a combination of four purposes: to disarm them, to make them an example to others, to impose revenge on behalf of victims, and to reform them. Similarly, in World War II, criminal regimes in Germany and Japan that destroyed massive numbers of other states' civilians and soldiers were disarmed. Both states were subjected to very destructive mass bombing that satisfied discernible urges for revenge and the creation of examples for the future. Finally, the political systems of both states were fundamentally reformed. In contrast, had the Soviet Union used nuclear weapons against the United States and its allies, any realistic pursuit of all four of these goals would have been unthinkable. A massive attack by the Soviet Union would have left the United States capable of little more than an angry and apocalyptic revenge. Limited use might have led to retaliation aimed more at disarmament, setting an example for the future, and a lesser revenge. However, pursuit of unconditional surrender and political reform of the Soviet Union would have risked escalation of the war and the annihilation of both sides.

Quester argues that renegade states with small to modest capabilities for mass destruction will be seen as too dangerous to live with once they use these weapons, but not too dangerous to defeat. Quester notes that counterproliferation programs that reduce the destruction a renegade could do would make it even clearer that the United States will not be bound by Cold War nuclear theory, but will follow its sense that seeking unconditional surrender of such a renegade is the "right thing to do."

Finally, in Chapter 9, "Rethinking How Wars Must End: NBC War Termination Issues in the Post–Cold War Era," Brad Roberts argues that in wars against NBC-armed regional challengers, the enormous advantages in power and overall survivability of the United States and its allies would allow them to choose how these conflicts will end. Roberts goes on to argue that stalemate would likely seem unacceptable, and that unconditional surrender and political reform of the opponent would seem necessary, if the opponent does substantial damage with NBC weapons. Perhaps most important, he argues that the United States will have to choose a course of action that addresses both the immediate problems posed by the war and the longer-term U.S. interests in the peace that follows.

Thus, the United States must not be seen as a "nuclear bully" that was overcome by rage and fear and used its weapons in impulsive, imperious, and excessively destructive ways. Such actions could lead to a widespread view that U.S. military power is too dangerous and needs to be counterbalanced, and could encourage further proliferation of NBC weapons. Alternatively, the United States must not be seen as a "nuclear wimp" whose fears of staying the course led it to appease an NBC-armed aggressor. This could encourage tests of U.S. willingness to defend other interests, undermine U.S. security guarantees, and thus spur further proliferation of NBC weapons by allies. It could also lead to a political backlash against those who had caused such a U.S. decline.

Instead, Roberts argues that the United States should seek to be seen as a responsible and just steward of the collective good. This requires ending the conflict in ways that resolve its underlying cause, that remove the threat posed by the aggressor, and that use nuclear weapons only to the extent that they appear needed to end the war and to save lives. Finally, Roberts argues that emerging from such a war as a just and responsible steward will be easier if the United States does three things. First, it must reduce its own and help to reduce its allies' vulnerabilities to NBC attack. Second, it must work with allies, the Congress, and the public to shape the political context for regional conflict involving NBC weapons. And, third, it is especially important to let potential aggressors understand how the United States would likely see its alternatives in such a conflict.

The main conclusion I draw from Part II of the book is that the United States seems likely to prove more resolute than much of the community that is expert on nuclear proliferation and its potential consequences seems to believe. Rosen argues that the United States would not see a net advantage in opposing regional aggression backed by nuclear weapons and thus that potential allies would not trust alliance with the United States to save them: all the following chapters argue the opposite.

In closing this introductory chapter, let me emphasize that I have presented only a few of the points made by each of my colleagues, and even fewer of the strong justifications they present in defense of their points. Moreover, my selections and renditions of their arguments are surely colored by my own views. Thus, I delay any further comments or interpretations of their work until Chapter 10, when the reader will have had a chance to read their work and to form his or her own opinions.

Chapter 10, which notes the different perspectives that the authors have brought to their analyses, argues that projections of the likely behavior of the United States and other nations can profit from bringing more consciously balanced mixtures of perspectives to bear. It then adds a strong measure of the strategic personality perspective to those of the other authors in an attempt to project likely U.S. behavior in response to several of the more important questions that would be posed when confronting nuclear-backed aggression.

*   *   *

Finally, I would like to thank Larry Welch, Bob Roberts, Phil Major, and Mike Leonard of the Institute for Defense Analyses for their steadfast support of this project. I also owe special thanks to Brad Roberts for his invaluable comments and suggestions on both my introductory and concluding chapters, and to Caroline Ziemke for her comments on the concluding chapter, which draws extensively on her research on strategic personalities. Rafael Bonoan was very helpful in providing supporting literature and analysis for Chapter 10. Richard Falkenrath and Leo MacKay deserve a salute for their frequent encouragement to create and especially to finish this book. Sean-Lynn Jones and Karen Motley made excellent suggestions on the organization and presentation of the book, and Miriam Avins's editorial suggestions have made the book far more readable and to the point. Kristina Cherniahivsky deserves great credit for her patient and long-suffering search for a suitable cover photo. Johnathan Wallis wins my thanks for his careful review of the entire text. Last, but not least, Olga Alvarado deserves special thanks for incorporating a myriad of changes into the text, and keeping the manuscript in its various forms organized and moving.

Chapter 2

Rethinking the Causes of Nuclear Proliferation: Three Models in Search of a Bomb

Scott D. Sagan

Why do states build nuclear weapons? An accurate answer to this question is critically important both for predicting the long-term future of international security and for current foreign policy efforts to prevent the spread of nuclear weapons. Yet given the importance of this central proliferation puzzle, it is surprising how little sustained attention has been devoted to examining and comparing alternative answers.

This lack of critical attention is not due to a lack of information: there is now a large literature on nuclear decision-making inside the states that have developed nuclear weapons, and a smaller but still significant set of case studies of states' decisions to refrain from developing nuclear weapons. Instead, the inattention appears to reflect a near consensus that the answer is obvious. Many U.S. policymakers and most international relations scholars have a clear and simple answer to the proliferation puzzle: states will seek to develop nuclear weapons when they face a significant military threat to their security that cannot be met through alternative means; if they do not face such threats, they will willingly remain non-nuclear states.[1]

An earlier version of this essay appeared as "Why Do States Build Nuclear Weapons? Three Models in Search of a Bomb," in International Security, Vol. 21, No. 3 (Winter 1996/97).

1. Among policymakers, John Deutch presents the most unadorned summary of the basic argument that "the fundamental motivation to seek a weapon is the perception that national security will be improved." John M. Deutch, "The New Nuclear Threat," *Foreign Affairs*, Vol. 71, No. 41 (Fall 1992), pp. 124–125. Also see George Shultz, "Preventing the Proliferation of Nuclear Weapons," *Department of State Bulletin*, Vol. 84, No. 2093 (December 1984), pp. 17–21. For examples of the dominant paradigm among scholars, see Michael M. May, "Nuclear Weapons Supply and Demand," *American Scientist*, Vol. 82, No. 6 (November–December 1994), pp. 526–537; Bradley A. Thayer, "The Causes of Nuclear Proliferation and the Nonproliferation Regime," *Security Studies*, Vol. 4, No. 3 (Spring 1995), pp. 463–519; Benjamin Frankel, "The

The central purpose of this chapter is to challenge this conventional wisdom about nuclear proliferation. I argue that the consensus view, which focuses on national security considerations as the cause of proliferation, is dangerously inadequate because nuclear weapons programs also serve more parochial and less obvious objectives. Like other weapons, nuclear weapons are more than tools of national security; they are political objects of considerable importance in domestic politics and bureaucratic struggles and also serve as international symbols of modernity and identity.

The chapter examines three alternative theoretical frameworks—what I call "models" in the very informal sense of the term—about why states decide to build or refrain from developing nuclear weapons: "the security model," according to which states build nuclear weapons to increase national security against foreign threats, especially nuclear threats; "the domestic politics model," which envisions nuclear weapons as political tools used to advance parochial domestic and bureaucratic interests; and "the norms model," under which nuclear weapons acquisition, or restraint in weapons development, is determined by the role of such weapons as a symbol of a state's modernity and identity. Although many of the ideas underlying these models appear in the case study and nonproliferation policy literature, they have not been adequately analyzed, nor placed in a comparative theoretical framework, nor properly evaluated against empirical evidence. When I discuss these models, therefore, I compare their explanations for nuclear weapons development, present alternative interpretations of the history of some major proliferation decisions, and contrast the models' implications for nonproliferation policy. The article concludes with an outline of a research agenda for future proliferation studies and an examination of the policy dilemmas illuminated by these models.

The crafting of effective nonproliferation policies will remain critical. Despite the successful 1995 agreement to permanently extend the Nuclear Non-Proliferation Treaty (NPT), there will be continuing NPT review conferences assessing the implementation of the treaty every five years; each member state can legally withdraw from the treaty, under the "supreme national interest" clause, if it gives three months' notice; and many

---

Brooding Shadow: Systemic Incentives and Nuclear Weapons Proliferation," and Richard K. Betts, "Paranoids, Pygmies, Pariahs, and Nonproliferation Revisited," both in Zachary S. Davis and Benjamin Frankel, eds., *The Proliferation Puzzle*, special issue of *Security Studies*, Vol. 2, Nos. 3/4 (Spring/Summer 1993), pp. 37–38 and pp. 100–124; and David Gompert, Kenneth Watman, and Dean Wilkening, "Nuclear First Use Revisited," *Survival*, Vol. 37, No. 3 (Autumn 1995), p. 39.

new states can be expected to develop, at a minimum, a "latent nuclear weapons capability" over the coming decade. Indeed, some fifty-seven states now operate or are constructing nuclear power or research reactors, and it has been estimated that about thirty countries today have the necessary industrial infrastructure and scientific expertise to build nuclear weapons on a crash basis if they chose to do so.[2] The NPT encourages this long-term trend by promoting the development of power reactors in exchange for the imposition of International Atomic Energy Agency (IAEA) safeguards on the resulting nuclear materials. This suggests that while most attention concerning proliferation in the immediate term has appropriately focused on controlling nuclear materials in the former Soviet Union, managing the new nuclear relationship between India and Pakistan, and preventing the small number of suspected active proliferators (such as Iraq, Iran, Libya, and North Korea) that currently appear to have vigorous nuclear weapons programs from getting the bomb, the longer-term and enduring proliferation problem will be to ensure that the larger and continually growing number of latent nuclear states maintain their non-nuclear weapons status. This underscores the policy importance of addressing the sources of the political *demand* for nuclear weapons, rather than focusing primarily on efforts to safeguard existing stockpiles of nuclear materials and to restrict the *supply* of specific weapons technology from the "haves" to the "have-nots."

If my arguments and evidence concerning the three models of proliferation are correct, however, any future demand-side nonproliferation strategy will face inherent contradictions. For, in contrast to the views of scholars who claim that a traditional realist theory focusing on security threats explains all cases of proliferation and nuclear restraint,[3] the historical evidence presented here suggests that each theory explains some past cases quite well and others quite poorly. Unfortunately, since the models provide different and often contradictory lessons for U.S. nonproliferation policy, policies designed to address one future proliferation

---

2. See Steve Fetter, "Verifying Nuclear Disarmament," Occasional Paper No. 29, Henry L. Stimson Center, Washington, D.C., October 1996, p. 38; and "Affiliations and Nuclear Activities of 172 NPT Parties," *Arms Control Today*, Vol. 25, No. 2 (March 1995), pp. 33–36. For earlier pioneering efforts to assess states' nuclear weapons latent capability and demand, see Stephen M. Meyer, *The Dynamics of Nuclear Proliferation* (Chicago: University of Chicago Press, 1984); and William C. Potter, *Nuclear Power and Nonproliferation* (Cambridge, Mass: Oelgeschlager, Gunn and Hain, 1982).

3. For example, May, "Nuclear Weapons Supply and Demand"; Thayer, "The Causes of Nuclear Proliferation and the Nonproliferation Regime"; and Frankel, "The Brooding Shadow: Systemic Incentives and Nuclear Weapons Proliferation."

problem will likely exacerbate others. As I discuss in more detail below, particularly severe tensions are likely to emerge in the future between U.S. extended deterrence policies designed to address security concerns of potential proliferators and U.S. nonproliferator policies designed to maintain and enhance international norms against nuclear use and acquisition.

The Security Model: Nuclear Weapons and International Threats

According to neorealist theory in political science, states exist in an anarchical international system and must therefore rely on self-help to protect their sovereignty and national security.[4] Because of the enormous destructive power of nuclear weapons, any state that seeks to maintain its national security must balance against any rival state that develops nuclear weapons by gaining access to a nuclear deterrent itself. Strong states will pursue a form of internal balancing by adopting the costly but self-sufficient policy of developing their own nuclear weapons. Weak states will join a balancing alliance with a nuclear power, utilizing a promise of nuclear retaliation by that ally as a means of extended deterrence. Acquiring a nuclear ally may be the only option available to weak states, but the policy inevitably raises questions about the credibility of extended deterrence guarantees, since the nuclear power would also fear retaliation if it responded to an attack on its ally.

Although states might also choose to develop nuclear weapons to serve either as deterrents against overwhelming conventional military threats or as coercive tools to compel changes in the status quo, a focus on states' responses to emerging nuclear threats is the most common and most parsimonious explanation for nuclear weapons proliferation.[5] George Shultz once nicely summarized the argument: "Proliferation be-

---

4. The seminal text of neorealism remains Kenneth N. Waltz, *Theory of International Politics* (New York: Random House, 1979). Also see Kenneth N. Waltz, "The Origins of War in Neorealist Theory," in Robert I. Rotberg and Theodore K. Rabb, eds., *The Origin and Prevention of Major Wars* (New York: Cambridge University Press, 1989), pp. 39–52; and Robert O. Keohane, ed., *Neorealism and Its Critics* (New York: Columbia University Press, 1986).

5. The Israeli, and possibly the Pakistani, nuclear weapons decisions might be the best examples of defensive responses to conventional security threats; Iraq, and possibly North Korea, might be the best examples of the offensive coercive threat motivation. On the status quo bias in neorealist theory in general, see Randall L. Schweller, "Bandwagoning for Profit: Bringing the Revisionist State Back In," *International Security*, Vol. 19, No. 1 (Summer 1994), pp. 72–107; and Richard Rosecrance and Arthur A. Stein, eds., *The Domestic Bases of Grand Strategy* (Ithaca, N.Y.: Cornell University Press, 1993).

gets proliferation."[6] Every time one state develops nuclear weapons to balance against its main rival, it also creates a nuclear threat to another state in the region, which then must initiate its own nuclear weapons program to maintain its national security.

From this perspective, the history of nuclear proliferation is a strategic chain reaction. During World War II, none of the major belligerents was certain that the development of nuclear weapons was possible, but all knew that other states were already working or could soon be working to build the bomb. This fundamental fear was the central impetus for the U.S., British, German, Soviet, and Japanese nuclear weapons programs. The United States developed atomic weapons first, not because it had any greater demand for the atomic bomb than these other powers but because it invested more heavily in the program and made the right set of technological and organizational choices.[7] After August 1945, the Soviet Union's program was reinvigorated because the U.S. atomic attacks on Hiroshima and Nagasaki demonstrated that nuclear weapons were technically possible, and the emerging Cold War made a Soviet bomb a strategic imperative. From the realist perspective, the Soviet response was perfectly predictable. Josef Stalin's reported request to Igor Kurchatov and B.L. Yannikov in August 1945 appears like a textbook example of realist logic: "A single demand of you comrades. . . . Provide us with atomic weapons in the shortest possible time. You know that Hiroshima has shaken the whole world. The balance has been destroyed. Provide the bomb—it will remove a great danger from us."[8]

Other states' decisions to develop nuclear weapons can also be explained within the same framework. London and Paris are thus seen to have built nuclear weapons because of the growing Soviet military threat and the inherent reduction in the credibility of the U.S. nuclear guarantee to NATO (North Atlantic Treaty Organization) allies once the Soviet Union was able to threaten retaliation against the United States.[9] China

---

6. Shultz, "Preventing the Proliferation of Nuclear Weapons," p. 18.

7. On the genesis of the atomic programs in World War II, see McGeorge Bundy, *Danger and Survival: Choices about the Bomb in the First Fifty Years* (New York: Random House, 1988) pp. 3–53; and Richard Rhodes, *The Making of the Atomic Bomb* (New York: Simon and Schuster, 1986).

8. A. Lavrent'yeva in "Stroiteli novogo mira," *V mire knig*, No. 9 (1970), in David Holloway, *The Soviet Union and the Arms Race* (New Haven, Conn.: Yale University Press, 1980), p. 20; also quoted in Thayer, "The Causes of Nuclear Proliferation," p. 487.

9. Important sources on the British case include Margaret Gowing, *Britain and Atomic Energy, 1939–1945* (London: Macmillan, 1964); Margaret Gowing, *Independence and Deterrence: Britain and Atomic Energy, 1945–1952*, vols. 1 and 2 (London: Macmillan,

developed the bomb because Beijing was threatened with possible nuclear attack by the United States at the end of the Korean War and again during the Taiwan Straits crises in the mid-1950s. Not only did Moscow prove to be an irresolute nuclear ally in the 1950s, but the emergence of hostility in Sino-Soviet relations in the 1960s further encouraged Beijing to develop, in Avery Goldstein's phrase, the "robust and affordable security" of nuclear weapons, since the border clashes "again exposed the limited value of China's conventional deterrent."[10]

After China developed the bomb in 1964, India, which had just fought a war with China in 1962, was bound to follow suit. India's strategic response to the Chinese test came a decade later, when its Atomic Energy Commission successfully completed the long research and development process required to construct and detonate what was called a "peaceful nuclear explosion" (PNE) in May 1974. According to realist logic, India maintained an ambiguous nuclear posture until 1998—building sufficient nuclear materials and components for a moderate-sized nuclear arsenal, but not testing or deploying weapons into the field—in a clever effort to deter the Chinese without encouraging nuclear weapons programs in other neighboring states.[11] After India's 1974 test, it was inevitable that Pakistan, which now faced a hostile neighbor with both a nuclear weapons capability and conventional military superiority, would seek to produce a nuclear weapon as quickly as possible.[12] By the late 1990s, Indian officials believed that it faced both a small nuclear

---

1974); and Andrew Pierre, *Nuclear Politics: The British Experience with an Independent Strategic Force, 1939–1970* (London: Oxford University Press, 1972). On the French case, see Lawrence Scheinman, *Atomic Energy Policy in France Under the Fourth Republic* (Princeton, N.J.: Princeton University Press, 1965), and Wilfred L. Kohl, *French Nuclear Diplomacy* (Princeton, N.J.: Princeton University Press, 1971).

10.   Avery Goldstein, "Robust and Affordable Security: Some Lessons from the Second-Ranking Powers During the Cold War," *Journal of Strategic Studies*, Vol. 15, No. 4 (December 1992), p. 494. The seminal source on the Chinese weapons program, which emphasizes the importance of U.S. nuclear threats in the 1950s, is John W. Lewis and Xue Litai, *China Builds the Bomb* (Stanford, Calif.: Stanford University Press, 1988).

11.   For important works on the history of the Indian nuclear program, see Itty Abraham, *The Making of the Indian Atomic Bomb: Science, Society, and the Postcolonial State* (London: Zed Books, 1998); George Perkovich, *India's Ambiguous Bomb* (Berkeley, Calif.: University of California Press, 1999); Ashok Kapur, *India's Nuclear Option: Atomic Diplomacy and Decision Making* (New York: Praeger, 1976); and T.T. Poulose, ed., *Perspectives of India's Nuclear Policy* (New Delhi: Young Asia Publications, 1978).

12.   Valuable sources on Pakistan's program include Ziba Moshaver, *Nuclear Weapons Proliferation in the Indian Subcontinent* (Basingstoke, U.K.: Macmillan, 1991); Ashok Kapur, *Pakistan's Nuclear Development* (New York: Croom Helm, 1987); and Samina Ahmed, "Pakistan's Nuclear Weapons Program," *International Security*, Vol. 23, No. 4 (Spring 1999), pp. 178–204.

Pakistani capability, developed with the assistance of China, and the continuing Chinese nuclear threat. These twin security problems forced the Indian government to move its nuclear capability out of the basement, testing five nuclear weapons on May 11 and May 13, 1998. "We live in a world where India is surrounded by nuclear weaponry," Prime Minister Atal Vajpayee exclaimed soon after he announced the Indian tests.[13] Predictably, Pakistan quickly followed suit: the government in Islamabad announced that it too had tested nuclear weapons on May 28 and May 30, less than a month after the Indian tests.

EXPLAINING NUCLEAR RESTRAINT

Given the strong deterrent capabilities of nuclear weapons, why would any state that is technically capable of developing the bomb voluntarily give up such powerful tools of security? Cases of nuclear weapons "reversal" can also be viewed through the lens provided by the security model if one assumes that external security threats can radically change or be reevaluated. The case of South Africa has most often been analyzed in this light, with the new security threats that emerged in the mid-1970s seen as the cause of South Africa's bomb program and the end of these threats in the late 1980s as the cause of its policy reversal. As President F.W. de Klerk explained in his speech to Parliament in March 1993, the Pretoria government saw a growing "Soviet expansionist threat to southern Africa"; "the buildup of the Cuban forces in Angola from 1975 onwards reinforced the perception that a deterrent was necessary, as did South Africa's relative international isolation and the fact that it could not rely on outside assistance should it be attacked."[14] Six atomic weapons were therefore constructed, but were stored disassembled in a secret location from 1980 to 1989, when the program was halted. It has been claimed that South African nuclear strategy during this period was designed to use the bomb both as a deterrent against the Soviets and as a

---

13. "We Have Shown Them That We Mean Business," interview with Prime Minister Vajpayee, *India Today*, May 25, 1998 (www.india-today.com/itoday/25051998/vajint.html).

14. F.W. de Klerk, March 24, 1993, address to the South African Parliament as transcribed in Foreign Broadcast Information Service (FBIS), JPRS-TND-93-009 (March 29, 1993), p. 1 (henceforth cited as de Klerk, "Address to Parliament"). For analyses that focus largely on security threats as the cause of the program, see Darryl Howlett and John Simpson, "Nuclearization and Denuclearization in South Africa," *Survival*, Vol. 35, No. 3 (Autumn 1993), pp. 154–173; and J.W. de Villers, Roger Jardine, and Mitchell Reiss, "Why South Africa Gave up the Bomb," *Foreign Affairs*, Vol. 72, No. 5 (November/December 1993), pp. 98–109. For a more detailed and more balanced perspective see Mitchell Reiss, *Bridled Ambition: Why Countries Constrain Their Nuclear Capabilities* (Washington, D.C.: Woodrow Wilson Center Press, 1995), pp. 7–44.

tool of blackmail against the United States. If Soviet or Soviet-supported military forces directly threatened South Africa, the regime reportedly planned to announce that it had a small arsenal of nuclear weapons, and dramatically test one or more of the weapons if necessary by dropping them from aircraft over the ocean, hoping that such a test would shock the United States into intervention on behalf of the Pretoria regime.[15] South Africa destroyed its small nuclear weapons arsenal in 1991, realist theory suggests, because of the radical reduction in the external security threats to the regime. By 1989, the risk of a Soviet-led or sponsored attack on South Africa was virtually eliminated. President de Klerk cited three specific changes in military threats in his speech to Parliament: a cease-fire had been negotiated in Angola; the tripartite agreement granted independence to Namibia in 1988; and most importantly, "the Cold War had come to an end."[16]

Although the details vary, the basic security model has also been used to explain other examples of nuclear restraint. For example, both Argentina and Brazil refused to complete the steps necessary to join the Latin American nuclear weapons–free zone (NWFZ) and began active programs in the 1970s that eventually could have produced nuclear weapons; however, their 1990 joint declaration of plans to abandon their programs is seen as the natural result of the recognition that the two states, which had not fought a war against one another since 1828, posed no fundamental security threat to each other.[17] Similarly, it has been argued that the non-Russian former states of the Soviet Union that were "born nuclear"—Ukraine, Kazakhstan, and Belarus—decided to give up their arsenals because of a mixture of two realist model arguments: their long-standing close ties to Moscow meant that these states did not perceive Russia as a major military threat to their security and sovereignty, and increased U.S. security guarantees to these states made their posses-

---

15. Military planners nonetheless developed nuclear target lists in their contingency military plans, and research was conducted on the development of the hydrogen bomb until 1985. See Reiss, *Bridled Ambition*, p. 16.

16. See de Klerk, "Address to Parliament," p. 2.

17. Thayer, "The Causes of Nuclear Proliferation," p. 497; and May, "Nuclear Weapons Supply and Demand," pp. 534–535. For analyses of the Argentine-Brazilian decision, see Monica Serrano, "Brazil and Argentina," in Mitchell Reiss and Robert S. Litwak, eds., *Nuclear Proliferation After the Cold War* (Washington, D.C.: Woodrow Wilson Center Press, 1994), pp. 231–255; Jose Goldemberg and Harold A. Feiveson, "Denuclearization in Argentina and Brazil," *Arms Control Today*, Vol. 24, No. 2 (March 1994), pp. 10–14; Reiss, *Bridled Ambition*, pp. 45–88; and John R. Redick, Julio C. Carasales, and Paulo S. Wrobel, "Nuclear Rapprochement: Argentina, Brazil, and the Nonproliferation Regime," *Washington Quarterly*, Vol. 18, No. 1 (Winter 1995), pp. 107–122.

sion of nuclear weapons less necessary.[18] In short, from a realist's perspective, nuclear restraint is caused by the absence of the fundamental military threats that produce positive proliferation decisions.

POLICY IMPLICATIONS OF THE SECURITY MODEL

Several basic predictions and prescriptions flow naturally from the logic of the security model. First, since states that face nuclear adversaries will eventually develop their own arsenals unless they secure credible alliance guarantees from a nuclear power, it is essential that the United States maintain its nuclear commitments to key allies, including some form of continued first-use policy.[19] Other efforts to enhance the security of potential proliferators—such as confidence-building measures or "negative security assurances" that the nuclear states will not use their weapons against non-nuclear states—can also be helpful in the short run, but will likely not be effective in the long term since the anarchic international system forces states to be suspicious of potential rivals.

Under the security model's logic, the NPT is seen as an institution permitting non-nuclear states to overcome a collective action problem. Non-nuclear states in the NPT would prefer to become the only nuclear weapons power in their regions, but since that is an unlikely outcome if one state develops a nuclear arsenal, each is willing to refrain from proliferation if, and only if, its neighbors remain non-nuclear. The treaty permits such states to exercise restraint with increased confidence that their neighbors will follow suit, or will at least give sufficient advance warning before a withdrawal from the regime. It follows, from this logic, that other elements of the NPT regime should be considered far less important. Specifically, the commitments that the United States and other nuclear states made under Article VI of the treaty—that they will pursue "negotiations in good faith on measures relating to cessation of the nuclear arms race at an early date and to nuclear disarmament"—are

---

18.   Sherman Garnett writes, for example, that "for many Ukrainian citizens—not just the ethnic Russians—it is difficult to conceive of Russia as an enemy to be deterred with nuclear weapons." Sherman W. Garnett, "Ukraine's Decision to Join the NPT," *Arms Control Today*, Vol. 25, No. 1 (January 1995), p. 8. Garnett also maintains that "the role that security assurances played in the creation of a framework for Ukrainian denuclearization is obvious. They were of immense importance." Sherman W. Garnett, "The Role of Security Assurances in Ukrainian Denuclearization," in Virginia Foran, ed., *Missed Opportunities? The Role of Security Assurances in Nuclear Non-Proliferation* (Washington, D.C.: Carnegie Endowment for International Peace, 1997).

19.   See Lewis Dunn, *Controlling the Bomb* (New Haven, Conn.: Yale University Press, 1982); May, "Nuclear Weapons Supply and Demand," p. 535; and Frankel, "The Brooding Shadow," pp. 47–54.

merely sops to public opinion in non-nuclear countries. The degree to which the nuclear states follow through on these Article VI commitments will not significantly influence the actual behavior of non-nuclear states, since it will not change their security status.

Under realist logic, however, U.S. nonproliferation policy can only slow, not eliminate, the spread of nuclear weapons. Efforts to slow the process may be useful, but they will eventually be countered by two very strong structural forces that create an inexorable momentum toward a world of numerous nuclear weapons states. First, the end of the Cold War created a more uncertain multipolar world in which U.S. nuclear guarantees will be considered increasingly less reliable; second, each time one state develops nuclear weapons, it will increase the strategic incentives for neighboring states to follow suit.[20]

PROBLEMS AND EVIDENCE

What's wrong with this picture? The security model is parsimonious; the resulting history is conceptually clear; and the theory fits our intuitive belief that important events in history (like the development of a nuclear weapon) must have equally important causes (like national security threats). A major problem exists, however, concerning the evidence. The realist history depends primarily on the statements of motivation by the key governmental leaders, who have a vested interest in explaining that the choices they made served the national interest, and on a correlation in time between the emergence of a plausible security threat and a decision to develop nuclear weapons. Indeed, many realist case studies observe a nuclear weapons decision and then work backwards, looking for the national security threat that "must" have caused the decision. Similarly, realist scholars too often observe a state decision not to have nuclear weapons and then work backwards to find the change in the international environment that "must" have led the government to believe that threats to national security were radically decreasing.

A deeper analysis opens up the black box of decision-making and reveals in more detail how governments actually made their nuclear decisions. Moreover, any rigorous attempt to evaluate the security model of proliferation also requires an effort to develop alternative explanations, and to assess whether they provide more or less compelling explanations for proliferation decisions. The following sections therefore develop a

---

20. See Kenneth N. Waltz, "The Emerging Structure of International Politics," *International Security*, Vol. 18, No. 2 (Fall 1993), pp. 44–79; and John J. Mearsheimer, "Back to the Future: Instability in Europe after the Cold War," *International Security*, Vol. 15, No. 1 (Summer 1990), pp. 5–56.

domestic politics model and a norms model of proliferation and evaluate the explanations that flow from their logic, versus the security model's arguments offered above, for some important cases of both nuclear proliferation and nuclear restraint.

The Domestic Politics Model: Nuclear Pork and Parochial Interests

A second model of nuclear weapons proliferation focuses on the domestic actors who encourage or discourage governments from pursuing the bomb. Whether or not the acquisition of nuclear weapons serves the national interests of a state, it is likely to serve the parochial interests of at least some individual actors within the state. Three kinds of actors are often important in proliferation decisions: the state's nuclear energy establishment (which includes officials in state-run laboratories as well as in civilian reactor facilities); important units within the professional military (often within the air force, though sometimes in navy bureaucracies interested in nuclear propulsion); and politicians in states where individual parties or the mass public strongly favor nuclear weapons acquisition. When such actors form coalitions that are strong enough to control the government's decision-making process—either through their direct political power or indirectly through their control of information—nuclear weapons programs are likely to thrive.

Unfortunately, there is no well-developed domestic political theory of nuclear weapons proliferation that identifies the conditions under which such coalitions are formed and become powerful enough to produce their preferred outcomes.[21] The basic logic of this approach, however, has been strongly influenced by the literature on bureaucratic politics and military technology decision-making in the United States and the Soviet Union during the Cold War.[22] In this literature, bureaucratic actors are not seen as passive recipients of top-down political decisions; instead,

---

21. This is a serious weakness shared by many domestic-level theories in international relations, not just theories of proliferation. On this issue, see Ethan B. Kapstein, "Is Realism Dead? The Domestic Sources of International Politics," *International Organization*, Vol. 49, No. 4 (Autumn 1995), pp. 751–774.

22. The best examples of this literature include Morton H. Halperin, *Bureaucratic Politics and Foreign Policy* (Washington, D.C.: The Brookings Institution, 1974); Matthew Evangelista, *Innovation and the Arms Race: How the United States and Soviet Union Develop New Military Technologies* (Ithaca, N.Y.: Cornell University Press, 1988); and Donald MacKenzie, *Inventing Accuracy: A Historical Sociology of Nuclear Missile Guidance* (Cambridge, Mass.: MIT Press, 1990). For a valuable effort to apply insights from the literature on the social construction of technology to proliferation problems, see Steven Flank, "Exploding the Black Box: The Historical Sociology of Nuclear Proliferation," *Security Studies*, Vol. 3, No. 2 (Winter 1993/94), pp. 259–294.

they create the conditions that favor weapons acquisition by encouraging extreme perceptions of foreign threats, promoting supportive politicians, and actively lobbying for increased defense spending. This bottom-up view focuses on the formation of coalitions within a state's scientific- military-industrial complex. The initial ideas for individual weapons innovations are often developed inside state laboratories, where scientists favor military innovation simply because it is technically exciting and keeps money and prestige flowing to their laboratories. Such scientists are then able to find, or even create, sponsors in the professional military whose bureaucratic interests and specific military responsibilities lead them also to favor the particular weapons system. Finally, such a coalition builds broader political support within the executive or legislative branches by shaping perceptions about the costs and benefits of weapons programs.

Of course, realists also recognize that domestic political actors have parochial interests, but they argue that such interests have only a marginal influence on crucial national security issues. The outcome of bureaucratic battles, for example, may well determine whether a state builds 900 or 1,000 ICBMs (intercontinental ballistic missiles) or emphasizes submarines or strategic bombers in its nuclear arsenal; but a strong consensus among domestic actors will soon emerge about the basic need to respond in kind when a potential adversary acquires nuclear weapons. In contrast, from the domestic politics perspective nuclear weapons programs are not obvious or inevitable solutions to international security problems; instead, nuclear weapons programs are solutions looking for a problem to justify their existence. In this model, international threats certainly exist, but are seen as more malleable and more subject to interpretation, and can therefore produce a variety of responses from domestic actors. Security threats are not the central cause of weapons decisions according to this model: they are merely windows of opportunity through which parochial interests can jump.

PROLIFERATION REVISITED: ADDRESSING THE INDIA PUZZLES

The historical case that most strongly fits the domestic politics model is the Indian nuclear weapons experience. A focus on domestic politics in New Delhi, rather than on the strategic threats posed by China and Pakistan, leads to a very different interpretation of India's nuclear weapons decisions. This is true both of India's first test of a "peaceful nuclear explosive" in 1974 and its May 1998 nuclear "weaponization" tests.

In contrast to the brief realist's account outlined above, a closer look at the genesis of the Indian bomb program reveals that there was no consensus among officials in New Delhi that it was necessary to have a nuclear deterrent as a response to China's 1964 nuclear test. According to realist logic, a consensus should have existed and produced one of two events. First, a crash weapons program could have been initiated; there is no evidence that such an emergency program was started, however, and indeed, given the relatively advanced state of Indian nuclear energy at the time, such an effort could have produced a nuclear weapon by the mid- to late 1960s, relatively soon after the Chinese test, instead of in 1974.[23] Second, leaders in New Delhi could have made a concerted effort to acquire nuclear guarantees from the United States, the Soviet Union, or other nuclear powers. Indian officials, however, did not adopt a consistent policy to pursue security guarantees: in diplomatic discussions after the Chinese test, officials rejected the idea of bilateral guarantees because they would not conform with India's nonaligned status; refused to consider allowing foreign troops in India to support a nuclear commitment; and publicly questioned whether any multilateral or bilateral guarantee could possibly be credible.[24]

Instead of producing a united Indian effort to acquire a nuclear deterrent, the Chinese nuclear test produced a prolonged bureaucratic battle, fought inside the New Delhi political elite and nuclear energy establishment, between actors who wanted India to develop a nuclear weapons capability as soon as possible and other actors who opposed an Indian bomb and supported global nuclear disarmament and later Indian membership in the NPT. Soon after the Chinese nuclear test, for example, Prime Minister Lal Bahadur Shastri argued against developing an Indian atomic arsenal, in part because the estimated costs of $42–84 million were deemed excessive; Homi Bhabba, the head of the Atomic Energy Commission (AEC), however, loudly lobbied for the development of nuclear weapons capability, claiming that India could develop a bomb in eighteen months and that an arsenal of fifty atomic bombs would cost less than $21 million (a figure that excluded the construction of reactors, separation plants, and the opportunity costs of diverting scientists from development projects).[25] Although Shastri continued to oppose weapons devel-

---

23. In 1963, U.S. intelligence agencies estimated that India could test a nuclear weapon in four to five years (1967 or 1968). By 1965, U.S. estimates were that it would take one to three additional years. See Peter R. Lavoy, "Nuclear Myths and the Causes of Proliferation," in Davis and Frankel, *The Proliferation Puzzle*, p. 202; and George Bunn, *Arms Control by Committee: Managing Negotiations with the Russians* (Stanford, Calif.: Stanford University Press, 1992), p. 68.

24. See A.G. Noorani, "India's Quest for a Nuclear Guarantee," *Asian Survey*, Vol. 7, No. 7 (July 1967), pp. 490–502.

25. Frank E. Couper, "Indian Party Conflict on the Issue of Atomic Weapons," *Journal of Developing Areas*, Vol. 3, No. 2 (January 1969), pp. 192–193. Also see Lavoy, "Nuclear Myths and the Causes of Proliferation," p. 201.

opment and rebuked legislators in congressional debates for quoting Bhabba's excessively optimistic cost estimates, he compromised with the pro-bomb members of the Congress party and the AEC leadership, agreeing to create a classified project to develop the ability to detonate a PNE within six months of any final political decision.[26] However, even this compromise was short-lived, as Bhabba's successor at the AEC, Vikram Sarabhai, opposed the development of any Indian nuclear explosives, whether they were called PNEs or bombs, and ordered a halt to the PNE preparation program.[27]

After Sarabhai's death in 1971, the pro-bomb scientists in the AEC began to lobby Prime Minister Indira Gandhi, and developed an alliance with defense laboratories whose participation was needed to fabricate the explosive lenses for a nuclear test.[28] Unfortunately, firm evidence on why Gandhi decided to approve the scientists' recommendation to build and test a "peaceful" Indian nuclear device does not exist: indeed, even nuclear scientists who pushed for the May 1974 test now acknowledge that it is impossible to know whether Gandhi was primarily responding to domestic motives—she neither asked questions at the critical secret meetings in early 1974 nor explained why she approved their PNE recommendations.[29] Three observations about the decision, however, do suggest that domestic political concerns, rather than international security threats, were paramount. First, the decision was made by Prime Minister Gandhi, with the advice of a very small circle of personal advisers and scientists from the nuclear establishment. Senior defense and foreign affairs officials in India were not involved in the initial decision to prepare the nuclear device, nor in the final decision to test it. The military services were not asked how nuclear weapons would affect their war plans and military doctrines; the Defense Minister was reportedly informed of, but not consulted about, the final test decision only ten days before the May 18 explosion; the Foreign Minister was merely given

26. See Shyam Bhatia, *India's Nuclear Bomb* (Ghaziabad: Vikas Publishing House, 1979), pp. 120–122. The director of the PNE study later wrote that "getting the Prime Minister to agree to this venture must have required great persuasion, as Shastriji was opposed to the idea of atomic explosions of any kind." Raja Ramanna, *Years of Pilgrimage: An Autobiography* (New Delhi: Viking, 1991), p. 74.

27. See Kapur, *India's Nuclear Option*, p. 195; Mitchell Reiss, *Without the Bomb: The Politics of Nuclear Nonproliferation* (New York: Columbia University Press, 1988), p. 221 and p. 325, n. 42; and Ramanna, *Years of Pilgrimage*, p. 75.

28. Ramanna, *Years of Pilgrimage*, p. 89.

29. See Perkovich, *India's Nuclear Bomb*; and Ramanna, *Years of Pilgrimage*, p. 89.

a forty-eight-hour notice of the detonation.[30] This pattern suggests that security arguments were of secondary importance, and at a minimum, were not thoroughly analyzed or debated before the nuclear test.

Second, the subsequent absence of a systematic program for either nuclear weapons or PNE development and testing, and New Delhi's lack of preparedness for Canada's immediate termination of nuclear assistance, suggest that the decision was taken quickly, even in haste, and thus may have focused more on immediate political concerns rather than on longer-term security or energy interests.

Third, domestic support for the Gandhi government had fallen to an all-time low in late 1973 and early 1974 due to a prolonged and severe domestic recession, the eruption of large-scale riots in a number of regions, and the lingering effects of the splintering of the ruling Congress Party. From a domestic politics perspective, it would be highly surprising for a politician with such problems to resist what she knew was a major opportunity to increase her standing in public opinion polls and to defuse an issue about which she had been criticized by her domestic opponents.[31] Indeed, the domestic consequences of the test were very rewarding: the nuclear detonation occurred during the government's unprecedented crackdown on striking railroad workers and contributed to a major, albeit short-lived, increase in support for the Gandhi government.[32]

---

30. See Neil H.A. Joeck, "Nuclear Proliferation and National Security in India and Pakistan," (Ph.D. dissertation, University of California, Los Angeles, 1986), p. 229; and Kapur, *India's Nuclear Option*, p. 198.

31. Although Gandhi denied, in a later interview, that domestic concerns influenced her 1974 decision, she did acknowledge that the nuclear test "would have been useful for elections." See Rodney W. Jones, "India," in Jozef Goldblat, ed., *Non-Proliferation: The Why and the Wherefore* (London: Taylor and Francis, 1985), p. 114.

32. For example, Indian public opinion polls taken in June 1974 reported that a full 91 percent of the literate adult population knew about the explosion and 90 percent of those individuals answered in the affirmative when asked if they were "personally proud of this achievement." The overall result was that public support for Gandhi increased by one-third in the month after the nuclear test, according to the Indian Institute of Public Opinion, leading the Institute to conclude that "both she and the Congress Party have been restored to the nation's confidence." The Institute's analysis was that the increase was the result of both "the demonstration of India's atomic capability and the decisive action on the Railway strike," though the data outlined above suggest that more emphasis should be placed on the weapons test. See "The Prime Minister's Popularity: June 1974," and "Indian Public Opinion and the Railway Strike," in *Monthly Public Opinion Surveys* (Indian Institute of Public Opinion), Vol. 19, No. 8 (May 1974), pp. 5–6 and pp. 7–11; and "Public Opinion on India's Nuclear Device," *Monthly Public Opinion Surveys*, Vol. 19, No. 9 (June 1974), Blue Supplement, pp. III–IV.

Similarly, the Indian government's decision to test five nuclear weapons in May 1998 should be seen in the light of domestic politics in New Delhi. For more than twenty years after the 1974 PNE explosion, Indian politicians had refrained from testing nuclear weapons, despite widespread public support for the Indian "nuclear option" and "weaponization" through further tests.[33] Throughout this period, many government officials and Indian defense analysts argued that India's national security would indeed be harmed by nuclear weapons tests, since that would simply compel Pakistan to conduct nuclear tests. How does one therefore explain the momentous shift of policy in 1998?

The best explanation focuses on important changes in domestic politics rather than developments in international relations in South Asia in the late 1980s. Although Indian political authorities often cited the need to deter the Chinese nuclear threat as the main purpose of the May 1998 tests, the degree to which Sino-Indian relations had improved during the 1990s raises doubts about this rationale.[34] Indeed, China and India signed a bilateral agreement in 1993 to maintain peaceful relations in disputed border areas, implemented a set of confidence-building measures, and negotiated a set of agreements in 1995 and 1996 for mutual reductions of military units in those regions.[35] There were certainly continuing tensions in Sino-Indian relations, but no major deterioration in the military balance or the political relationship had occurred prior to the May 1998 nuclear tests.

What had occurred just prior to the tests was the March 1998 election, which brought into office an eighteen-party coalition government led by the Hindu nationalist party, the Bharatiya Janata Party (BJP). This government was highly unstable, however, and a number of coalition partner parties threatened to leave the coalition soon after Prime Minister Atal Vajpayee took office. The nuclear tests were tremendously popular throughout India: 86 percent of those polled in May 1998 supported the test and 44 percent stated that they were now more likely to vote for the

---

33. See David Cortright and Amitabh Matoo, *India and the Bomb: Public Opinion and Nuclear Options* (Notre Dame, Ind.: University of Notre Dame Press, 1996).

34. See, for example, the comments of Defense Minister George Fernandes in "China and Pakistan Present a Collaborated Threat," *The Times of India*, April 12, 1998 (www.timesofindia.com/120498/12edit4.htm).

35. See the discussion in Hua Han, "Sino-Indian Relations and Nuclear Arms Control," in Eric Arnett, ed., *Nuclear Weapons and Arms Control in South Asia After the Test Ban*, SIPRI Research Report No. 14, Stockholm International Peace Research Institute (Oxford: Oxford University Press, 1998), pp. 35–52; and Gaurav Kampani, "From Existential Deterrence to Minimum Deterrence," *Nonproliferation Review*, Vol. 6, No. 1 (Fall 1998), pp. 12–24.

BJP in any forthcoming election.[36] The resulting shift in public opinion silenced the BJP's internal critics, at least temporarily, and may have broadened the base of support for the BJP over the longer term, by adding the appeal of scientific progress or "nuclear nationalism" to its less popular Hindu nationalist platform.[37]

This domestic politics explanation is predicated on the existence of a widespread belief among Indian voters that the state's ability to build a nuclear weapon is a sign of its scientific prowess. Yet, as noted before, it is estimated that some thirty countries today could quickly develop nuclear weapons should they choose to do so. Ironically, the very success of the NPT has thus produced one counterproductive result: New Delhi officials can proudly proclaim that India's scientists have made India one of only eight nuclear weapons states, but only because the NPT has helped reduce the incentives to "go nuclear" in many other states that have more advanced technological capabilities.

DEVELOPMENT AND DENUCLEARIZATION: SOUTH AFRICA REVISITED

The domestic politics model predicts that reversals of weapons decisions occur not when external threats are diminished, but rather when there are major internal political changes. There are a number of reasons why internal changes could produce restraint: a new government has an opportunity to change course more easily because it can blame the failed policies of the previous regime; actors with parochial interests in favor of weapons programs may lose internal struggles to newly empowered actors with other interests; and the outgoing government may fear that the incoming government would not be a reliable custodian of nuclear weapons. Each of these domestic pathways to restraint can be relatively independent of changes in international security threats.

Such forces can be seen when one reexamines the history of South Africa's weapons program with a focus on domestic political interests rather than national security. For example, President de Klerk's public explanation for the program stressed that it was caused by the need to deter "a Soviet expansionist threat to Southern Africa," especially after Cuban military forces intervened in Angola in October 1975. Yet the preliminary research needed to develop nuclear devices was started inside South Africa's Atomic Energy Board in 1971, on the independent authority of the Minister of Mines. A non-nuclear scale model of a gun-

---

36. The polling data are reported in "Solid Support," *India Today*, May 25, 1998.

37. A thorough analysis of the domestic political factors influencing the Indian decision appears in Perkovich, *India's Nuclear Bomb.* Also see Kalpana Sharma, "The Hindu Bomb," *Bulletin of Atomic Scientists*, Vol 54, No. 4 (July/August 1998), pp. 30–33.

type explosive device was secretly tested in May 1974, and later in 1974, after the results of this test were known, Prime Minister John Voster approved plans to construct a small number of explosive devices and to build a secret testing site in the Kalahari desert.[38] Such evidence strongly supports the claims of South African scientists that the nuclear program was originally designed to produce PNEs, and that it was championed within the government by the South African nuclear power and mining industries to enhance their standing in international scientific circles and to be utilized in mining situations.[39]

This explanation for the origin of the nuclear program helps to explain South African nuclear doctrine, which otherwise appears so strange, as a post hoc development used to exploit devices that were originally developed for other purposes. (Testing a nuclear device in the event of a Soviet invasion might *reduce* the likelihood of U.S. intervention and would raise great risks that the Soviet Union would use nuclear weapons.) Senior officials in the program have stated, for example, that the military was not consulted about the bomb design and that operational considerations, such as the size and weight of the devices, were not taken into account.[40] As a result, the first South African nuclear device was actually too large to be deliverable by an aircraft and had to be redesigned because it did not meet the safety and reliability standards set by Armscor, the engineering organization run by the South African military, which took over the nuclear program in 1978.[41]

The timing and details of actions concerning South Africa's decision to dismantle and destroy its bomb stockpile also suggest that domestic political considerations were critical. In September 1989, before the Cold War was unambiguously over (the Berlin Wall fell in November 1989), de Klerk was elected president and immediately requested a high-level report on the possibility of dismantling the six nuclear devices. Officials in South Africa considered de Klerk's request as a sign that he had already decided to abandon the weapons program. Although possible

---

38. See the chronology in Reiss, *Bridled Ambition*, p. 8 and p. 27; and Waldo Stumpf, "South Africa's Nuclear Weapons Program: From Deterrence to Dismantlement," *Arms Control Today*, Vol. 25, No. 10 (December 1995/January 1996), p. 4. Also see David Fischer, "South Africa," in Reiss and Litwak, *Nuclear Proliferation After the Cold War*, p. 208; and David Albright, "South Africa's Secret Nuclear Weapons," *ISIS Report* (Washington, D.C.: Institute for Science and International Security, May 1994), pp. 6–8.

39. See Mark Hibbs, "South Africa's Secret Nuclear Program: From a PNE to a Deterrent," *Nuclear Fuel* (May 10, 1993), pp. 3–6; and Stumpf, "South Africa's Nuclear Weapons Program," p. 4.

40. See Reiss, *Bridled Ambitions*, p. 12.

41. Albright, "South Africa's Secret Nuclear Weapons," p. 10.

concerns about who would inherit nuclear weapons are rarely discussed in the public rationales for the dismantlement decision, the de Klerk government's actions spoke more loudly than its words: the weapons components were dismantled *before* IAEA inspections could be held to verify the activities, and all the nuclear program's plans, history of decisions, and approval and design documents were burned prior to the public announcement of the program's existence. This highly unusual step strongly suggests that fear of African National Congress control of nuclear weapons (and perhaps also concern about possible seizure by white extremists) was critical in the decision.[42]

Domestic politics can also be seen as playing a critical role in other cases of nuclear restraint. In Argentina and Brazil, for example, the key change explaining the shift from nuclear competition to cooperative restraint in the 1980s could not have been a major reduction of security threats, since there was no such reduction. Indeed, a traditional realist view would predict that the experience of the 1982 Falklands War—in which Argentina was defeated by a nuclear power, Great Britain—would have strongly encouraged Argentina's nuclear ambitions. Instead, the important change was the emergence of liberalizing domestic regimes in both states, governments supported by coalitions of actors—such as banks, export-oriented firms, and state monetary agencies—that value unimpeded access to international markets and oppose economically unproductive defense and energy enterprises. Nuclear programs that were run as fiefdoms and served the interests of the atomic industry bureaucrats and the military were therefore abandoned by new civilian regimes with strong support by liberalizing coalitions.[43]

POLICY IMPLICATIONS OF THE DOMESTIC POLITICS MODEL

For U.S. nonproliferation policy, the domestic politics approach both cautions modest expectations about U.S. influence and calls for a broader set of diplomatic efforts. Modest expectations are in order, since the domestic factors that influence decisions are largely outside the control of U.S. policy. Nevertheless, a more diverse set of tools could be useful

---

42. A rare public hint that concerns about domestic stability played a role in the decision is the acknowledgment by the head of the Atomic Energy Corporation that the government discussed issuing an immediate announcement revealing the existence of the weapons and thus permitting the IAEA to dismantle them. They rejected this plan, however, because "the state of the country's internal political transformation was not considered conducive to such an announcement at the time." See Stumpf, "South Africa's Nuclear Weapons Program," p. 7.

43. The best analysis is Etel Solingen, "The Political Economy of Nuclear Restraint," *International Security*, Vol. 19, No. 2 (Fall 1994), pp. 126–169.

to help create and empower domestic coalitions that oppose the development or maintenance of nuclear arsenals.

A variety of activities could be included in such a nonproliferation strategy focused on influencing domestic debates. International financial institutions are already demanding that cuts in military expenditures be included in conditionality packages for aid recipients. More direct conditionality linkages to nuclear programs—such as deducting the estimated budget of any suspect research and development program from the International Monetary Fund or U.S. loans to a country—could heighten domestic opposition to such programs.[44] Providing technical information and intellectual ammunition for domestic actors—by encouraging more accurate estimates of the economic and environmental costs of nuclear weapons programs and highlighting the risks of nuclear accidents—could bring new members into antiproliferation coalitions.[45] In addition, efforts to encourage strict civilian control of the military, through educational and organizational reforms, could be productive, especially in states where the military has the capability to create secret nuclear programs to serve its parochial interests (like Brazil in the 1980s). Finally, U.S. attempts to provide alternative sources of employment and prestige to domestic actors who might otherwise find weapons programs attractive could decrease nuclear incentives. To the degree that professional military organizations support nuclear proliferation, encouraging their involvement in other military activities (such as Pakistani participation in peacekeeping operations or the Argentine Navy's role in the Persian Gulf) could decrease such support. Where the key actors are laboratory officials and scientists, assistance in non-nuclear research and development programs (as in the U.S.-Russian "lab-to-lab" program) could decrease personal and organizational incentives for weapons research.

A different perspective on the role of the NPT also emerges from the domestic politics model. The NPT regime is not just a device to increase states' confidence about the limits of their potential adversaries' nuclear programs; it is also a tool that can help empower domestic actors who are opposed to nuclear weapons development. The NPT negotiations

---

44. Etel Solingen, *The Domestic Sources of Nuclear Postures*, Institute of Global Conflict and Cooperation, Policy Paper No. 8 (October 1994), p. 11.

45. On these costs and risks, see Kathleen C. Bailey, ed., *Weapons of Mass Destruction: Costs Versus Benefits* (New Delhi: Manohar Publishers, 1994); Stephen I. Schwartz, "Four Trillion and Counting," *Bulletin of the Atomic Scientists*, Vol. 51, No. 6 (November/December 1995); Bruce G. Blair, *The Logic of Accidental Nuclear War* (Washington, D.C.: The Brookings Institution, 1993); and Scott D. Sagan, *The Limits of Safety: Organizations, Accidents, and Nuclear Weapons* (Princeton, N.J.: Princeton University Press, 1993).

and review conferences create a well-placed elite in the foreign and defense ministries with considerable bureaucratic and personal interests in maintaining the regime. The IAEA creates monitoring capabilities and enforcement incentives against unregulated activities within a state's own nuclear power organizations. The network of nongovernmental organizations built around the treaty supports similar antiproliferation pressure groups in each state.

According to this model, progress in arms control agreements among the five nuclear powers recognized by the NPT regime is important for nonproliferation because it can provide political cover at home for actors in potential and emerging nuclear powers to practice restraint. For example, the U.S. commitment under Article VI of the NPT to work for the eventual elimination of nuclear weapons is important because of the impact that the behavior of the United States and other nuclear powers can have on the domestic debates in non-nuclear states. Whether or not the United States originally signed Article VI merely to placate domestic opinion in non-nuclear states is not important; what is important is that the loss of this pacifying tool could influence outcomes in potential proliferators. In future debates inside such states, the arguments of antinuclear actors—that nuclear weapons programs do not serve the interests of their states—can be more easily countered by pro-bomb actors whenever they can point to specific actions of the nuclear powers that highlight these states' continued reliance on nuclear deterrence such as the maintenance of nuclear first-use doctrines.

The Norms Model: Nuclear Symbols and State Identity

A third model of nuclear weapons proliferation focuses on norms concerning weapons acquisition. In this model, nuclear decisions serve important symbolic functions, both shaping and reflecting a state's identity. State behavior is determined not by leaders' cold calculations about the national security interests or their parochial bureaucratic interests, but rather by deeper norms and shared beliefs about what actions are legitimate and appropriate in international relations.

Given the importance of the subject, and the large normative literature in ethics and law concerning the use of nuclear weapons, it is surprising that so little attention has been paid to "nuclear symbolism" and the development of international norms concerning the acquisition of nuclear weapons.[46] Sociologists and political scientists have studied

---

46. On nuclear ethics, see Joseph S. Nye, Jr., *Nuclear Ethics* (New York: Free Press, 1986); and Steven P. Lee, *Morality, Prudence, and Nuclear Weapons* (New York: Cam-

the emergence and influence of international norms in other substantive areas, however, and their insights can lead to a valuable alternative perspective on proliferation. Within sociology, the "new institutionalism" literature suggests that modern organizations and institutions often come to resemble each other (what is called institutional isomorphism) not because of competitive selection or rational learning but because institutions mimic each other.[47] These scholars emphasize the importance of roles, routines, and rituals: individuals and organizations may well have "interests," but such interests are shaped by the social roles actors are asked to play, are pursued according to habits and routines as much as through reasoned decisions, and are embedded in a social environment that promotes certain structures and behaviors as rational and legitimate and denigrates others as irrational and primitive.

From this sociological perspective, military organizations and their weapons can be envisioned as serving functions similar to those of flags, airlines, and Olympic teams: they are part of what modern states believe they have to possess to be legitimate, modern states. Air Malawi, Royal Nepal Airlines, and Air Myanmar were not created because they are cost-effective ways to travel nor because domestic pressure groups pushed for their development, but because government leaders believed that a national airline is something that modern states must have to have to be modern states. Very small and poor states, without a significant number of scientists, nevertheless have official government-sponsored science boards. From a new institutionalist perspective, such similarities are not the result of functional logic (actions designed to serve either international or domestic goals); they are the product of shared beliefs about what is legitimate and modern behavior.[48]

---

bridge University Press, 1993). For a recent analysis of legal restraints on the use of nuclear weapons, see Nicholas Rostow, "The World Health Organization, the International Court of Justice, and Nuclear Weapons," *Yale Journal of International Law*, Vol. 20, No. 1 (Winter 1995), pp. 151–185. For a rare analysis of the symbolism of nuclear weapons, see Robert Jervis, "The Symbolic Nature of Nuclear Politics," in Jervis, *The Meaning of the Nuclear Revolution* (Ithaca, N.Y.: Cornell University Press, 1989), pp. 174–225.

47. Among the most important sources are the essays collected in Walter W. Powell and Paul J. DiMaggio, eds., *The New Institutionalism in Organizational Analysis* (Chicago: University of Chicago Press, 1991); and John W. Meyer and W. Richard Scott, *Organizational Environments: Ritual and Rationality*, 2nd ed. (Newbury Park, Calif.: Sage Publications, 1992).

48. The most influential work developing new institutional approaches for international security issues is Peter J. Katzenstein, ed., *The Culture of National Security: Norms and Identity in World Politics* (New York: Columbia University Press, 1996). Also see

Within political science, a related literature has evolved concerning the development and spread of norms within international regimes. Although this norms perspective has rarely been applied to the proliferation problem, scholars have studied such important phenomena as the global spread of anticolonialism, the abolition of the African slave trade, the near-total elimination of piracy at sea, and constraints against the use of chemical weapons.[49] A diverse set of ideas emerging in this field is producing a valuable debate about the role of global norms, but not a well-developed theory about their causal influence. Still, as one would expect of political scientists, coercion and power are seen to play a more important role in spreading norms than in the sociologists' literature. Normative pressures may begin with the actions of entrepreneurial non-state actors, but their beliefs only have significant influence once powerful state actors join the cause. Religious and liberal opposition to slavery, for example, was clearly important in fueling U.S. and British leaders' preferences in the nineteenth century, but such views would not easily have become an international norm without the bayonets of the Army of the Potomac at Gettysburg or the ships of the British Navy patrolling the

Marc C. Suchman and Dana P. Eyre, "Military Procurement as Rational Myth: Notes on the Social Construction of Weapons Proliferation," *Sociological Forum*, Vol. 7, No. 1 (March 1992), pp. 137–161; Martha Finnemore, "International Organizations as Teachers of Norms: UNESCO and Science Policy," *International Organization*, Vol. 47, No. 4 (Autumn 1993), pp. 565–598; Francisco O. Ramirez and John Boli, "Global Patterns of Educational Institutionalization," in George M. Thomas, John W. Meyer, Francisco O. Ramirez, and John Boli, eds., *Institutional Structure: Constituting State, Society, and the Individual* (Newbury Park, Calif.: Sage Publications, 1987), pp. 150–172. For an excellent survey and critique, see Martha Finnemore, "Norms, Culture, and World Politics: Insights from Sociology's Institutionalism," *International Organization*, Vol. 50, No. 2 (Spring 1996), pp. 325–348.

49. For rare applications of the norms perspective to proliferation, see Harald Müller, "The Internalization of Principles, Norms, and Rules by Governments: The Case of Security Regimes," in Volker Rittberger, ed., *Regime Theory and International Relations* (Oxford: Clarendon Press, 1995), pp. 361–390; and Müller, "Maintaining Nonnuclear Weapon Status," in Regina Cowen Karp, ed., *Security With Nuclear Weapons?* (New York: Oxford University Press, 1991), pp. 301–339. Also see Robert H. Jackson, "The Weight of Ideas in Decolonization: Normative Change in International Relations," in Judith Goldstein and Robert O. Keohane, eds., *Ideas and Foreign Policy* (Ithaca, N.Y.: Cornell University Press, 1993), pp. 111–138; Neta C. Crawford, "Decolonization as an International Norm," in Laura W. Reed and Carl Kaysen, eds., *Emerging Norms of Justified Intervention* (Cambridge, Mass.: American Academy of Arts and Sciences, 1993), pp. 37–61; Ethan A. Nadelmann, "Global Prohibition Regimes: The Evolution of Norms in International Society," *International Organization*, Vol. 44, No. 4 (Autumn 1990), pp. 479–526; and Richard Price, "A Genealogy of the Chemical Weapons Taboo," *International Organization*, Vol. 49, No. 1 (Winter 1995), pp. 73–104.

high seas between Africa and Brazil.[50] Similarly, normative beliefs about chemical weapons were important in creating legal restrictions against their use in war; yet, the norm was significantly reinforced at critical moments by the fear of retaliation in kind and by the availability of other weapons that were believed by military leaders to be more effective on the battlefield.[51]

The sociologists' arguments highlight the possibility that nuclear weapons programs serve symbolic functions reflecting leaders' perceptions of appropriate and modern behavior. The political science literature reminds us, however, that such symbols are often contested and that the resulting norms are spread by power and coercion, and not by the strength of ideas alone. Both insights usefully illuminate the nuclear proliferation phenomenon. Existing norms concerning the nonacquisition of nuclear weapons (such as those embedded in the NPT) could not have been created without the strong support of the most powerful states in the international system, who believed that the norms served their narrow political interests. Yet, once that effort was successful, these norms shaped states' identities and expectations and even powerful actors became constrained by the norms they had created.[52] The history of nuclear proliferation is particularly interesting in this regard because a major discontinuity—a shift in nuclear norms—has emerged as the result of the NPT regime.

Although many individual case studies of nuclear weapons decisions mention the belief that nuclear acquisition will enhance the international prestige of the state, such prestige has been viewed simply as a reasonable, though diffuse, means used to enhance the state's international influence and security. What is missing from these analyses is an understanding of why and how actions are granted symbolic meaning: why are some nuclear weapons acts considered prestigious, while others produce opprobrium, and how do such beliefs change over time? Why, for example, was nuclear testing deemed prestigious and legitimate in the

---

50. Ethan Nadelman, who stresses this point about power, nonetheless adds that "even among the laggards, indeed especially among the laggards, the consciousness of being perceived as primitive and deviant surely weighed heavily in the decisions of local rulers to do away with slavery." Nadelman, "Global Prohibition Regimes," p. 497.

51. See Price, "A Genealogy of the Chemical Weapons Taboo"; and Jeffrey Legro, *Cooperation Under Fire: Anglo-German Restraint During World War II* (Ithaca, N.Y.: Cornell University Press, 1995), pp. 144–216.

52. For an excellent analysis of how such a process can work in other contexts, see Michael Byers, "Custom, Power, and the Power of Rules," *Michigan Journal of International Law,* Vol. 17, No. 1 (Fall 1995), pp. 109–180.

1960s, but is today considered illegitimate and irresponsible? The answer is that the NPT regime appears to have shifted the norm concerning what acts grant prestige and legitimacy from the 1960s notion of joining "the nuclear club" to the 1990s concept of joining "the club of nations adhering to the international nuclear agreements." Moreover, the salience of the norms that were made explicit in the NPT treaty has shifted over time. These arguments are perhaps best supported by contrasting two cases— France's decision to build and test nuclear weapons and the Ukraine's decision to give up its nuclear arsenal—in which perceptions of legitimacy and prestige appear to have had a major influence, albeit with very different outcomes.

PROLIFERATION REVISITED: FRENCH GRANDEUR AND WEAPONS POLICY

According to realist theory, the French decision to develop nuclear weapons has a very simple explanation: in the 1950s, the Soviet Union was a grave military threat to French national security, and the best alternative to building an independent arsenal—reliance on the U.S. nuclear guarantee to NATO—was ruled out after the Soviet development of a secure second-strike capability reduced the credibility of any U.S. nuclear first-use threats. According to this explanation, the need for a French arsenal was driven home by the 1956 Suez Crisis, when Paris was forced to withdraw its military intervention forces from Egypt after a nuclear threat from Russia and under U.S. economic pressure. "The Suez humiliation of 1956 was decisive," writes David Yost. "It was felt that a nuclear weapons capability would reduce France's dependence on the U.S. and her vulnerability to Soviet blackmail."[53] The central realist argument for French nuclear weapons was clearly expressed in the rhetorical question President Charles de Gaulle posed to President Dwight Eisenhower in 1959: "Will they [future U.S. presidents] take the risk of devastating American cities so that Berlin, Brussels and Paris might remain free?"[54]

This explanation of French nuclear policy does not stand up very well against either evidence or logic. Indeed, the two most critical decisions initiating the weapons program—Prime Minister Pierre Mendes-France's December 1954 decision to start a secret nuclear weapons research program inside the Commissariat à l'énergie atomique (CEA) and the May

---

53. David S. Yost, "France's Deterrent Posture and Security in Europe, Part I: Capabilities and Doctrine," Adelphi Paper No. 194 (London: International Institute for Strategic Studies [IISS], Winter 1984/85), p. 4. Also see Kohl, *French Nuclear Diplomacy,* p. 36.

54. Jean Lacouture, *De Gaulle: The Ruler 1945–1970* (New York: W.W. Norton, 1993), p. 421, as quoted in Thayer, "The Causes of Nuclear Proliferation," p. 489.

1955 authorization by the Ministry of Defense for funds to be transferred to the CEA for the development of a prototype weapon—predated the 1956 Suez Crisis.[55] In addition, it is by no means clear why French leaders would think that the traumatic Suez experience could have been avoided if there had been an independent French nuclear arsenal, since Great Britain had also been forced to withdraw from the intervention in Egypt under U.S. and Soviet pressure, despite its possession of nuclear weapons.[56] A simple exercise in comparative logic also raises doubts about the security model. If the critical cause of proliferation in France was the lack of credibility of U.S. nuclear guarantees given the growing Soviet threat in the mid-1950s, why then did not other nuclear-capable states in Europe, faced with similar security threats at the time, also develop nuclear weapons?[57] Of all the nuclear-capable states in Europe that were both threatened by Soviet military power and had reasons to doubt the credibility of the U.S. first-use pledge, France was the only state to develop a weapon; West Germany, the Netherlands, Italy, Switzerland, Belgium, Norway, and Sweden all restrained their nuclear programs. This presents a puzzle for the security model, since the Soviet Union's conventional and nuclear threat to most of these states' security was at least as great as its threat to France; the U.S. nuclear guarantee should not have been considered more credible by those states that had been U.S. enemies or neutrals in World War II, than by France, a U.S. ally of long standing, and one which the United States had strongly aided once it entered the war in 1941.

A stronger explanation for the French decision to build nuclear weapons emerges when one focuses on French leaders' perceptions of the bomb's symbolic significance. The belief that nuclear power and nuclear weapons were deeply linked to a state's position in the international system was present as early as 1951, when the first French five-year plan was put forward with its stated purpose being "to ensure that in 10 years' time France will still be an important country."[58] France emerged from World War II in an unusual position: it was a liberated victor whose military capabilities and international standing were not at all comparable to the power and status it had before the war. It should therefore not

---

55. See Bertrand Goldschmidt, *The Atomic Complex* (La Grange Park, Ill.: American Nuclear Society, 1982), p. 131; and Scheinman, *Atomic Energy Policy in France Under the Fourth Republic*, pp. 120–122.

56. Scheinman, *Atomic Energy Policy in France Under the Fourth Republic*, pp. 171–173.

57. The British acquisition of nuclear weapons in 1952 predated the Soviet development of a secure second-strike capability.

58. The document is quoted in Goldschmidt, *The Atomic Complex*, p. 126.

be surprising that the governments of both the Fourth and the Fifth Republics vigorously explored alternative means to return France to its historical great power status.[59] After the war, the initial French effort to restore its tarnished prestige focused on the fight to hold onto an overseas empire, yet as Michel Martin has nicely put it, "as the curtain was drawn over colonial domination, it became clear that the country's *grandeur* had to be nourished from other sources."[60]

After 1958, the Algerian crisis contributed greatly to Charles de Gaulle's obsession with nuclear weapons as the source of French grandeur and independence. In contrast, de Gaulle appeared less concerned about whether French nuclear forces could provide adequate deterrence against the Soviet military threat. For example, during both the Berlin crisis of 1958 (before the 1960 French nuclear weapons test) and the 1962 Cuban crisis (after the test, but before French nuclear forces were operational), de Gaulle expressed great confidence that the Soviets would not risk an attack on NATO Europe.[61] Wilfred Kohl also reports on a revealing incident in which a French military strategist sent de Gaulle a copy of a book on French nuclear doctrine and de Gaulle replied, "thanking the man for his interesting analysis of strategic questions, but stressing that for him the central and clearly the only important issue was: 'Will France remain France?'"[62] For de Gaulle, the atomic bomb was a dramatic symbol of French independence and was thus needed for France to continue to be seen, by itself and others, as a great power. He confided to Dwight Eisenhower in 1959:

A France without world responsibility would be unworthy of herself, especially in the eyes of Frenchmen. It is for this reason that she disapproves of NATO, which denies her a share in decision-making and which is confined to Europe. It is for this reason too that she intends to provide herself with an atomic armament. Only in this way can our defense and foreign policy be independent, which we prize above everything else.[63]

---

59. For detailed analyses of the French nuclear weapons decision that focus on political prestige as the central source of policy, see Scheinman, *Atomic Energy Policy in France Under the Fourth Republic;* and Kohl, *French Nuclear Diplomacy.* Also see Bundy, *Danger and Survival*, pp. 472–487, 499–503.

60. Michel L. Martin, *Warriors to Managers: The French Military Establishment Since 1945* (Chapel Hill, N.C.: University of North Carolina Press, 1981), p. 21.

61. See Philip H. Gordon, "Charles de Gaulle and the Nuclear Revolution," *Security Studies,* Vol. 5, No. 1 (Autumn 1995), pp. 129–130.

62. Kohl, *French Nuclear Diplomacy*, p. 150, quoted in Bundy, *Danger and Survival*, p. 502.

63. Charles de Gaulle, *Memoirs of Hope: Renewal and Endeavor* (New York: Simon and

When the French nuclear weapons arsenal is viewed as primarily serving symbolic functions, a number of puzzling aspects of the history of French atomic policy become more understandable. The repeated Gaullist declarations that French nuclear weapons should have world-wide capabilities and must be aimed in all directions ("*tous les azimuts*") are not seen as the product of security threats from all directions, but rather as the only policy that is logically consistent with global grandeur and independence. Similarly, the French strategic doctrine of "proportional deterrence" against the Soviet Union during the Cold War—threatening more limited destruction in a retaliatory strike than did the United States under its targeting doctrine—is not a product of France's geographical position or limited economic resources, but rather an indication that deterrence of the Soviet Union was a justification, and never the primary purpose of its arsenal. Finally, the profound French reluctance to stop nuclear testing in the mid-1990s is seen as being produced not only by the stated concerns about weapons modernization and warhead safety, but also because weapons tests were perceived by Parisian leaders as potent symbols of French identity and status as a great power.

RESTRAINT REVISITED: THE NPT AND THE UKRAINE CASE

Stark contrasts exist between French nuclear decisions in the 1950s and Ukrainian nuclear decisions in the mid-1990s. When the Soviet Union collapsed in 1991, an independent Ukraine was "born nuclear" with more than 4,000 nuclear weapons on or under its soil. In November 1994, the Rada—the Ukrainian parliament in Kiev—voted overwhelmingly to join the NPT as a non-nuclear state, and all weapons were removed from Ukrainian territory by June 1996.

This decision to give up a nuclear arsenal is puzzling from the realist perspective: a number of prominent realist scholars, after all, maintained that given the history of Russian expansionist behavior and continuing tensions over Crimea and the treatment of Russian minorities, Ukraine's independence was seriously threatened. They further argued that nuclear weapons were the only rational solution to this security threat.[64] The disarmament decision is also puzzling from a traditional domestic politics perspective. Despite the tragic consequences of the Chernobyl accident,

---

Schuster, 1971), p. 209, quoted in Yost, *France's Deterrent Posture and Security in Europe,* pp. 13–14.

64. See John J. Mearsheimer, "The Case for a Ukrainian Nuclear Deterrent," *Foreign Affairs,* Vol. 72, No. 3 (Summer 1993), pp. 50–66; and Barry R. Posen, "The Security Dilemma and Ethnic Conflict," *Survival,* Vol. 35, No. 1 (Spring 1993), pp. 44–45.

public opinion polls in Ukraine showed rapidly growing support for keeping nuclear weapons in 1992 and 1993: polls showed support for an independent arsenal increasing from 18 percent in May 1992 to 36 percent in March 1993, to as much as 45 percent in the summer of 1993.[65] In addition, well-known retired military officers, such as Rada member General Volodomyr Tolubko, vigorously lobbied to maintain an arsenal and senior political leaders, most importantly Prime Minister (then President) Leonid Kuchma, came from the Soviet missile-building industry and would not therefore be expected to take an antinuclear position.[66]

An understanding of Ukraine's decision to eliminate its nuclear arsenal requires that more attention be focused on the role that emerging NPT nonproliferation norms played in four critical ways. First, Ukrainian politicians initially adopted antinuclear positions as a way of buttressing Kiev's claims to national sovereignty. In one of its first efforts to assert a foreign policy independent of Moscow, Ukraine tried to accede to the NPT as a non-nuclear state in early 1990, attempting to use NPT membership to separate itself from the Soviet Union.[67] In July 1990, this policy was underscored when the Parliament in Kiev issued its Declaration of Sovereignty. Embedded in declarations about Ukraine's right to participate as a full member in all agreements concerning "international peace and security" was the proclamation that Ukraine would "become a neutral state that does not participate in military blocs and that adheres to three non-nuclear principles: not to maintain, produce, or acquire nuclear weapons." This extraordinary statement was an expedient designed to buttress Kiev's claim to independence from the Soviet Union, rather than a blueprint laying out Ukraine's long-term strategy: indeed, it was adopted by a vote of 355–4, without extensive debate, by the parliament in which conservative communists (many of whom would later take pro-nuclear positions) still held the majority of seats.[68] Nevertheless, the declaration placed the onus of reneging on an international commitment on the politicians and scholars who afterwards called for keeping an arsenal, and it is revealing that even many of the more hawkish analysts

---

65. See William C. Potter, "The Politics of Nuclear Renunciation: The Cases of Belarus, Kazakhstan, and Ukraine," Henry L. Stimson Center, Occasional Paper No. 22 (April 1995), p. 49.

66. For a detailed analysis, see Bohdan Nahaylo, "The Shaping of Ukrainian Attitudes Toward Nuclear Arms," *RFE/RL (Radio Free Europe/Radio Liberty) Research Report*, Vol. 2, No. 8 (February 19, 1993), pp. 21–45.

67. Potter, "The Politics of Nuclear Renunciation," p. 19.

68. See Nahaylo, "The Shaping of Ukrainian Attitudes," pp. 21–22.

thereafter defensively advocated keeping the arsenal on a temporary basis until other sources of security could be found.[69] Second, the strength of the NPT regime created a history in which the most recent examples of new or potential nuclear states were so-called "rogue states" such as North Korea, Iran, and Iraq. This was hardly a nuclear club whose new members would receive international prestige, and during the debate in Kiev, numerous pro-NPT Ukrainian officials insisted that the renunciation of nuclear weapons was now the best route to enhance Ukraine's international standing.[70] Third, economic pressures were clearly critical to the Ukrainian decision: the United States and NATO allies encouraged Kiev to give up the arsenal not by convincing officials that nuclear weapons could never serve as a military deterrent against Moscow, but by persuading them that not following the NPT norm would result in very negative economic consequences.[71] The ability of the West to coordinate such activities, and credibly to threaten collective sanctions and promised inducements for disarmament, were significantly heightened by the NPT norm against the creation of new nuclear weapons states. Fourth, the Kiev government and the Ukrainian public could more easily accept the economic inducements offered by the United States—such as Nunn-Lugar payments to help transport and destroy the weapons—with the belief that they were enabling Ukraine to keep an international commitment.

As with all counterfactuals, it is impossible to assess with certainty whether Ukraine would have divested itself of nuclear weapons had the NPT norms not been in existence. Still, it is valuable to try to imagine how much more difficult a disarmament outcome would have been in the absence of the NPT and its twenty-five-year history. Without the NPT, a policy of keeping a nuclear arsenal would have placed Ukraine in the category of France and China; with the NPT, it would place Ukraine in the company of dissenters like India and Pakistan and pariahs like Iraq and North Korea. International threats to eliminate economic aid and suspend political ties would be less credible, since individual states would be more likely to defect from an agreement. Finally, without the NPT norm, U.S. dismantlement assistance would have been seen in Kiev as the crass purchase of Ukrainian weapons by a foreign government,

---

69. Potter, "The Politics of Nuclear Renunciation," pp. 21–23; and Nahaylo, "The Shaping of Ukrainian Attitudes."

70. See Potter, "The Politics of Nuclear Renunciation," p. 44; and Garnett, "Ukraine's Decision to Join the NPT," p. 12.

71. An excellent analysis, of U.S. policy appears in Garnett, "Ukraine's Decision to Join the NPT," pp. 10–12.

instead of being viewed as friendly assistance to help Kiev implement an international agreement.

POLICY IMPLICATIONS OF THE NORMS MODEL

If the norms model of proliferation is correct, the key U.S. policy challenges are to recognize that such norms can have a strong influence on other states' nuclear weapons policies, and to adjust U.S. policies to increase the likelihood that norms will push others toward policies that also serve U.S. interests. Recognizing the possibility that norms can influence other states' behavior in complex ways should not be difficult. After all, the norms of the NPT have already influenced U.S. nuclear weapons policy in ways that few scholars or policymakers predicted ahead of time: in January 1995, for example, the Clinton administration abandoned the long-standing U.S. position that the Comprehensive Test Ban Treaty (CTBT) must include an automatic escape clause permitting states to withdraw from the treaty after ten years. Despite the arguments made by Pentagon officials that such a clause was necessary to protect U.S. security, the administration accepted the possibility of a permanent CTBT because senior decision-makers became convinced that the U.S. position was considered illegitimate by non-nuclear NPT members, due to the Article VI commitment to eventual disarmament, and might thereby jeopardize the effort to negotiate a permanent extension of the NPT treaty.[72]

Adjusting U.S. nuclear policies in the future to reinforce emerging nonproliferation norms will be difficult, however, because many of the recommended policies derived from the norms perspective directly contradict recommendations derived from the other models. A focus on NPT norms raises especially severe concerns about how the U.S. nuclear first-use doctrine influences potential proliferators' perceptions of the legitimacy or illegitimacy of nuclear weapons possession and use.[73] To the degree that such first-use policies create beliefs that making nuclear threats is what great powers do, they will become desired symbols for states that aspire to that status. The norms argument against the U.S. nuclear first-use doctrine, however, contradicts the policy advice derived from the security model, which stresses the need for continued nuclear

---

72. Douglas Jehl, "U.S. in New Pledge on Atom Test Ban," *New York Times,* January 31, 1995, p. 1; and Dunbar Lockwood, "U.S. Drops CTB 'Early Out' Plan; Test Moratorium May Be Permanent," *Arms Control Today,* Vol. 25, No. 2 (March 1995), p. 27.

73. On this issue, see Barry M. Blechman and Cathleen S. Fisher, "Phase Out the Bomb," *Foreign Policy,* No. 97 (Winter 1994–95), pp. 79–95; and Wolfgang K.H. Panofsky and George Bunn, "The Doctrine of the Nuclear-Weapons States and the Future of Non-Proliferation," *Arms Control Today,* Vol. 24, No. 6 (July/August 1994), pp. 3–9.

guarantees for U.S. allies. Similarly, the norms perspective suggests that current U.S. government efforts to maintain the threat of first use of nuclear weapons to deter the use of biological or chemical weapons would damage the nuclear nonproliferation regime.[74] Leaders of non-nuclear states are much less likely to think that their own acquisition of nuclear weapons to deter adversaries with chemical and biological weapons is illegitimate and ill-advised if the greatest conventional military power in the world can not refrain from making such threats.

Other possible policy initiatives are less problematic. For example, if norms concerning prestige are important, then it would be valuable for the United States to encourage the development of other sources of international prestige for current or potential proliferators. Thus, a policy that made permanent UN Security Council membership for Japan or Germany conditional upon the maintenance of non-nuclear status under the NPT might further remove nuclear weapons possession from considerations of international prestige.

Finally, the norms model leads to a more optimistic vision of the potential future of nonproliferation. Norms are sticky: individual and group beliefs about appropriate behavior change slowly, and over time norms can become rules embedded in domestic institutions.[75] In the short run, therefore, norms can be a brake on nuclear chain reactions: in contrast to more pessimistic realist predictions that "proliferation begets proliferation," the norms model suggests that normative constraints can deter or at least delay proliferation. The long-term future of the NPT regime is also viewed with more optimism, for the model envisions the possibility of a gradual emergence of a norm against all nuclear weapons possession. The development of such a norm may well have been inadvertent—for quite understandable reasons, the United States did not take its Article VI commitment to work in good faith for complete nuclear disarmament seriously during the Cold War. But to the degree that other states believe that such commitments are real and legitimate, their perceptions that the United States is backsliding away from Article VI will influence their behavior over time.

---

74. For contrasting views on this policy, see George Bunn, "Expanding Nuclear Options: Is the U.S. Negating its Non-Use Pledges?" *Arms Control Today*, Vol. 26, No. 4 (May/June 1996), pp. 7–10; and Gompert, Watman, and Wilkening, "Nuclear First Use Revisited."

75. For useful discussions, see Abram Chayes and Antonia Handler Chayes, *The New Sovereignty: Compliance with International Regulatory Agreements* (Cambridge, Mass.: Harvard University Press, 1995); and Andrew P. Cortell and James W. Davis, Jr., "How Do International Institutions Matter? The Domestic Impact of International Rules and Norms," *International Studies Quarterly*, Vol. 40, No. 4 (December 1996), pp. 451–478.

This emphasis on emerging norms therefore highlights the need for the nuclear powers to reaffirm their commitments to global nuclear disarmament, and suggests that it is essential that the United States and other governments develop a public, long-term strategy for the eventual elimination of nuclear weapons.[76] The norms model cannot, of course, predict whether such efforts will ever resolve the classic risks of nuclear disarmament: that states can break treaty obligations in crises, that small arsenals produce strategic instabilities, and that adequate verification of complete dismantlement is exceedingly difficult. But the model does predict that there will be severe costs involved if the nuclear powers are seen to have failed to make significant progress toward nuclear disarmament.

Causal Complexity and Policy Tradeoffs

The ideas and evidence presented in this chapter suggest that the widely held security model explanation for nuclear proliferation decisions is inadequate. A realist might well respond to this argument by asserting that evidence is always ambiguous in complex historical events, and that I underestimate foreign threats and thus provide a poor measure of the effects of security concerns on decision-makers. Moreover, it could be argued that the best theories are those that explain the largest number of cases and that the largest number of positive nuclear weapons decisions in the past (the United States, the Soviet Union, China, Israel, and perhaps Pakistan) and the majority of the most pressing proliferation cases today (Iraq, Libya, and possibly North Korea and Iran) appear to be best explained by the basic security model.

I have no quarrel with the argument that the largest number of past proliferation cases, and even many emerging ones, are best explained by the security model. But explaining the majority of cases should not be satisfying. The evidence presented above strongly suggests that multi-causality, rather than measurement error, lies at the heart of the nuclear proliferation problem. Nuclear weapons proliferation and nuclear restraint have occurred in the past, and can occur in the future, for more than one reason: different historical cases are best explained by different causal models.

---

76. For important efforts to rethink the elimination issue, see "An Evolving U.S. Nuclear Posture," Report of the Steering Committee of the Project on Eliminating Weapons of Mass Destruction, Henry L. Stimson Center, Washington, D.C., December 1995; and Donald MacKenzie and Graham Spinardi, "Tacit Knowledge, Weapons Design, and the Uninvention of Nuclear Weapons," *American Journal of Sociology*, Vol. 101, No. 1 (July 1995), pp. 44–100.

This argument has important implications for future scholarship on proliferation as well as for U.S. nonproliferation policy. The challenge for scholars is not to produce increasing numbers of detailed but atheoretical case studies of states' nuclear proliferation and restraint decisions; it is to produce theory-driven comparative studies to help determine the conditions under which different causal forces produced similar outcomes. Predicting the future based on such an understanding of the past will still be problematic, since the conditions that produced the past proliferation outcomes may themselves change. But future scholarship focusing on how different governments assess the nuclear potential and intention of neighbors, on why pro-bomb and anti-bomb domestic coalitions form and gain influence, and on when and how NPT norms about legitimate behavior constrain statesmen will be extremely important.[77]

For policymakers, the existence of three different reasons why states develop nuclear weapons suggests that no single policy can ameliorate all future proliferation problems. Fortunately, some of the policy recommendations derived from the models are compatible: for example, many of the diplomatic tools suggested by the domestic politics model, which attempts to reduce the power of individual parochial interests in favor of nuclear weapons, would not interfere with simultaneous efforts to address states' security concerns. Similarly, efforts to enhance the international status of some non-nuclear states need not either undercut deterrence or promote pro-nuclear advocates in those countries.

Unfortunately, other important recommendations from different models are more contradictory. Most importantly, a security-oriented strategy of maintaining a major role for U.S. nuclear guarantees to restrain proliferation among allies will eventually create strong tensions with a norms-oriented strategy seeking to delegitimize nuclear weapons use and acquisition. The final outcome of these alternative strategies, of course, is not under the control of the United States, as leaders of potential proliferators will decide for themselves whether to pursue or reject nuclear weapons programs.

U.S. policy will not be without influence, however, and intelligent decisions will not emerge if U.S. officials refuse to recognize that painful tradeoffs are appearing on the horizon. U.S. decision-makers will eventually have to choose between the difficult nonproliferation task of weaning allies away from nuclear guarantees without producing new nuclear states, and the equally difficult task of maintaining a norm against nuclear proliferation without the U.S. government's facing up to its logical consequences.

---

77. See the essays in Peter R. Lavoy, Scott D. Sagan, and James J. Wirtz, eds., *Planning the Unthinkable* (Ithaca, N.Y.: Cornell University Press, 2000).

Chapter 3

Universal Deterrence or Conceptual Collapse? Liberal Pessimism and Utopian Realism

Richard K. Betts

Weapons of mass destruction (WMD) tend to turn typical thinking upside down. For example, many strategic pundits argued during the Cold War that for nuclear weapons, "offense is defense, defense is offense. Killing people is good, killing weapons is bad."[1] More generally, the very existence of WMD makes those who are otherwise optimistic about the course of history cringe in horror, while it leaves some who are otherwise pessimistic quite cheerful. The former fear that no good can come from a growing number of bad things; the latter argue that a half-century of peace among nuclear powers proves the reverse.

Most U.S. citizens view international relations through the prism of the liberal tradition, which emphasizes the idea of progress from primitive conflict toward enlightened cooperation, a view reinforced by the end of the Cold War. Progress should obviate the quest for instruments of mass killing, not spur it. The spread of such instruments is a wrench in the works of the end of history. Thus most normal people in the United States see the diffusion of WMD as a dire threat to be contained and rolled back. On this issue, ironically, liberals prove profoundly conservative and elitist. They enshrine the status quo, seeking to freeze the international hierarchy indefinitely through treaties to prevent others from obtaining the weapons the United States has held for half a century and will continue to hold indefinitely, despite lofty rhetoric about abolition.

---

1. Axiom quoted in John Newhouse, *Cold Dawn: The Story of SALT* (New York: Holt, Rinehart and Winston, 1973), p. 176. The rationale was that offense dominance would ensure mutual assured destruction and thus prevent either side from striking; the development of defenses would mislead them to believe that war could be tolerable; the ability to target only population centers would make a first strike useless; and the capacity to target military forces would be a temptation to strike first. See also Robert Jervis, "Cooperation Under the Security Dilemma," *World Politics*, Vol. 30, No. 2 (January 1978).

The most Panglossian view, with equal irony, comes out of international relations theory's realist tradition. For the most part, realism is pessimistic about the international human condition. From Thucydides, through Machiavelli, Hobbes, Carr, and Morgenthau, this school has viewed international conflict as inevitable as long as there is no world government to prevent it. History is not progress toward world order and the rule of international law, but a cycle of conflict among states or groups. Hopes for collective security and lions lying down with lambs are utopian.

Those who identify with classical realism, with its emphasis on human fallibility, worry about mass destruction as much as liberals do. Out of the realist tradition, however, comes a schismatic coterie of academics led by Kenneth Waltz. More than classical realists, this group emphasizes the constraining effects of the international power structure and holds that nuclear weapons can produce the permanent peace that liberals have always believed in and realists have always said is impossible. In the paradoxical tradition of theories on nuclear weapons this minority might be dubbed utopian realists. They foresee the universalization of vulnerability compelling the universalization of military restraint.[2] The spread of murderous weapons will ensure that victims can destroy their attackers in retaliation, which in turn will mean that resorting to force will not serve anyone's interests. Nuclear weapons offer a technological fix to neutralize the traditional natural dynamics of the international political jungle. The specter of apocalypse becomes the guarantor of peace. This benign view derives from standard Cold War deterrence theory, and faith that national leaders are only irresponsible when stakes are not astronomical.

The utopian realists are right about the statistically probable effect of proliferation in any specific case, but the liberal pessimists are right about the ultimate general effect. The first exception to the utopian rule may upset all confidence in stable order and set off dangerous behavior. But while the liberal diagnosis is a better guide to the danger, realist norms—relying on incentives based on interest rather than on law or preachment—are a better guide to policy. Neither school, however, has consid-

---

2. The principal works making this argument are Kenneth N. Waltz, *The Spread of Nuclear Weapons: More May Be Better*, Adelphi Paper No. 171 (London: International Institute for Strategic Studies [IISS], Autumn 1981); Shai Feldman, *Israeli Nuclear Deterrence: A Strategy for the 1980s* (New York: Columbia University Press, 1982); John J. Mearsheimer, "The Case for a Ukrainian Nuclear Deterrent," *Foreign Affairs*, Vol. 72, No. 3 (Summer 1993); Bruce Bueno de Mesquita and William H. Riker, "An Assessment of the Merits of Selective Nuclear Proliferation," *Journal of Conflict Resolution*, Vol. 26, No. 2 (June 1982).

ered very carefully how much the effects of spreading chemical and biological WMD will parallel those of nuclear weapons.

How Soon and How Big a Problem?

WILL THE FUTURE BE LIKE THE PAST?

Not all fear the spread of capabilities for mass destruction, but few wish to spur it. Most believe that much greater proliferation will occur soon if some extraordinary solution is not found. Is that expectation warranted? Confidence in predicting proliferation should be tempered by the poor track record of past estimates.

At the beginning of the nuclear era, hubris caused most U.S. officials to underestimate how quickly the Soviet Union would get its own atom bomb. Since then, however, it has been more common to overestimate the rate of spread. John F. Kennedy's oft-cited remark about a world of twenty-five nuclear-armed states by the 1970s is typical. In later years it was also common to predict the rapid spread of chemical weapons (CW), the "poor man's atomic bomb." Proliferation of chemical weapons went into reverse after World War II. Only five or six countries still kept them by the early 1960s.[3] They did spread more widely than nuclear arms, however, in the latter part of the century. Unlike nuclear weapons, which were only ever used in one week in 1945, they are not bound up in a taboo, since they have actually been used a number of times over the years.

Since the 1950s, few observers have underestimated the rate at which nuclear and chemical weapons would spread (although the prospects for biological weapons programs have been underestimated in recent years). Past overestimates of "horizontal" nuclear proliferation (the spread of weapons to additional countries) do not mean there is no problem, but they raise a question about whether there is any consistent trend.[4] Though slower than anticipated, nuclear spread has occurred at the rate of one net entrant every six or seven years: as of 1998 there were eight

---

3. Edward M. Spiers, *Chemical and Biological Weapons: A Study of Proliferation* (London: Macmillan, 1994), p. 11.

4. Even "vertical" proliferation of nuclear weapons (the increase in numbers within the superpowers' inventories) was less consistent than folklore had it. Rather than an ever-upward spiral of procurement, the number of U.S. weapons, their aggregate megatonnage, and their total destructive power ("equivalent" megatonnage) actually declined significantly over the course of the 1960s. Albert Wohlstetter, "Racing Forward or Ambling Back?" in *Defending America* (New York: Basic Books, 1977), pp. 135–141. Much of this reflected retirement of air defense warheads, though, and deployment of multiple warheads on offensive missiles after the 1960s reversed the trend.

known nuclear weapons states fifty-three years after the first nuclear detonation. More interesting, the rate has not been accelerating. Six of the eight entered the club in the first half of the post-1945 period. The latest spate of hand-wringing that came with the Indian and Pakistani nuclear tests of 1998 obscured the fact that these were not new entrants to the club (India originally tested in 1974, and Pakistan had the capability at least since 1990), and that they are the *only* net entrants since Israel in the 1960s (South Africa entered but then left). For the second half of the nuclear era, the net rate of spread has been one every fifteen years.

Concern about the issue has also varied significantly. During the Cold War few hawkish strategists got exercised about proliferation; they were too busy worrying about Soviet capabilities and the East-West competition. Except for China's accession in 1964, they often pooh-poohed concern with proliferation, perhaps because all the other entrants were U.S. allies. Those who used to wring their hands the most about the proliferation of weapons of mass destruction were doves, who were viscerally disposed to regard any such weapons as bad, whoever had them, and who tended to focus on international legal solutions or export controls rather than on the strategic dimension of the issue.[5]

The track record of predictions suggests three questions: Why did most analysts think that the spread of WMD would be faster and wider than it has been? Will the reasons that the spread was slower than anticipated persist? Does the rate matter?

There are at least five reasons for overestimation. First, those interested in the question were those inclined to be worried about it. Second, because good news is no news, pessimists focus on the reasons that bad things could happen, rather than on the reasons that they might not. Third, at least in regard to nuclear weapons, the technical ease of procurement was sometimes exaggerated.[6] Fourth, in the 1970s it was mistakenly expected that commercial nuclear facilities for generating electricity would burgeon around the world, providing options to build weapons as a byproduct. Fifth, two strategic considerations were underestimated. During the Cold War most of the countries that could produce nuclear weapons were protected by a nuclear superpower ally who did not want them to do so. In addition, there have always been quite powerful disadvantages to acquiring nuclear weapons, costs that countries would

---

5. There were exceptions—well-known hawks such as Albert Wohlstetter and Fred Iklé—who worried about proliferation during the Cold War, but even they did most of their campaigning on the subject in the 1970s, the trough between intense phases of the East-West conflict.

6. A prime example is the popularity of John McPhee's *The Curve of Binding Energy* (New York: Ballantine, 1975).

not wish to bear unless they felt extremely vulnerable or extremely cocky. Apart from the alienation of a superpower patron, the strategic costs include aggravating relations with neighbors, and getting one's country put on some great power's nuclear targeting list. It is conceivable that some whose incentives for nuclear weapons came more from concern with prestige than with safety might have been deterred by fear of international opprobrium. That is doubtful, however, since it is not evident that countries that *did* acquire such weapons ever suffered for doing so. (It is no accident that the five permanent members of the United Nations Security Council are the five declared nuclear powers of the half-century since World War II.)

The wave of concern about proliferation in the 1990s is the third we have seen in the nuclear era. The first was in the 1960s, when debates about the credibility of the U.S. commitment to deliberate escalation in a NATO–Warsaw Pact War, the development of the British and French independent deterrent forces, and Gaullism led Americans to worry that other industrial nations would follow suit. The second was in the 1970s, when the energy crisis led many to worry that the spread of nuclear power for generation of electricity would enable Third World countries to enter the club. The third wave is different, for both technical and political reasons.

Previously, almost all concern focused on the spread of nuclear weapons. Now there is appreciable attention to chemical and biological arms as well.[7] The end of the Cold War also means that hawks and defense experts now focus on proliferation, because without the Soviet Union and Warsaw Pact to worry about, it is where the action is. This affects the tenor of debates and proposals for how to deal with the question. More of the discussion of proliferation now has a strategic flavor—that is, terms of reference oriented to balance of power, national security objectives, and military strategy—than it did when the Cold War was at its height. "*Counter*proliferation" became a new watchword, now respectable in its connotations of muscular policies. During the Cold War, proposals for military solutions were not favored by those most worried about the problem, who looked to diplomacy, propaganda, and arms control treaties.

The main effect of the end of the Cold War in respect to proliferation is on incentives to acquire WMD. To the countries for whom the end of that war brings more peace and security, the attractiveness of WMD should be lower. But who are those countries? For the most part they are

---

7. See Richard K. Betts, "The New Threat of Mass Destruction," *Foreign Affairs,* Vol. 77, No. 1 (January/February 1998).

members of NATO, but their incentives had already been controlled during the Cold War by the U.S. umbrella. Since France became a nuclear power more than three decades ago none of those countries has moved in that direction. The new peace does little to change the odds in western Europe, although it may increase incentives for vulnerable states in the east. The two weightiest of those countries, however, show no signs of moving for WMD: Poland has gotten into NATO, and Ukraine, which is more vulnerable to reincorporation in Moscow's orbit, bartered away the nuclear weapons it already had for Western economic aid.

The end of the Cold War produced at least as much insecurity as it resolved. Most of the new disarray involves not interstate violence but conflict over the formation or disintegration of states and political communities. For those countries or groups whose security problems are internal contests for control within a state, WMD have less strategic value than in interstate conflict. Large area weapons are not attractive to intermingled contending populations and forces that want to preserve their own groups' lives and the property they are fighting over. For civil wars in which the contestants and their home territories are clearly separated, on the other hand, this may be less true. In either case, chemical weapons, which I classify as only borderline WMD for reasons discussed below, can remain militarily useful.

Those countries whose security problems are external and more pressing than during the Cold War, who now feel more alone or less protected by allies, have stronger incentives to acquire WMD. Along with prestige, security self-reliance is the biggest incentive for a deterrent, and has figured in the decisions of all countries that have acquired nuclear weapons to date.[8] The most likely candidates to seek a serious WMD capability have long been countries that have strong enemies and that lack reliable great power protectors, and they will continue to be the most likely candidates—although the end of the Cold War changes some of the membership in that category.[9] Although insecurity has been an ingredient in all proliferation so far, it is not a sufficient condition—not all isolated states have sought an independent deterrent.

---

8. South Africa's decision to reverse its nuclear weapons program and disarm was consistent with the change in its security situation. Once the apartheid government decided to accept majority rule, external military threats were largely neutralized, and the whites of the country had no interest in bestowing such weapons on a black government. North Korea is now more politically isolated, and Israel less so, than they were some years ago.

9. See Richard K. Betts, "Paranoids, Pygmies, Pariahs and Non-Proliferation Revisited," *Security Studies*, Vol. 2, Nos. 3/4 (Spring/Summer 1993), which updates an article originally published in *Foreign Policy*, No. 26 (Spring 1977).

Changes in incentives are half of the problem; changes in capabilities are the other half. Here trends cut across each other. For nuclear options, the dramatic slowdown in dispersion of peaceful nuclear power generation facilities limits the infrastructures that countries could choose to divert toward weapons programs. It is unclear, however, how much this matters. Every known nuclear weapons program, up to and including those of Iraq and North Korea, has been a *dedicated* program, with facilities largely designed and procured for the purpose of developing a weapon. No country with nuclear weapons has gotten them by lurching absentmindedly out of a peaceful nuclear power program. Cutting in the other direction, however, is the "loose nukes" problem in the former Soviet Union. There, controls on existing weapons, fissionable material, and trained scientists have been under tremendous stress.

Does the rate at which WMD spread matter? The answer must be yes if the costs or benefits vary with the number of members of the club. In the long run we are all dead, as John Maynard Keynes said, so the longer we can delay the costs, or the sooner we can gain the benefits, the better. The probability that WMD will be used varies directly by some degree with the number of potential users, and a panicky scramble could encourage half-baked, poorly safeguarded programs. It is also widely believed that there is some threshold of membership beyond which restraints will collapse and all but microstates will rush to avoid being left behind. Similarly, a phase of rapid acceleration in which several members are added in a brief time could provoke other states that might be willing to watch and wait as long as additions occurred only every half-dozen years or so.

Detonations by India and Pakistan in 1998 raised fears of a chain reaction. Such fears in past waves of concern did not pan out. Several countries bear watching (especially Iraq, North Korea, Iran, and Libya), but their ambitions existed before the tests. Unless Pakistan writes off the United States and deliberately aids Iran's nuclear program, it is doubtful that events on the subcontinent will spur emulation elsewhere.

WHICH WEAPONS?

Posing "weapons of mass destruction" as a category obscures the important differences among nuclear, biological, and chemical arms.[10] Yet it

10. Some observers lump ballistic missiles with WMD. This is awkward since they are delivery systems rather than weapons themselves, and missiles should not be especially significant apart from their role in carrying WMD—and for a country that has WMD, the mode of delivery is far less significant than having the weapons, which can be delivered by other means such as aircraft. Despite the psychological effects of missile attacks with high explosives, missiles—especially ballistic missiles, which un-

remains important to think of WMD generically, in terms of what distinguishes them from "normal" weapons: their primary function or strategic comparative advantage. The main purpose of most normal weapons is to defeat or destroy military forces in combat, or to attack a specific war-supporting installation. The comparative advantage of WMD is to inflict maximum injury on a society and its population. WMD should be capable of producing tens of thousands of casualties in a single attack. (In these terms, conventional high explosives were WMD in World War II when they were used in incendiary attacks against residential areas by the Royal Air Force over Germany and by the U.S. Army Air Force against Japan. Since 1945 there has been an unbroken taboo not only against the use of nuclear weapons, but also against the conventional bombing of large areas of dense civilian population on a scale comparable to what was done in the 1940s. The closest exceptions—the U.S. bombing of Hanoi in 1972, and the Iran-Iraq "war of the cities"—were not in the same class.)

Functions do overlap. Nuclear weapons can be used to strike other nuclear weapons, or for small attacks on clumps of conventional forces; biological weapons could be used against large concentrations of troops in staging areas far from a front; and chemicals are well recognized as weapons for the battlefield. Major nuclear powers during the Cold War also devoted much planning and a significant portion of their nuclear resources to tactical applications, although they did so only after they had fielded hundreds of weapons for "strategic" use against the most vital targets in each other's interiors. Of the weaker countries likely to seek WMD in the next few decades, few are likely to have so many that they can be "wasted" on the battlefield (although Pakistani officials have

---

like air-breathing cruise missiles are expensive and must carry all their own energy—are extremely inefficient means for delivering conventional ordnance. Aircraft can carry far greater payloads, and are reusable. The "war of the cities" between Iraq and Iran in 1988 "ranks as one of the smallest strategic bombing campaigns in history." Hanoi absorbed 500 times as much ordnance from U.S. aircraft in the Linebacker II campaign in eleven days as Tehran did from Iraqi missiles in seven weeks. The attacks on Tehran had great effect on public morale, while those on Baghdad did not, because the Iraqis were winning. See Thomas L. McNaugher, "Ballistic Missiles and Chemical Weapons," International Security, Vol. 15, No. 2 (Fall 1990), pp. 11, 13; and Steve Fetter, "Ballistic Missiles and Weapons of Mass Destruction," International Security, Vol. 16, No. 1 (Summer 1991), pp. 9, 13. Ballistic missiles have also not been the subject of efforts at treaty constraint in the same way nuclear, biological, and chemical weapons have. Strategic arms limitation treaties and the Intermediate-range Nuclear Force treaty limits U.S. and Russian missiles, but no one else's. The "Missile Technology Control Regime" is a voluntary arrangement, not a treaty. This fact led the Bush administration to oppose the Gore-McCain bill in 1989, which sought mandatory sanctions on foreigners for transfers of missile-related items that would not be illegal outside the United States. See Senate Committee on Foreign Relations, National Security Implications of Missile Proliferation, 101st Cong., 1st sess., 1989, pp. 43, 47, 66.

spoken of plans for tactical use of nuclear weapons). The destructive effects and operational comparative advantage of these agents are greatest when used against undefended civilians and economies. The relative importance of the three types of WMD may be misjudged, however, in the popular imagination.

Nuclear weapons have long been seen as the principal WMD problem, ever since they became the essential military ingredient in the titanic Cold War struggle between the superpowers, the main symbol of strategic independence and world-class clout. But even though the technical barriers to acquiring fissionable material and weaponizing nuclear ordnance have eroded over time, they are still substantial, and still far greater than the obstacles to getting either chemical or biological weapons.

Chemical weapons are most ambiguous. In the past they have been unpopular among militaries as counterforce weapons, ever since the mixed results of World War I (when winds sometimes turned chemical salvos against the side that launched them), and because it is comparatively easy for modern forces to defend against CW given some tactical warning. Protection against chemicals does impose burdens on a defense, which could prove decisive where the local balance of power was close to even. So chemical weapons are tactically significant, but pound for pound, they are most lethal when used against unprepared civilians. Dominated by pictures of dead Kurdish women and children, the image of chemical ordnance in recent times has been of a terror weapon. Chemicals certainly count as WMD if used in high volume against concentrated populations, but that is hard to do.[11] The amount of weaponized chemical agent necessary for killing large numbers of people is hard to stockpile, transport, and deploy. In some cases aircraft can deliver these weapons effectively. Otherwise, the best way to deliver large quantities of CW is by artillery, which can only be done within very close range. An army that can get that close to its enemy's cities is usually winning the war anyway. On the other hand, chemical weapons can be very effective against military forces if used in a surprise attack, before the defenders can suit up. The unfavorable view of CW's battlefield effectiveness also changed in the course of the Iran-Iraq War.[12] Observers have long recog-

---

11. With favorable weather conditions and no civil defense, chemical weapons can be up to 500 times more deadly than conventional explosives. Fetter, "Ballistic Missiles and Weapons of Mass Destruction," p. 23.

12. In World War I, chemical weapons produced "1.6 to five times as many casualties as a similar amount of high explosive," but this average came from cases of use against unprotected troops. McNaugher, "Ballistic Missiles and Chemical Weapons," p. 19. In the last year of the war, in contrast, chemical barrages accounted for 20 percent of the artillery munitions used, but caused only 15 percent of casualties and 1.4 percent of

nized chemicals as more appropriate for tactical use than biological weapons because chemicals can be used with more precision and discrimination and take quicker tactical effect.[13] Given the limitations of logistics and delivery capability for covering large areas, CW offer the least efficient mass destruction of all WMD.

Biological weapons (BW) received far less attention than either of the other WMD until the late 1990s. They were banned by the superpowers in 1972, and were largely forgotten by policymakers until recent revelations that the Soviet Union and Russia had cheated and continued their program. Inattention was also abetted by a widely held but incorrect notion that, unlike chemical weapons, biological weapons have not been used in warfare. The purposeful spreading of disease to enemies occurred in a number of cases in premodern times, and the Japanese conducted numerous field tests of biological weapons during World War II, killing substantial numbers of Chinese.[14]

Biological weapons may be the greatest WMD threat of all, for two simple reasons. First, they are far more effective for mass killing than are chemical weapons. Second, they are far easier to acquire than are nuclear weapons. To illustrate the first point, the U.S. Office of Technology Assessment estimated that *a single airplane* delivering 100 kilograms of anthrax spores by aerosol on a clear, calm night could kill more than 300 times as many people as one delivering 1,000 kilograms of Sarin nerve gas—*between one and three million fatalities* if the target was the Washington, D.C. area.[15] On the second point, earlier technical obstacles to making

---

deaths. Effective Iraqi use of chemicals against Iranian human wave assaults changed attitudes. Brad Roberts, *Chemical Disarmament and International Security*, Adelphi Paper No. 267 (London: IISS, Spring 1992), pp. 6, 16.

13. Victor Utgoff, *The Challenge of Chemical Weapons: An American Perspective* (London: Macmillan, 1990), p. 98.

14. The Japanese attacks included the contamination of reservoirs, ponds, and wells; aerial spraying against wheat crops; and the use of infected fleas to spread typhus rickettsia, cholera, and plague. The campaign was limited because delivery systems worked poorly (for example, pathogens sent by artillery burned up on impact). Sheldon Harris, "Japanese Biological Warfare Research on Humans: A Case Study of Microbiology and Ethics," *Annals of the New York Academy of Sciences*, Vol. 666 (December 1992), p. 32. Had the Japanese "been able to develop an efficient aerosol dissemination system, the history of World War II would read quite differently today." Col. Randall J. Larsen and Robert P. Kadlec, *Biological Warfare: A Post–Cold War Threat to America's Strategic Mobility Forces*, Viewpoints No. 95-4 (Ridgway Center, University of Pittsburgh, n.d.), p. 5.

15. U.S. Congress, Office of Technology Assessment [OTA], *Proliferation of Weapons of Mass Destruction: Assessing the Risks*, OTA-ISC-559 (Washington, D.C.: U.S. Government Printing Office [GPO], August 1993), p. 54.

Table 3.1. Incentives for Choice Among WMD

		Destructiveness	
		High	*Low*
Availability	*High*	Biological	Chemical
	Low	Nuclear	

biological agents effective have largely been overcome.[16] Problems in stability, controlling effects, keeping agents alive over long periods, delivery, and the danger of infecting the user had made most strategists uninterested in them. Modern fermentation procedures, aerosol dispensing systems, genetic engineering, and other innovations in biotechnology have abated many of the problems. Neither knowledge nor equipment for the mass production of BW agents are esoteric or contained by export controls. As a result, as long ago as 1988 the U.S. Director of Central Intelligence reported that at least ten countries were developing biological weapons; up to one hundred countries have the capability to do so, and some estimated at the beginning of the 1990s that more than twenty had a program.[17]

WHO WANTS THE WEAPONS?

Rational strategic reasons to want WMD include: to deter the use of WMD against one's own country; to redress inferiority in conventional military capabilities by threatening to escalate in retaliation against an

16.   At the end of the Korean War, the U.S. Joint Strategic Plans Committee concluded that there were "major technical obstacles" to producing and disseminating biological weapons effectively. John Ellis Van Courtland Moon, "Biological Warfare Allegations: The Korean War Case," *Annals of the New York Academy of Sciences*, Vol. 666 (December 1992), p. 68.

17.   Kathleen C. Bailey, *Doomsday Weapons in the Hands of Many: The Arms Control Challenge of the '90s* (Urbana: University of Illinois Press, 1991), pp. 87–90; and Larsen and Kadlec, *Biological Warfare*, pp. 1, 5. Until recently access to pathogens was easy. Larsen and Kadlec reported that "Virtually any agent can be legally acquired from organizations such as the America Type Culture Collection (ATCC) in Rockville, Maryland. ATCC is a legitimate corporation that serves the world-wide research community. . . . One of its customers, however, was the Iraqi government, which purchased anthrax and four other agents" (p. 7). It is possible to tighten restraints on legal commerce, but it is doubtful that controls can ever be strict enough to prevent determined buyers from acquiring pathogens through intermediaries, theft, or collection from natural sources.

enemy's conventional attack; and to coerce an adversary into political concessions. The first two reasons make WMD weapons for the weak, the third makes them weapons for the adventurous. The difference in these motives matters for determining how dangerous proliferation will prove to be, and what great powers ought to do about it.

People in the United States often lose sight of the defensive motives for WMD, or assume that the desire of poor countries to obtain them is less legitimate than the reasons that the United States itself has wanted them. But is there any rationale for why the United States ever needed nuclear weapons that is not an even stronger rationale for India or Pakistan? India can hardly be deemed alarmist for wanting an option to deter China—a superior power, on its border, with whom it fought a war in 1962. And whatever is sufficient to deter China is more than enough to frighten Pakistan. The probability of war between Pakistan and India has always been higher than it ever was between the United States and Soviet Union; India demonstrated its nuclear capability more than two decades ago; and the imbalance of conventional military power in India's favor is greater than was the imbalance between NATO and the Warsaw Pact.[18]

Cuba, Vietnam, North Korea, and Taiwan are the most obviously vulnerable states today, since they face enemies much stronger than they are, and lack any assured alliance. By the same token, however, they also have an incentive to avoid provoking their enemies into action—which a nascent nuclear program might do. Other WMD programs are less likely to be so risky—chemical and biological weapons have never energized as much discussion of preventive attack.

The American tendency to depreciate defensive motives is greatest in regard to countries considered bad or irresponsible, ones now commonly called "rogue states." Some of these are considered rogues because they mount what are seen as unprovoked attacks on Western interests, or because they are run by wild and crazy despots. However, it is not necessary to make excuses for such governments to recognize that they have good reason to worry about being put out of business by a superpower. The more ostracized a regime, the more warranted is that fear. The more that outside powers do to confront the regime, for however many good reasons, the more they do to confirm the fear.

At the same time, if an effort to develop WMD is met with diplomatic fulmination but acquiescence, the wisdom of the effort is confirmed. Take

---

18. See Richard K. Betts, "Incentives for Nuclear Weapons," and "Nuclear Defense Options: Strategies, Costs, and Contingencies," in Joseph A. Yager, ed., *Nonproliferation and U.S. Foreign Policy* (Washington, D.C.: The Brookings Institution, 1980).

North Korea. It has been one of the most grotesque and odious of governments. But who has any better reason for wanting a deterrent than a country alone, one of the last anachronistic holdouts from the defeated Second World of communist states, whose legitimacy is challenged by the vastly richer and more populous southern half of the divided country, which is in turn supported by the world's only superpower? And what penalty did Pyongyang suffer for violating its Nuclear Non-Proliferation Treaty obligations and moving toward a weapon capability? It was not attacked. It did not even have to bear economic sanctions. Rather, the penalty was that it had to accept billions of dollars in aid and the construction of new reactor facilities for generating energy, and it achieved a breakthrough in diplomatic exchange with Washington that it had been seeking for forty years. All of this bolsters the Pyongyang regime, rather than promoting its collapse. Moreover, North Korea was not required to rectify its treaty violation immediately by surrendering or even accounting for the fissionable material it had that was sufficient for one or two weapons, and was required only to begin moving in the direction of neutralizing its nuclear option.[19] While the agreement may have been the best that Washington could do, the point is simply that Pyongyang was given no reason to regret its nuclear gambit.

Effects on Stability and Strategy

NOT TO WORRY?
Liberal pessimism requires little exposition, since its logic strikes most Westerners as self-evident. Utopian realism argues, in contrast, that we should learn from the long peace between the superpowers during the Cold War that nuclear weapons have been good for children and other living things. They stabilized international relations by enforcing caution, making the danger of attacking another nuclear power so obvious that none dared take a chance on challenging the status quo by force.

---

19. The "Agreed Framework Between the United States of America and the Democratic People's Republic of Korea," which many in the West believe has brought North Korea's nuclear option under control, did not require prompt destruction of the plutonium reprocessing plant, nor even the immediate transfer of spent fuel to foreign custody. The new reactors to be supplied will not be completed or turned over until these matters are resolved to Washington's satisfaction, but the new reactors, while less easily applicable to production of fissionable material, are not incapable of being used for it. The agreement commits the United States to reverse more than four decades of policy—to "move toward full normalization of political and economic relations," and to "provide formal assurances to the DPRK, against the threat or use of nuclear weapons by the U.S."

This argument must be taken seriously, at least by anyone who believes that the mutual deterrence relationship between the United States and Soviet Union during the Cold War was stable and reduced their willingness to risk war, or who believes that the prospect of mutual assured destruction rendered all the elaborate schemes for counterforce targeting fanciful and made the piling up of tens of thousands of weapons superfluous. As Waltz says, "Miscalculation causes wars. One side expects victory at an affordable price, while the other side hopes to avoid defeat. Here the differences between conventional and nuclear worlds are fundamental. In the former, states are too often tempted to act on advantages that are wishfully discerned and narrowly calculated."[20] He also argues that the reasons that nuclear deterrence kept the Cold War from turning hot will be even more applicable and evident in the post–Cold War world:

Waltz

Nuclear weapons restore the clarity and simplicity lost as bipolar situations are replaced by multipolar ones. . . .

Deterrent strategies offer this great advantage: Within wide ranges neither side need respond to increases in the other side's military capabilities. . . . This should be easier for lesser nuclear states to understand than it was for the United States and the Soviet Union. Because most of them are economically hard-pressed, they will not want to have more than enough. . . .

States can safely shrink their borders because defense in depth becomes irrelevant. . . .

The problem of stretching a deterrent [to cover allies], which agitated the western alliance, is not a problem for lesser nuclear states. Their problem is not to protect others but to protect themselves. . . . Weak states easily establish their credibility.[21]

The Waltz argument cannot be brushed off, but surprisingly few academic strategists besides Scott Sagan have refuted it in detail. Although most intellectuals as well as normal people oppose proliferation, writings arguing the benefits are more obtrusive in the literature of international relations theory. Why this difference between conventional wisdom and some currents of academic fashion? One reason is that outside of political science departments, people in the United States do not approach the question as detached analysts shorn of national identity.

---

20. Kenneth Waltz, "More May Be Better," in Scott D. Sagan and Kenneth N. Waltz, *The Spread of Nuclear Weapons: A Debate* (New York: W.W. Norton, 1995), p. 6.

21. Waltz, in Sagan and Waltz, *The Spread of Nuclear Weapons*, pp. 14, 27, 30, 31, 33.

They do not abstract themselves from the policy interests of the United States, and do not care nearly as much about resolving the security anxieties of non-nuclear countries as they do about minimizing the chances that nuclear weapons will ever be used anywhere. The world is an uncertain place, where parsimonious theories about stability may or may not prove correct, but where our own country would have less to worry about if we were the only ones to have weapons of mass destruction. It is easier for officials than for analysts to apply shameless double standards and recommend policies that are better for the United States than for other countries that want strategic independence.

High-quality theory is not necessarily a direct guide to good policy. In the scientifically rickety world of social science, any theory that predicts, say, 90 percent of outcomes on some important matter is an amazingly good theory. The Waltz argument may be in that category. In the overwhelming majority of cases, new nuclear states may be more cautious and remain deterred by each other. In the world of policy, on the other hand, people do not marvel at all the cases where nuclear weapons will make the world safer, but worry about the exceptions where things will go wrong. Those wrapped up in policy also take more seriously the prospects for nonrational or accidental action associated with complex organizations, problems that Sagan poses as the main grounds for greater pessimism than Waltz derives from looking at the broader logic of the international balance of power.[22]

In most cases, the *logic* of deterrence theory that became the bedrock of U.S. strategic thought in the course of the nuclear era suggests that the acquisition of nuclear weapons should have a stabilizing effect—that is, they should make it hard to change the status quo by force. Those who have a powerful deterrent will be less coercible or conquerable. It is less clear whether they will coerce non-nuclear neighbors. "Rogue" states that start brandishing nuclear threats risk bringing down an international consensus—and more significantly, U.S. power and countercoercion—on themselves. On the other hand, they may sometimes find that nuclear capability makes the outside powers more amenable to negotiation than they might otherwise be (as in the case of North Korea's diplomatic coup with the United States).

If nuclear spread enhances stability, this is not entirely good news for the United States, since it has been accustomed to attacking small countries with impunity when it felt justified and provoked. The United States is not accustomed to being deterred by anyone but the Russians or

---

22. Scott Sagan, "More May Be Worse," and "Sagan Responds to Waltz," in Sagan and Waltz, *The Spread of Nuclear Weapons*, pp. 47–92, 115–136.

Chinese. This is not the main reason, however, that the Waltz argument fails to command enthusiasm. The main reason is the worry that real statesmen may not always have the courage of Waltz' convictions, that one exception to the rule may be too many, and that the ramifications of the first breakage of the half-century taboo on nuclear use are too unpredictable to tempt us to run the experiment. If the probability that nothing will go seriously wrong in any one case of proliferation is a reassuring 90 percent, the odds that nothing will go seriously wrong in *any* of them decline steadily as the number of cases grows. In short, when it comes to nuclear weapons, "very" stable in "almost all" cases is great for purposes of theoretical clarification, but not good enough for purposes of policy prescription.

POTENTIAL EFFECTS

Nevertheless, if logic suggests that the most common effect of the spread of WMD is to increase security and caution, and that more countries should be able to take care of their own security, there should be declines in expansive conceptions of core security; the disposition of states armed with WMD to meddle with each other; the importance of alliances; and the importance of international legal regulations for WMD.

Most countries' security policies are only about their core security—territorial integrity and sovereignty over the country itself. Only great powers define their security more broadly, to include buffer zones or the security of friends and allies. But all of the countries historically considered great powers (except Germany and Japan, which still recall the experience of their own states' extinction) have already had a nuclear capability for some time. Some small countries, such as Israel, have been impelled by geography to include control of adjacent territory within their concept of core security. Shai Feldman, a student of Waltz, argued that reliance on nuclear deterrence could allow Israel to relinquish occupied territory.[23] Yet Israel has had nuclear weapons for over a quarter of a century, long before Feldman's analysis, and long before it proved willing to cede territory conquered in 1967.

The notion that widespread nuclear capability would inhibit aggression by creating a world of porcupines or a "unit veto system" of omnilateral deterrence is an old one.[24] The suppression of military interventionism, however, could simply channel impulses to meddle into covert

---

23. Feldman, *Israeli Nuclear Deterrence.*

24. Morton Kaplan, *System and Process in International Politics* (New York: Wiley, 1957), pp. 50–52, 69; and George Liska, *Nations in Alliance* (Baltimore, Md.: Johns Hopkins University Press, 1962), pp. 271–274.

political action or other less direct methods. These in turn could increase diplomatic tension and the chances of miscalculation, especially since many of the political systems of the potential proliferators are likely to be weak, permeable, and praetorian, unlike the stable institutionalized governments of the developed world.[25] Internal political weakness and externally deployable military strength (via WMD) are a volatile combination. It was reckless enough for the Argentine junta in 1982 to divert public attention from internal economic problems by grabbing the Falkland (or Malvinas) Islands—one of only two cases of a non-nuclear state initiating combat against a nuclear power (the other being Egypt and Syria against Israel in 1973). Would that conflict have unfolded differently if Argentina had had its own nuclear weapons?

To the extent that the spread of deterrence capability makes more countries strategically independent, it should depreciate the importance of alliance guarantees, by making them both less necessary and less credible. This was the rationale behind the Gaullist *force de frappe* and critique of NATO in the 1960s—a rationale that made much more sense than U.S. policymakers would ever admit. Indeed, the attraction of an independent deterrent should vary inversely with confidence in alternative means of protecting one's security. States have not acquired nuclear weapons during periods of growing support from allies. Britain acquired its nuclear capability in a period of no change in its alliance relationship with the United States; France got its nuclear force after the Suez crisis weakened its tie to Washington; China as relations with the Soviet Union worsened; Israel and India when they were not members of dependable military alliances; and North Korea's nuclear program took off when first U.S.-Soviet détente and Sino-U.S. rapprochement, and then the end of the Cold War, weakened external guarantees to Pyongyang.[26] (The lure of strategic independence of course is a matter of degree.[27] Despite with-

---

25. On praetorianism, which makes the dynamics of domestic politics in poorly developed states analogous to those of the international state of nature, see Samuel P. Huntington, *Political Order in Changing Societies* (New Haven, Conn.: Yale University Press, 1968), chap. 4. For reflections on the problem of controlling nuclear weapons held by unstable governments, see Lewis A. Dunn, "Military Politics, Nuclear Proliferation, and the 'Nuclear Coup d'Etat'," *Journal of Strategic Studies*, Vol. 1, No. 1 (May 1978).

26. Pakistan is a mixed case, since it brought its nuclear program to fruition in the 1980s when, motivated by the need for Pakistani support in the Afghan guerrillas, the United States was supplying more foreign aid than before. The program was initiated and accelerated, however, much earlier, when Washington was refusing to supply even A-7 aircraft.

27. Waltz exaggerated the extent to which Cold War bipolarity reduced the importance of alliances. His only explanation for why the United States and Soviet Union

drawing from the integrated military command—an institutionalized mechanism of alliance historically unique in peacetime—France remained a steadfast member of the Atlantic alliance itself.) The spread of independent deterrents should also further diminish the already tattered hopes for collective security through multilateral enforcement by international organizations like the United Nations. If deterrence by threat of mass destruction works against a regular alliance, there is no reason it should not work against a coalition representing the UN.

Even after WMD have spread much more, there will be areas of conflict among countries or groups that do not have their own such weapons. In those cases the incentives and inhibitions on multilateral involvement will not necessarily change much. What will be more common and more awkward are cases in which one local contestant has WMD, the others do not, and outside powers support these others. We have already run one experiment of that sort, after Iraq invaded Kuwait. Baghdad's demonstrated chemical and suspected biological capability did not deter U.S. attack. Was this because U.S. forces had the equipment with which to defend against chemicals and had vaccinated troops against anthrax? Would knowledge that Baghdad had a few nuclear weapons have deterred the Western coalition from invading Iraqi territory? Is a superpower just not likely to be intimidated by a country that is no more than a medium power apart from its WMD arsenal? Or will leaders think of chemical and biological weapons in significantly different ways from how they have understood nuclear arms, and thus limit the transferability of logic from the highly developed deterrence theory of the Cold War nuclear competition? (See the chapters in this volume by Stephen Rosen and Barry Posen.)

DUBIOUS RESTRAINTS

Liberals opposed to proliferation have put most of their hope into promoting a nonproliferation "regime" based on a bevy of treaties, especially the Nuclear Non-Proliferation Treaty (NPT), the Comprehensive Test Ban Treaty (CTBT), and international conventions banning chemical and biological arms. Much energy has been expended in conferences and speeches about reinforcing nonproliferation norms. This emphasis overlooks the essence of the problem and ignores the nontrivial

---

scrambled so furiously to line up allies and keep them in line was that they were overreacting to each other. Similarly, he does not offer a compelling reason for why, if both superpowers possessed nuclear deterrents that were as solid and dependable for ensuring their security as Waltz maintained, they did not consider isolationist foreign policy options more seriously. Kenneth N. Waltz, *Theory of International Politics* (Reading, Mass.: Addison-Wesley, 1979), chap. 8.

costs of making treaty adherence the centerpiece of nonproliferation strategy. What are the downsides of treaties aimed at proscribing weapons?

First, as useful as treaties are, it is a misconception to see them as a solution. They are effects of nonproliferation, not causes of it. The NPT and CTBT *reflect* the intent of their adherents to abjure nuclear weapons. To date, the countries considered problematic—those that might acquire nuclear weapons—simply did not join the NPT. (South Africa stayed out while it had a nuclear weapons program and joined when it decided to get rid of it.) Or else they joined and cheated (Iraq and North Korea). The inspection obligations of treaty membership did not reveal or reverse the weapons programs of Iraq and North Korea. Iraq passed International Atomic Energy Agency (IAEA) inspections, and it was unilateral U.S. intelligence collection, not the IAEA, that detected the illegal North Korean activity. *If the NPT or CTBT themselves prevent proliferation, one should be able to name at least one specific country that would have sought nuclear weapons or tested them, but refrained from doing so, or was stopped, because of either treaty.* None comes to mind.

Second, however useful the treaties are, some of them—the chemical and biological weapons conventions—are not verifiable with any confidence. Facilities needed to develop such weapons are sufficiently compact and similar to legitimate commercial facilities that inspectors would need much more luck to overcome a dedicated concealment effort than they would in looking for forbidden nuclear installations. It is better to have countries under legal inhibitions than not, but such efforts should not take precedence over other measures to limit countries' incentives and capabilities.

Third, selling the treaties on nuclear limitations in terms of legal and moral norms is politically counterproductive. When Indians and Pakistanis with a handful of nuclear devices hear sanctimonious demands to give them up coming from Americans—who face far fewer uncertainties about their core territorial security, and who will continue to possess thousands of well-tested nuclear weapons even if stockpiles come down by 90 percent or more—they understandably react with exasperation. The charge that the NPT attempts to establish an international caste system is not hyperbole; that is the very essence of the treaty. The NPT is based on an irrevocable division between the few rich countries who had these weapons in 1968, and all others who were expected to remain without them forever. This division is in the interest of the United States and the other nuclear "haves," but it undermines the pursuit of that interest to pose it as a moral principle. It is less quixotic to deal in terms of a bargain, offering things in the interest of the targets to convince them that the benefits of giving up the capabilities exceed the costs in security or prestige.[28]

Fourth, attempting to coerce countries to observe treaty regimes, even when they have not signed the treaties, risks damaging other important interests without achieving the nonproliferation objective. The legislation that required imposing economic sanctions on India and Pakistan in 1998 put the United States in the bizarre position of being on better economic terms with and more politically supportive of China, an authoritarian power that abuses human rights, poses potential problems for U.S. security interests, and has many well-tested nuclear weapons, than it is in regard to India, the world's largest democracy, which has few and less frequently tested nuclear weapons, and has never threatened the United States. It also left Pakistan weakened by sanctions and with little left to lose from sharing its nuclear technology with richer Islamic states.

In any event, if WMD spread widely, international treaty restraints will by definition wane to irrelevance. This will matter symbolically, and for the morale of antiproliferation crusaders. These are not minor considerations, and treaty restraints are useful in other important respects. For example, the inspection provisions of the NPT can provide warning of the diversion of fissionable material, or at least complicate the diverter's job of deception. The NPT, as well as treaties on chemical and biological weapons, do have value as rallying points around which great powers might mobilize a coalition to move against violators. (This has not happened yet beyond the case of Iraq, where detected violations reinforced the case for sanctions.) Without the invasion of Kuwait and the additional evidence uncovered after the Iraqi surrender, it is unlikely that sanctions would have been imposed because of the NPT violation alone. The NPT did provide a legal pretext on which to demand international sanctions against North Korea, but Washington did not press for them. Instead it moved to buy off the North Koreans.

BACK TO OPTIMISM?

If the nonproliferation regime unravels, should we count on the traditional deterrence logic of utopian realism? It is based more on logic than

---

28. The NPT originally did this, by offering a quid pro quo. Article IV of the treaty guarantees the have-nots access to "the fullest possible exchange of technology" for peaceful nuclear energy generation in exchange for not building weapons. This was undermined by the development of the nuclear suppliers cartel, which functioned to cut off trade in technology for reprocessing plutonium or enriching uranium. And since the nuclear energy industry is near collapse now anyway, technology transfer is not much of an inducement anymore. In contrast to its original rationale, the NPT now constitutes a simple demand to the nuclear weapons have-nots to remain so.

evidence. Empirical validation for the logic rests on the one experience of long-routinized deterrence, that between the United States and Soviet Union in the second half of the Cold War. The early phase of the Cold War, before crises over Berlin and Cuba worked out the limits to probes and provocations, is a less reassuring model. Only with hindsight is it easy to assume that because the superpowers did not go over the edge, it was foreordained by deterrence that they could not. Only because China did not take Quemoy and Matsu, the Soviet Union did not take West Berlin, and the United States did not take Cuba does it appear inherent in the logic of deterrence that they would not. But then why did the Soviet Union and China—both nuclear powers in 1969—come to blows on the Ussuri River? Why did Egypt and Syria attack Israeli-occupied territory in 1973? Why did Argentina attack territory held by a nuclear power in 1982? Waltz answers that nuclear deterrence only works to protect core security—home territory—not allies, occupied territory, or faraway possessions.[29] But the territory in dispute on the Ussuri was not a colony or an ally. The U.S.-Soviet model of Cold War deterrence does not match the geographic conditions of many other conflicts. Unlike China, India, and Pakistan, the United States and Soviet Union did not share a land border, and had no territorial claims against each other's homelands. And if nuclear deterrence cannot be expected to cover allied territory, why did Waltz denigrate anxiety about NATO's reliance on U.S. extended nuclear deterrence, and pooh-pooh U.S. efforts to shift toward reliance on conventional defense? There are too many contortions necessary to square all these exceptions and contradictions.

If nuclear deterrence makes countries less worried about their core security, why should it necessarily make them less expansive or adventurous? They may feel less compulsion to rely on territorial buffers, but they may also feel more able to throw their weight around outside their borders. If U.S. and Soviet nuclear power coincided with their global competition and expansive conceptions of security, why should nuclear capability now make smaller powers more content to look no further than their most fundamental interests? Might it not offer them more options for wider-ranging foreign policy than they had before?

This should be especially true where mutual deterrent balances do not emerge immediately, and new nuclear powers are the only ones in their neighborhoods. If local powers start pushing their neighbors around, the risks increase that a superpower accustomed to thinking and acting as a global policeman will get involved. In few such situations would the rules and limits of the competitive game already have been

---

29. Waltz, in Sagan and Waltz, *The Spread of Nuclear Weapons*, p. 16.

worked out, as it took Moscow and Washington more than fifteen years to do.

Perhaps a completed Iraqi nuclear capability would have deterred the United States from rolling back the conquest of Kuwait. This is far from certain, however, since the other Iraqi WMD capability did not. There was much public discussion of the danger posed by Iraq's chemical weapons, and somewhat less of its biological weapons, but negligible argument that they made it too risky to attack Iraq.[30] Perhaps this just reflects the weakness of chemical weapons as WMD. Or perhaps it reflects the extent to which people in the United States have trouble thinking about being deterred by a country that is not a great power. On the other hand, the extent to which Washington rewarded North Korea for violating the NPT points in the opposite direction: it may be easy not only for a second-rate power to deter, but also for it to coerce.

There are many technical and political counterarguments to Waltzian optimism about nuclear spread: poor countries' nuclear forces may be both small and vulnerable, creating the incentive for preventive attack; conflict dyads in many regions are between contiguous adversaries, precluding the tactical warning time of surprise attack on which U.S.-Soviet deterrence rested; animosities among contestants in some parts of the world (for example, Hindus vs. Muslims, Arabs vs. Israelis, Serbs vs. Albanians) are more intense and visceral than the tensions between Americans and Russians ever were, and the difference threatens cool calculation during crises; cognitive orientations, value systems, and modes of calculation in some non-Western cultures are not conducive to the economistic logic of Western deterrence theory; and so forth.

The optimists' answers boil down to two points. First, deterrence does not require either the overkill or the elaborate technical aspects of stability associated with the evolution of the U.S.-Soviet balance. Simple uncertainty about the ability to escape even a few weapons from a victim's retaliation is enough to deter a country from mounting a nuclear attack, since the potential loss of a few cities is out of all proportion to possible gains. It is also easy for even the poorest country to hide a few weapons and deliver them inefficiently, slowly, but on target nonetheless. The aficionados' lore of technical stability, counterforce duels, and whiz-bang command and control systems is beside the point, and mistakes the details that preoccupy developed nations' militaries and professional strategists for the concerns that govern the decisions of political leaders

---

30. The Chairman of the Joint Chiefs of Staff at the time reports that biological weapons were a bigger concern internally than was manifested in the press. General Colin L. Powell, USA (Ret.) with Joseph Persico, *My American Journey* (New York: Random House, 1995), pp. 494, 503–504. See also Barry Posen's chapter in this volume.

with common sense. Second, worries about irrational and reckless leaders represent a crypto-racist double standard. Even the wildest and nastiest prove quite sober when contemplating the extinction of their societies and of themselves. The magic of nuclear reality is that its risks are so perfectly stark that no one can goof.

There are two main rebuttals, and they are convincing. First, the fact of a half-century of nuclear peace between the superpowers leads the optimists to assume that what was, had to be, and to overestimate how intrinsically safe the confrontation was. Although U.S. and Soviet leaders meant to be cautious, there were numerous accidents that raised the risk of inadvertent escalation.[31] Moreover, the tendency of military elites to consider preventive war as a solution more readily than civilian politicians do manifested itself even in the United States; in newly nuclear countries with military *governments*, these tendencies would not be as reliably constrained as they have been. Second, irrationality and idiocy do occur in the behavior of political leaders. What is too risky to contemplate is often clear only in hindsight. Waltz himself attributes the mistaken overarmament and anxiety about stability in the superpower relationship to "fuzzy thinking," yet does not worry that fuzzy thinking will lead new nuclear states to build vulnerable forces or make other dangerous mistakes. As Sagan puts it:

Waltz and other nuclear proliferation optimists have confused prescriptions of what rational states *should* do with predictions of what real states *will* do. . . . This is an error that Waltz avoided in *Theory of International Politics*, where he noted that "the theory requires no assumptions of rationality . . . the theory says simply that if some do relatively well, others will emulate them or fall by the wayside . . . " Adding this element of natural selection to a theory of international relations puts less of a burden on the assumption of rationality. My approach is consistent with this vision. Many nuclear states may well behave sensibly, but some will not and will then "fall by the wayside." Falling by the wayside, however, means using their nuclear weapons in this case.[32]

Finally, experts in both government and academia tend to forget that the lore of deterrence taken for granted in the United States after decades of facing the issue on the front burner of policy—lore that seems so quintessentially logical, transcultural, and, to the optimists, self-evident—is not so obvious to all responsible individuals in countries to whom the

---

31. Scott D. Sagan, *The Limits of Safety* (Princeton, N.J.: Princeton University Press, 1993).

32. Sagan, in Sagan and Waltz, *The Spread of Nuclear Weapons*, pp. 86–87.

questions are new. Perhaps the implications of vulnerability to mass destruction are not subject to ethnocentric interpretation, and will have forced themselves into the consciousness of nuclear neophytes by the time they get far down the road to deployment. But cultural or other subjective reasons for differences in assumptions or sensitivities are not always evident before some crisis forces them into view.

Two anecdotes heard in conversation years ago illustrate this point. One came from a U.S. consultant who described a meeting with a group of the Brazilian political elite (business leaders and military officers) in the mid-1970s, when a deal to acquire full-cycle nuclear power capabilities from Germany was in the works. The American told the Brazilians that they should think carefully before venturing toward a nuclear weapons program. "Oh, why?" he said they asked. "Well," he responded, "right now Brazil is the most powerful country in the region and has no external security problems at all. But if you get nuclear weapons, Argentina will get them too. Then for the first time in Brazilian history all your cities will be at risk." Some hesitant mumbling among the participants followed, and one said, "You know, we never thought of that." The second story is of a conversation in the 1980s. An American was spouting standard arguments about reciprocal fear of surprise attack, survivable second-strike capability, and the delicate balance of terror, to try to convince a Pakistani air force officer of the danger of deploying a small but vulnerable force. The American emphasized that a few weapons on aircraft on soft airfields would be a magnet to Indian attack. He insisted that at the least the Pakistanis should refrain from deploying nuclear weapons until they conformed to responsible standards of technical stability. Throughout this monologue, the officer kept shaking his head. Finally the Pakistani said, with great vehemence, "You just don't understand. Once we have it, they won't dare attack."[33]

Of course these stories do not prove how countries would develop strategy or deploy forces. But they remind us that there is no reason to assume that the deterrence theory that became second nature to U.S. strategists in the Cold War will govern future developments either, any more than the theory of laissez-faire capitalism that seemed self-evident to many Western economists governed economic policies in India and in

---

33. Sagan notes that military organizations, manifesting the typical tendency of organizations toward goal displacement, resist diverting funds from procurement of weapons systems to security measures. For example, the U.S. Air Force refused until long into the nuclear era to allocate substantial resources to reducing the vulnerability of Strategic Air Command bases, and China left its developing nuclear force vulnerable for a full decade. Sagan, in Sagan and Waltz, *The Spread of Nuclear Weapons*, pp. 67–68, 72.

other non-Western countries. Jaswant Singh, a former military officer and principal spokesman on security policy for the new government in India in 1998, noted in several fora that U.S. obsessions with command and control arrangements or technical configurations of forces were not a concern for his government.[34]

The perceived stability of superpower mutual deterrence took years to evolve. Most would estimate that if nuclear spread encourages preventive war, it is most likely to happen in the early phase of force development, before redundancy or protective measures secure a striking force against destruction, or better yet, in the phase when states are developing fissionable material, when the critical targets (reactor, reprocessing, or enrichment facilities) cannot be easily hidden. There is a natural contradiction, however, between timing that makes sense in strategic terms (the sooner the better, when the odds of disabling the capability without suffering retaliation are highest) and the timing that makes sense in political terms (the later the better, when the danger has become real and unambiguous, when forming a government consensus for action is easiest, but when there is more doubt about escaping retaliation).

As of 1998, about eleven countries had built or come close to producing nuclear weapons (the United States, Russia, Great Britain, France, China, Israel, India, Pakistan, South Africa, Iraq, and North Korea). Some others have ventured toward a nuclear weapons program but not gotten as far. *Only once has an adversary mounted a preventive attack:* Israel destroyed Baghdad's Osirak reactor in 1981. It is debatable whether that strike did more to retard Iraqi moves toward nuclear weapons than to spur them.[35] In any case, Iraq was far closer to getting nuclear weapons a decade later than it was when the Israelis struck. The United States attacked a much wider range of Iraqi nuclear facilities in 1991, but only because a war triggered by events having nothing to do with the nuclear problem provided the excuse.

Nothing in the record so far suggests that preventive war will be a frequent choice of countries facing a developing nuclear capability. It is least improbable in the combination of circumstances that came together in the case of the Israeli attack on Iraq: the nuclearizing adversary appears aggressive and risk prone; it lacks significant conventional military means to retaliate for a strike on nuclear facilities; the attacker has an expansive

---

34. "'We are not replicating the experience of the West,' Singh says. 'Therefore, what the West constructed in the management of their arsenals is not what India requires'." Kenneth J. Cooper and John Ward Anderson, "Walking on Thin Ice," *Washington Post National Weekly Edition,* June 8, 1998, p. 6.

35. Richard K. Betts, "Nuclear Proliferation After Osirak," *Arms Control Today,* Vol. 11, No. 7 (September 1981), p. 1.

conception of its own core security; and attack with conventional ordnance can effectively destroy the existing nuclear capability. However,
since the incentives for preventive war are long-term vulnerability rather
than immediate crisis, there is always time to continue assessing the
question, mulling options, searching for other solutions. There is no
moment at which a decision seems necessary, so the path of least resistance is to wait.

Preemptive attack should be less unlikely than preventive war.[36] In a
crisis, with ambiguous warnings or short mobilization times (especially
for a nuclear striking force), the distinction between preventive and
preemptive attack (which is obscure to some observers anyway) may
blur.[37] During the Cold War there was only one near-war crisis, in October
1962, and the United States prepared its nuclear striking forces for preemption by moving to DEFCON-2 (readiness just short of wartime)
before there was any evidence of Soviet preparation to attack.[38] Future
crises among newer nuclear nations may involve more volatile political
interactions and more vulnerable forces to tempt attack. Preemption
becomes most tempting if it also appears possible to accomplish with
non-nuclear forces.

All of these arguments and judgments are oriented to nuclear weapons. There has been less consideration of how deterrence theory applies
to chemical weapons, and far less in regard to biological. These questions

---

36. This may not be saying much. See Dan Reiter, "Exploding the Powder Keg Myth:
Preemptive Wars Almost Never Happen," *International Security*, Vol. 20, No. 2 (Fall
1995).

37. Preventive attack is undertaken against a potential and growing threat, lest the
target country become too strong to defeat at a later date. Preemptive attack is spurred
by strategic warning, evidence that the enemy is already preparing an imminent
attack. See Richard K. Betts, *Surprise Attack* (Washington, D.C.: The Brookings Institution, 1982), pp. 145–147, and Betts, "Surprise Attack and Preemption," in Graham T.
Allison, Albert Carnesale, and Joseph S. Nye, Jr., eds., *Hawks, Doves, and Owls* (New
York: W.W. Norton, 1985), pp. 56–58.

38. Consistent with readying the option to strike first, the U.S. alert served the
purpose of diplomatic coercion. Thus the preparations were not hidden from the
Soviets. See Marc Trachtenberg, "The Influence of Nuclear Weapons in the Cuban
Missile Crisis," *International Security*, Vol. 10, No. 1 (Summer 1985), pp. 157, 159, 161;
and Richard K. Betts, *Nuclear Blackmail and Nuclear Balance* (Washington, D.C.: The
Brookings Institution, 1987), pp. 118–120. The DEFCON-3 (for "Defense Condition" 3)
alert in October 1973, and especially the interaction of U.S. and Soviet ships in the
Eastern Mediterranean, was more dangerous than political authorities in Washington
realized. See Joseph F. Bouchard, *Command in Crisis: Four Case Studies* (New York:
Columbia University Press, 1991), chap. 6. In that case, however, there was no serious
belief within the military establishment that preemption might occur.

become more important as the United States faces less danger from hostile great powers and more from aggrieved subnational groups.

Preventive or preemptive first strikes become most probable when WMD capability is deployed by nongovernmental terrorists. By definition, terrorists have the greatest interest in using terror weapons.[39] To date this possibility has commanded a great deal of attention, but while nearly a dozen states have acquired nuclear weapons and many more have acquired other forms of WMD, no non-state terrorist organization has yet used WMD for blackmail.[40]

This restraint is ironic, because terrorist groups represent the best potential exception to the logic of Waltzian optimism. They should be harder to deter than states. Their location—the address upon which retaliation can be visited—is often hard to pin down. If terrorists can be pinned down, they are likely to be targeted anyway, irrespective of whether they have WMD. Because they are hard to find and target, at least without inflicting tremendous collateral damage, terrorists should have more leeway to use WMD for coercion (it would be easy to kill certain known cliques if counterterrorists were willing to wipe out all of Damascus or Tehran). Holding a city hostage beats holding the passengers of an airliner.

The danger that terrorists might obtain nuclear weapons has been the focus of most concern, but the more probable danger is their use of chemical or biological weapons, given the easier access to them. This was evident in 1995 when the Aum Shinrikyo cult released chemical agents in the Tokyo subway and elsewhere.[41] But why have terrorists not done

---

39. I define terrorism in two variants, which overlap: individualistic anomic violence designed to punish the target for past sins (for example, Unabomber murders); and a strategy of coercion aimed to induce compliance with demands by threatening to kill the target's civilian population. (Making the threat credible may require demonstrating the capability.) The second variant is most likely to be a strategy of the weaker side in a conflict dyad, since the stronger can more easily achieve objectives by using superior conventional power and defeating its opponent's military forces.

40. A borderline case occurred late in 1995, when Chechens reportedly placed a small amount of radioactive material in a public area in Moscow, and claimed to have smuggled three other parcels of cesium into Russia. Rebel leader Shamil Basayev was quoted as saying, "People these days . . . think we can no longer hurt the Russians. So we will give them a little sign . . . that we are completely prepared to commit acts of terrorism." Michael Specter, "Chechen Insurgents Take Their Struggle to a Moscow Park," *New York Times*, November 24, 1995, pp. A1, A17. The Aum Shinrikyo chemical incidents in Tokyo do not count as coercion; no demands preceded or followed the attacks.

41. See the thorough review of these incidents in Ron Purver, *Chemical and Biological*

more of this? For one thing, any use of WMD might overcome inhibitions of great powers to attack terrorists whose address *is* ascertainable, or the states who tolerate their presence. But perhaps the greatest reason for optimism is that there is yet little evidence of terrorists who have an interest in inflicting true mass destruction (tens of thousands of casualties). The closest things have been hijackings involving a few hundred potential victims. For those promoting a political cause, it may seem that really impressive mass murder would be counterproductive.

The biggest danger would be from a group or state that did not want to be recognized and negotiated with, one that wanted only to hurt its victim without getting credit (as with the Lockerbie bombing). And the biggest danger within this category would be a group or state that sees its victim's whole society or culture as a threat, rather than a particular government, and could thus want mass destruction for its own sake—for example, religious or cultural radicals of the sort who bombed the World Trade Center, those who see liberal Western culture as a threat to their preferred social order. Since even fanatical governments are not usually suicidal, deterrence by threat of annihilation in retaliation should still help the United States.

Other Options

ALTERNATIVES TO DETERRENCE AND COMPELLENCE

The realist tradition offers the best guidance to security policy, on balance, but this does not include the utopian realist faction's approval of WMD. The best world for the United States—especially while it is the only superpower in conventional military terms—is one in which the smallest possible number of other countries have WMD. Better that some weak countries be forced to live with more insecurity than that they have the capacity to catalyze events that injure the United States. But what if more proliferation nevertheless occurs? Two approaches that have been neglected or rejected in U.S. policy discussions should become less unthinkable as WMD spread: passive defenses and political appeasement.

There has been much discussion of active defenses, even continuing proposals for investment in ballistic missile defense despite the end of the Cold War. An attempt at such an active defense system makes sense for threats in the middle range, rather than threats at either extreme. That is, a capability to counter a major power that can deliver more than a handful of weapons, but remains unable to deliver thousands (as the

Terrorism: The Threat According to the Open Literature (Canadian Security Intelligence Service, June 1995), pp. 153–189.

Soviet Union could), might make sense. At the extremes, in contrast to this middle range, anti-missile systems offer too little or too much—too little against a sophisticated saturation attack, too much against weaker enemies that do not rely on ballistic missiles. For poor and weak countries, other means of delivery will be far more logical than ballistic missiles for a long time to come. Commercial aircraft, ground-hugging cruise missiles, or clandestine insertion are easier to use and just as effective as ballistic missiles for purposes less ambitious than destroying dozens of targets. Clandestine means are not absolutely easy to use, but they are *comparatively* easy to use.

Passive civil defenses should have a higher priority than ballistic missile defense. They pose a far smaller claim on resources and (at least against nuclear and chemical arms) offer more "anti-bang" for the buck. Civil defenses that would offer little protection against an attack by hundreds of megatons could be much more effective in coping with a dozen or so low-yield fission weapons. It would not help near ground-zero, but could reduce blast casualties in adjacent areas and fallout casualties in outlying areas.

Until recently it was hard to imagine how politicians could summon the public to embrace a civil defense program now when it could not be sold successfully during the Cold War. In the past, arguments for major efforts to anticipate and block WMD threats from small enemies were attacked by some as alarmist and by others as offering false hope. As the Cold War fades, however, so do old prejudices. President Clinton's 1998 announcement of initiatives to improve home defense preparations against biological attack reflect the opening in this direction. Policy should now build momentum to develop various measures for blunting WMD effects: contingency plans for crisis relocation; simple broadcastable instructions on how to build hasty shelters; stockpiling of protective masks, suits, decontamination equipment, chemical antidotes, and antibiotics; standby vaccination programs; and so forth.

Such measures are desirable, especially because they are cheap compared to most programs in the defense budget. But they will only reduce the problem, not solve it. Biological weapons pose the greatest challenge. Because they will be easier to manufacture than nuclear weapons and easier to use for mass killing, they could become a weapon of choice for a state or group determined to inflict mass casualties on the United States. Defense against BW is possible, but least feasible in the most likely situation. If the government knows where a BW attack will come, if it knows exactly what agent will be used, and if it has sufficient advance notice, it may be possible to produce and distribute appropriate vaccines or antibiotics in the right communities; these are too many "ifs" to allow much confidence.[42] It is possible to deploy potentially effective protective masks, and given their comparatively low cost this appears advisable.[43] But would such masks be effective against a surprise night attack? How long would wearers have to keep the masks on to avert infection? Would they be able to drink and eat? Our current inability to defend against the common cold shows the limitations even under ideal conditions of intelligence, and "the use of recombinant DNA technologies to alter viruses limits the possibilities of vaccination even against specific strains."[44]

This is a reminder that the "mass" element in WMD is why such weapons give an intrinsic advantage to the attacker. It is why they are associated with apocalyptic consequences, rather than just lots of death, as conventional warfare is. It is a reminder of why any who have such weapons have a powerful deterrent, and thus why anyone who faces them—such as the United States—should be deterred from meddling with them.

One way to avoid meddling is to retreat from political and strategic involvement in the affairs of other countries and their enemies. Would the World Trade Center have been attacked in 1993 had not the radicals who perpetrated the crime seen the United States as responsible for the frustration of their compatriots' cause in the Middle East? To draw the lesson that our best defense is retreat would be outrageous and dishonorable to many Americans. It would certainly repeal the dominant ethos of righteous world power that the United States has cultivated since Pearl Harbor. But as Waltz says, "A big reason for America's resistance to the

---

42. Some conclude that overall, vaccines are more useful to aggressors (for protecting themselves against the agents they disseminate) than to defenders. For example, Erhard Geissler notes, "Immunological prophylaxis is highly specific. Protection is provided only for a particular type of pathogen." He also notes that so many variations of biological and toxin weapons are possible that the defender needs exact intelligence in time to research, produce, and disseminate vaccines or antidotes (even assuming that the particular biological threat in question is one that can be countered). Erhard Geissler, "A New Generation of Vaccines Against Biological and Toxin Weapons," in Geissler, ed., *Biological and Toxin Weapons Today*, Stockholm International Peace Research Institute (New York: Oxford University Press, 1986), pp. 63–64.

43. See Karl Lowe, Graham Pearson, and Victor Utgoff, *Potential Values of a Simple BW Protective Mask*, IDA P-3077 (Alexandria, Va.: Institute for Defense Analyses, September 1995).

44. Harlee Strauss and Jonathan King, "The Fallacy of Defensive Biological Weapon Programmes," in Geissler, ed., *Biological and Toxin Weapons Today*, pp. 70–71. The capacity to outrace defensive counters may be great in large WMD establishments associated with hostile governments, but small in terrorist groups with limited resources and repertoires. For the latter, some measures that could be circumvented in principle—such as vaccinations against standard anthrax strains—might raise the hurdle high enough to frustrate the effort.

spread of nuclear weapons is that if weak countries have some they will cramp our style. Militarily punishing small countries for behavior we dislike would become much more perilous."[45]

Has the United States been willing to take responsibility for righting wrongs or preventing them in faraway places in part because it faced few serious risks in doing so? In the future, which will be the greater risk: failing to confront a country or group that is acquiring WMD and thus allowing our vulnerability to grow, or confronting them, asking for a test?

Options if the Taboo is Broken

Theorizing about deterrence, arms control, and nuclear strategy was a cottage industry for decades. The emphasis, however, is on "cottage" and "was." Those accustomed to bandying these ideas easily forget that they were never on the tips of most people's tongues. Moreover, the corps of experts who promoted these terms of reference in the United States is aging and dwindling.

Common sense and conventional wisdom about WMD remain abstract, because neither nuclear nor biological weapons have been used against populations since the end of World War II. The common sense and conventional wisdom is also all very simple—and almost entirely about nuclear weapons. The closest thing to wide agreement about these matters in society at large is that these weapons are unusable. We may need them, but for the purpose of ensuring that no one else uses them against us. The fine-grained scenarios and schemes for tactical first-use, counterforce duels, or graduated escalation and limited nuclear war that dominated defense policy debates inside the Washington beltway during the Cold War never resonated outside. What is most impressive about nuclear and biological weapons is that the unthinkability of their use seems confirmed by the lengthening tradition of non-use. *The taboo has reinforced itself.*[46]

As time goes on, and the number of independent decision centers capable of unleashing the arms grows, some weapons of mass destruction are likely to be used somewhere. What then? Utopian realists say it will be too bad, but the effects need not extend beyond whatever faraway

---

45. Waltz, in Sagan and Waltz, *The Spread of Nuclear Weapons*, p. 111.

46. In this sense the taboo might be thought of as constituting a "regime" as the term has been used in international relations theory since the early 1980s—an informal rule that takes on a life of its own as the tradition of observance and mutual interest in its functioning accumulates. On the notion of a nuclear taboo, see Peter Gizewski's forthcoming Columbia University dissertation.

backwater suffers the tragedy. It is hard to believe, however, that anyone elsewhere will continue to view the issues and options as they did during a half-century or more of non-use. If ten thousand or more people are killed by a nuclear or biological attack, this is likely to alter attitudes overnight. But which way?

Quite opposite reactions are imaginable. The shock might jar sluggish statesmen into taking the danger seriously, cutting through diplomatic and military red tape, and undertaking dramatic actions to push the genie back in the bottle. Or the shock might prompt panic and a rush to stock up on WMD, as the possibility of use underlines the need for deterrent capability, or the effectiveness of such weapons as instruments of policy.

One seldom-noticed danger is that breakage of the taboo could demystify the weapons and make them look more conventional than our post-Hiroshima images of them. It helps to recall that in the 1930s, popular images of conventional strategic bombing were that it would be apocalyptic, bringing belligerent countries to their knees quickly. The apocalyptic image was fed by the German bombing of Guernica, a comparatively small city in Spain. When World War II came in Europe, both British and Germans initially refrained from bombing attacks on cities. Once city bombing began and gathered steam, however, it proved to be far less decisive than many had expected. British and German populations managed to adjust and absorb it. Over time, however, the ferocity of Allied bombing of Germany and Japan did approach the apocalyptic levels originally envisioned. In short, dire assumptions about the awesomeness of strategic bombing deterred its initiation, but once initiated did not prevent gradual escalation to the devastating level originally envisioned.

Nuclear weapon inventories of countries like India and Pakistan are likely to remain small in number and yield for some time. According to press reports, by some U.S. estimates the yields of the 1998 tests were only a few kilotons. If the first weapon detonated in combat is a low-yield device in a large city with uneven terrain and lots of reinforced concrete, it might only destroy a small part of the city. A bomb that killed 10,000 to 20,000 people would be seen as a stunning catastrophe, but there are now many parts of the world where that number would be less than 1 percent of a city's population. The disaster could seem surprisingly limited, since in the popular imagination (underwritten by the results in the small and flimsy cities of Hiroshima and Nagasaki), nuclear weapons mean "one bomb, one city." Awful destruction that yet seems surprisingly limited could prompt revisionist reactions among lay elites in some countries about the meaning of nuclear ordnance. (It is notable that reaction in the United States to the Aum Shinrikyo attacks in Japan was quite mild, probably because casualties were surprisingly low.)

If potential results range from shoring up the dam to breaking it, there is no clear guidance about the consequences of breaking the nuclear taboo. What seems most improbable is that the willingness to rely on old concepts of mutual deterrence and stability as solutions to living with WMD would survive unshaken. If the first use of WMD is against a country not comparably armed, the main lesson may seem the validation of deterrence; doubters or the previously complacent will be more likely to bank on utopian realism and tank up on their own WMD to prevent being victimized. (It is notable, though, that Japan did not react this way to its own experience in August 1945 once it regained independence.) If the first use is against a country with comparable arms, however, it will probably appear to discredit reliance on deterrence. In that case, it is hard to believe that the first lapse in the long nuclear peace will not trigger stampedes in at least one of three directions: toward pacifist foreign policies, however out of character and costly they may be; toward more energetic development of defensive measures, however limited their effectiveness may be; or toward brutal interventionist head-cracking by great powers to get the genie back in the bottle, however risky such activism could be.[47]

Another wild card is uncertainty about whether the logic of deterrence concepts that developed almost entirely around nuclear weapons will apply as well to biological and chemical weapons, neither of which has been enshrouded in the same taboo or mystique as have nuclear weapons. History—in the form of plagues across the millennia or gas warfare in this century—has given people reason to think of chemical and biological weapons as more normal and more usable than nuclear arms.

---

47. Many would hope in addition for a fourth possibility, that the shock might make feasible more effective multilateral regulation by the United Nations or some other international organization and collective security mechanism. No one can prove this is impossible, but there is no precedent by which to expect it. As Waltz says in explaining why so many injunctions for states to subordinate their immediate interests to the needs of the global ecosystem have gone unheeded: "The problems are found at the global level. Solutions to the problems continue to depend on national policies. What are the conditions that would make nations more or less willing to obey the injunctions . . . for the sake of the system? No one has shown how that can be done, although many wring their hands and plead for rational behavior. The very problem, however, is that rational behavior, given structural constraints, does not lead to the wanted results. With each country constrained to take care of itself, no one can take care of the system. . . . Necessities do not create possibilities." *Theory of International Politics,* p. 109.

It is also easier for an attacker to conceal responsibility for an epidemic caused by biological attack. If WMD proliferation takes the form of quests for more easily obtainable or more anonymous CW and BW, how certain can it be that the solid deterrence celebrated by utopian realists fixated on the Cold War model will be seen to apply to these other vehicles of mass destruction?

Utopian realism should not convince us to promote the spread of WMD, but it may hold out some hope if nonproliferation fails. At the least, like the rest of the realist tradition, it calls attention to a dimension of the problem that is woefully neglected by the most ardent opponents of proliferation: the fact that weapons do not proliferate by accident, but because countries or groups believe they need them. If they need weapons for prestige, there may be little the great powers can do other than threaten sanctions and hope that the price will strike the targets as greater than the prestige is worth, or at the other extreme, attempt to buy them off (for example, give a permanent UN Security Council seat to India).

If countries need the weapons for security, there may be more the great powers can do. If nonproliferation is truly a vital interest for the WMD "haves," they should be willing to assuage the security incentives of weaker states by providing security guarantees. It is hard to believe, however, that they are more likely to do this after the Cold War than during it. Today, nonproliferation must compete with incentives to re-trench military commitments. For committed internationalists, and those who want to keep the United States a global power for whatever reason, nonproliferation is a vehicle for resisting isolationism. Robert Art, for example, argues, "Taking out additional insurance against nuclear weapons spread, then, is the prime, *indeed the only*, security rationale for a continuing military global role for the United States."[48] In a world where the United States is the only superpower, however, few others will want or dare to challenge its core security. The interest of foreigners in hurting the United States is more likely to be driven by desire to deter the United States from acting against *their* core interests. Thus, it is not as easy as it once was to refute the anti-interventionist argument that the best way to keep others from trying to hurt the United States is to avoid meddling in their affairs. Leaving such governments or movements alone would often comport with utopian realism, but it would negate the liberal impulse to reform illiberal regimes.

For most of the nuclear era the priority that the United States placed on nonproliferation was high in principle but low in practice. Washington

---

48. Robert J. Art, "A Defensible Defense: America's Grand Strategy After the Cold War," *International Security,* Vol. 15, No. 4 (Spring 1991), p. 30 (emphasis added).

was always willing to promote nonproliferation when it did not have to short-change some other objective, but seldom did it prove willing to sacrifice other interests for the cause. The most obvious example was Pakistan, where Washington applied the Symington Amendment and cut the country off from U.S. aid in the late 1970s, until the Soviet Union invaded Afghanistan and made Pakistan a necessary ally in another cause—at which point Washington turned around and gave even more aid, including F-16 combat aircraft even more advanced than the models it had earlier refused to sell. Then, when India and Pakistan tested in 1998, sanctions were imposed, but only because legislation allowed no presidential waiver and contrary interests in relations with the two countries did not appear as compelling at the moment as during the Cold War.

New solutions for proliferation are not obvious: the problem is not new, and thinking about solutions has gone on for a long time. What is new is not the problem, but the amount of concern about it (which is entirely by default, due to the disappearance of the Soviet threat). As the essential problem remains the same, so too do the two main barriers to solution. One is that as long as world politics revolves around sovereign states, those states (or the groups that seek to create new states or supplant old ones) will seek means to protect themselves and exert power. The other is that countries have interests that compete with nonproliferation for their diplomatic, military, and political capital. Until either of those conditions change, we are likely to continue sliding, but slowly, toward a world where more countries and groups have more means to hurt each other.

For some of these actors, getting a strategic deterrent from WMD may improve their security, and sometimes the spread may even avert conventional wars that might otherwise occur in unstable regions. But for traditionally pessimistic realists as well as liberals, what may sometimes be good for weak and vulnerable states will not ultimately be good for those whose strength and security can only be compromised by proliferation. The United States should act as if the utopian realists are wrong, but hope that they are right.

Chapter 4

The National Myth and Strategic Personality of Iran: A Counterproliferation Perspective

Caroline F. Ziemke

The proliferation of nuclear weapons is undoubtedly the most complex and pressing strategic challenge facing the United States and the international community in the twenty-first century. Many of those seeking to acquire nuclear weapons are states with which the United States has strained relations and major cultural differences, making the task of forging effective counterproliferation approaches even more difficult. This chapter focuses on one such state—the Islamic Republic of Iran—to demonstrate the value of national myth as a tool to understand and predict state conduct.

There are several reasons for the focus on Iran. First, it tops most lists of states seeking and likely to acquire nuclear weapons in the not-too-distant future. Second, it is one of the states that the United States has singled out in its national strategy as a likely regional troublemaker, one of what former National Security Advisor Anthony Lake called "backlash states."[1] Third, opinions vary widely, especially between the United States and many of its European allies, as to how great a danger Iran poses to the international system and how best to cope with it. Finally, Iran was formerly one of the United States' most important regional allies and is now a sworn enemy, suggesting that perhaps the 1979 Revolution marked a dramatic change in Iran's strategic personality. Not so. Through a brief survey of Iran's national myth, this chapter demonstrates the continuities in Iran's strategic personality that provide the best basis for predicting its likely future conduct and national strategies. In the conclusion, I suggest how decision-makers might use such insight into Iran's strategic personality to shape future counterproliferation approaches.

---

1. Anthony Lake, "Confronting Backlash States," *Foreign Affairs*, Vol. 73, No. 2 (March/April 1994).

Policymakers and analysts thinking about strategies for responding to the proliferation of nuclear weapons must begin by trying to answer three basic questions. First, what might motivate any particular state to devote substantial and often scarce national resources to developing or acquiring a nuclear weapons capability? Second, where would nuclear weapons fit in that state's national strategy, and how and under what circumstances might it employ them? Third, how can the United States and the international community persuade such a state not to acquire nuclear weapons or, in the event that counterproliferation fails, deter their use? Counterproliferation and deterrence are both essentially forms of persuasion—attempts to modify a state's conduct using rational and, preferably, noncoercive means. But effective persuasion requires some understanding of how the target state makes its own strategic decisions— how it sees its relationship to the outside world, to what situations and events its pays the most attention, and what standards it uses for assessing circumstances and arriving at a decision as to how to proceed.

The problem with standard social science models for analyzing strategic decision-making in specific states is that, in measuring a state's conduct against a general theoretical model, they inevitably pluck events out of their historical context. But a state's conduct *always* follows from a whole complex of motives, preferences, beliefs, prejudices, and ways of thinking that have deep roots in history. As a result, the rationality that informs the strategic conduct of a real, live, flesh-and-blood state with a past is unlikely to conform precisely to what a hypothetical scientific and mathematical model would predict, and the policy community tends to dismiss the state's behavior as a willful attempt to flaunt international norms, or somehow irrational. But to draw such conclusions gets us no closer to answering the important questions: Why would a particular state want nuclear weapons? What might it do with them if it gets them? And what can be done to prevent it from acquiring or using nuclear capabilities?

Nations are the product of their unique historical experience in the same way that individuals are, in a biological sense, the product of their genetic background. A state's historical experience shapes how it sees itself, how it views the outside world, and how it makes its strategic decisions. To make use of their historical experience, nations tend to focus most on those aspects of their history that have the most meaning and tell them the most about who they are and what they aspire to be. Those aspects of a state's historical experience are "stored" in its national myth: the remembered history of how the nation came to be and what heroes, demons, traumas, golden ages, and symbols of national identity are most important in defining the boundaries between "us" and "the others."

National myth is not necessarily factual history; rather, it is a metaphorical representation of a state's history that a state's people believe to be true and use as their guide for social and political life. In the realm of national myth, historical accuracy is less important than motivational power. National myth presents an exaggerated view of the strengths, virtues, triumphs, and traumas that make up a state's collective self-image and provides the blueprint for its strategic personality: how it sees its relationship to the outside world, assesses its options and national interests, and makes decisions about how to act.

From National Myth To Strategic Personality

THE IRANIAN CASE

This section uses national myth as a tool to decipher Iran's rational model and derive a broad strategic personality profile that provides insight into how Iran perceives reality, makes judgments about its interests and goals, and draws conclusions concerning how best to advance those interests.

Three aspects of national myth are of particular utility in assessing Iran's strategic personality: Iran's remembered history (including the history of the ancient Persian Empire), its civic philosophy (including the civic aspects of Islamism), and its shared culture (including past and present religious doctrines). The influence of Shi'a myth is explicit in the present Iranian theocracy. Still, the contemporary Iranian national myth has much deeper roots in the pre-Islamic Zoroastrian and Persian secular myths than are readily apparent. In this respect, the mythic tradition of Iran differs fundamentally from that of its Arab neighbors. The Arabs look to the High Caliphate of 750 to 1258 CE as the "golden age" of Islam and the high water mark of Arab (and Islamic) cultural and political power. But the "golden age" of Islam was a period of defeat and decline for the Persian Empire, and an era of persecution for the Shi'a. Persia's glory days predated Islam, and while Islam is now an integral part of Iranian culture, even the Shi'a clerics have not managed to purge the influence of the religious, cultural, and political traditions that came before it.

ZOROASTRIANISM AND PRE-ISLAMIC MYTH

The dominant faith of pre-Islamic Persia was Zoroastrianism, a monotheistic faith founded on the teachings of the prophet Zarathustra in the sixth century BCE and the official religion of the Persian Empire until the rise of the Safavid Dynasty in the fifteenth century CE. Zoroastrian cosmology centered on the celestial dual between *Ahura Mazda* (the Lord of Light) and *Angra-Mainyu* (the Lord of Darkness) that would culminate in a final judgment day, when the dead would be resurrected and the wicked banished to eternal suffering. Ahura Mazda would thereafter preside over a purified universe and endless age of peace and prosperity on earth in the land of Turan (Tehran).[2]

Zoroastrianism was not a universalist faith and hence did not spread much outside Persia, but it had a profound impact on the evolution of later monotheism: the roots of the doctrine of redemptive suffering in early Judaism, Christianity, and Shi'a Islam can be traced directly to the influence of Zoroastrian teaching.[3] This cosmology left an indelible mark on the Persian worldview: the scale of evil in the outside world is alarming and the surest route to safety is to hew to one's community, and thus remain close to Ahura Mazda and the forces of good.[4] Iranian Shi'ism retains elements of Zoroastrian beliefs and rituals that, in large part, account for its theological distinctions from the Sunni Islamic mainstream.

Pre-Islamic Persia also left important secular nationalist legacies. The *Shahnemeh* is the mythic saga of the reign of Jamshid, who liberated Persia from the rule of Zahhak, a Faust-like Arab prince who became king of Persia through a pact with the Lord of Darkness. Zahhak's three-hundred-year reign is a period of suffering and evil that ends when Fariydun, a Moses-like figure, defeats the forces of evil to usher in five hundred years of peace and prosperity. Persia eventually collapses into civil war when Fariydun's sons make alliances with central Asian warlords, allowing Persia, once again, to fall to the mercy of foreign interlop-

---

2. *The Encyclopedia of Philosophy* (New York: Macmillan, 1967), Vol. 8, p. 380, s.v. "Zoroastrianism"; and Hans-Joachim Klimkeit, *Gnosis on the Silk Road* (San Francisco: Harper, 1993), p. 202.

3. Joseph Campbell, *Myths to Live By* (New York: Penguin Arkana, 1993), pp. 182–183.

4. The Puritanism of early settlers in New England instilled a similarly dualist vein in the U.S. national myth that today is most commonly expressed as the "shining city on the hill." The Puritans fervently believed that their mission in the New World was divinely decreed, that its achievement required strict discipline and single-minded pursuit of definite ends (establishing the "shining city on the hill"), and that deviation from those ends was tantamount to mortal sin. The Puritans, like the Shi'a, saw themselves as the outpost of righteousness surrounded by a dangerous, heathen world (state-sponsored Anglo-Catholicism in England; and the Native Americans in the New World). The Puritan myth instilled a strict spiritual and communal discipline—not unlike that of modern Shi'ism—as a means to keep the atomistic inclinations of individuals under control and the community focused on its shared objective. Just as the Iranians fear temptation from the materialism and individualism of the West, the Puritans saw in Native American culture—with its emphasis on individual liberty and freedom—a potentially divisive influence and a dire threat to their quest for righteousness. See Richard Slotkin, *Regeneration through Violence: The Mythology of the American Frontier, 1600–1860* (New York: Harper Perennial, 1996), pp. 36–37, 45.

ers until Jamshid ends the civil war and reunifies Persia. Again, the xenophobic lesson of the myth is clear: Persian culture thrives when its leaders use their power to keep it isolated and suffers when they pursue alliances and power-sharing with the outside world. The historical basis of the *Shahnemeh* is probably the reign of Cyrus the Great (550–529 BCE), who established the first great Persian Empire. Cyrus is remembered as a benevolent ruler and founder of the world's first "superpower": a Persian empire that at its peak extended from Anatolia to China. Persia thereafter became the economic and cultural crossroads of the ancient world, the center of the "Silk Road" that connected the Turkish and Arab lands to the west with Central Asia and India to the east. The current Iranian regime has gone to great pains to revive this aspect of Persia's legacy, steeping its efforts to develop economic ties with Central and South Asia in the mythic image of the Silk Road.

Persian secular history, like the mythic *Shahnemeh*, is peppered with court intrigues, coups, counter-coups, and bureaucratic power struggles. Persian politics, then and now, is obsessed with conspiracies. The secular national myth that has grown around the remembered history of the Persian empire, along with the spiritual and cultural superiority that follow from Zoroastrianism, have instilled the conviction that Iran's present weakness and vulnerability are unnatural and follow from Persian attempts to compromise with the outside world. The pre-Islamic Persian myth imprinted the Iranian national psyche with an "us-versus-them" mentality and a preoccupation with conspiracy and externalized evil. In interpreting events, Iran takes a comprehensively external and fatalist view: failures are not the fault of any weakness on Iran's part; therefore, changes in behavior are unnecessary and, at any rate, would bring no change in outcomes. The Shi'a myth, with its strong sense of martyrdom, gave Iran's fatalism a clear victim mentality. Once the Islamic revolution radicalized Shi'a myth, the sense of victimization led to rhetoric that looked to the outside world like a dangerous thirst for revenge.

SHI'ISM AND IRANIAN NATIONAL MYTH

The importance of pre-Islamic myth notwithstanding, it is impossible to separate the Iranian national character from Iranian Shi'ism, the polestar of modern Iranian nationalism. If Islam is poorly understood in the West, Shi'ism is an even greater enigma. This is ironic, since the theology and cosmology of Shi'ism and Christianity share common ancestry and are in many ways more similar to each other than either is to Sunni Islam. The ecclesiastical structure of Shi'ism, like Christianity and unlike Sunni Islam, consists of a formal and highly educated clerical hierarchy. The Shi'a clergy also share with their Christian counterparts a tradition of abstract theology and a broad exposure to comparative religion, something that had little appeal for the more pragmatic Arabs. It was, in large part, the erudition of the Shi'a clergy that ultimately made them a threat to the established political and religious order in Iran and, later, the broader regional order.[5] The Shi'a take very seriously the Koranic injunction that Allah gave man not only eyes and ears to receive his word, but also the power of judgment by which to discern the path of righteousness.[6]

Islam probably first came to Persia in the eleventh and twelfth centuries CE, but it did not become the dominant faith until the rise of the Safavid dynasty in the mid-sixteenth century. The Safavids were initially followers of Sufism, a school of Islamic mysticism that drew heavily on local mythologies and cultural identity.[7] Because the messianic Sufis proved difficult to control, the early Safavid shahs saw in Shi'ism the basis of a version of Islam that would promote social cohesion and build the foundation of a Persian national identity, and hence enhance their political power. Once they established "Twelver" Shi'ism as the official religion of Iran, the Safavids conducted a brutal campaign of persecution and conversion against the Sunni, Sufi, and Zoroastrians but buffered the social trauma by incorporating elements of the suppressed faiths.[8] The Safavids further isolated Iran by encouraging Shi'a alternatives to Sunni symbols of veneration: shrines at Mashad and Qom to rival those at Mecca and Medinah, the pilgrimage to Karbala (the site of the martyrdom of the Imam Husayn), and the establishment of an annual observance of the life and death of Husayn. By the late seventeenth century, Shi'ism had become a distinct and comprehensively Iranian alternative to the predominantly Arab Sunni Islam. The Shi'a clerics, however, proved less malleable than the shahs had hoped and, over time, became increasingly independent and critical of state authority. The clerics gradually withdrew from their association with the decreasingly popular shahs and became a parallel authority, often acting as the ringleaders of growing opposition to the power and abuses of the secular leaders.

Two mythic archetypes clearly distinguish Iranian Shi'ism from mainstream Sunni Islam: the martyrdom of the Imam Husayn, and the expec-

5. Olivier Roy, *The Failure of Political Islam*, trans. Carol Volk (Cambridge, Mass.: Harvard University Press, 1994), p. 136.

6. Abdullah Yusuf Ali, *The Meaning of the Holy Qur'an* (Brentwood, Md: Amana Corporation, 1992), n. 6138 to Surah 90: 6–12. This translation and edition is the source for all Koranic references in this chapter.

7. Karen Armstrong, *A History of God* (New York: Alfred A. Knopf, 1994), p. 260.

8. Ira M. Lapidus, *A History of Islamic Societies* (New York: Cambridge University Press, 1988), pp. 295–296.

tation of a messiah, the *Mahdi*. The martyrdom of the Imam Husayn—the grandson of the Prophet Mohammed—is the central archetype for Shi'a devotion and piety. The myth of Husayn has its historical roots in the second Arab civil war (682–991 CE). The urbanized Umayyad Caliphate sought to solidify its political and religious authority over the Islamic world, but it met resistance from the rural, nomadic Muslim followers of 'Ali Ibn Abi Talib—the cousin and son-in-law of the Prophet. Ali's followers believed that the caliphs had sacrificed their spiritual responsibility in the quest for temporal power and were, hence, corrupt and illegitimate rulers. 'Ali's son, Husayn, was ambushed and murdered by the forces of the Caliph Yazid at Karbala (Iraq) in 680 CE. The murder of Husayn provides Shi'ism its origin myth: the struggle against the corrupt followers of the assassin Yazid is the first skirmish in the apocalyptic struggle between goodness and corruption. The open display of mourning for Husayn is an essential element of Shi'a piety, and is ritually observed in two public rites: the *ta'ziyah majalis,* an annual passion play reenacting the assassination, and the *ziyarah,* the pilgrimage to Karbala.

Husayn set the standard for Shi'a piety, idealism, nobility of character, and aesthetic detachment from worldly concerns, but at the same time he became a key political role model. Husayn died because he refused to compromise the principles of Islam for mere political gain, and his martyrdom established a precedent: defending the purity of the faith and the Muslim community, no matter what the cost, is the highest religious responsibility.

Western culture and social and political philosophies are, in modern Iranian national myth, the sort of corrupting influences that Husayn died to resist. To modern Iran, the United States and its Arab allies are the contemporary embodiment of the caliphate; they seek to weaken and persecute Iran and the Shi'a while holding out to the Iranian people the promise of material comfort if they stray from the path of righteousness and submit to the authority of corrupt, temporal powers. Two heroes of modern Iranian nationalism personify Husayn's example of incorruptibility—Muhammed Mossadegh and the Ayatollah Khomeini.

The Husayn myth has not remained static, however. For centuries, the mourning for Husayn focused Iranian national myth on the historical injustice, persecution, and temporal defeat of the Shi'a faithful, contributing mightily to the Iranian tendency to externalize misfortune.[9] Modern Islamist revolutionaries, led by the Ayatollah Khomeini, revitalized and

---

9. Haggay Ram and Galia Sabar-Friedman, "The Political Significance of Myth: The Case of Iran and Kenya in a Comparative Perspective," *Cultural Dynamics,* Vol. 8, No. 1 (1996), p. 55.

radicalized the Husayn myth into a call to action. Radical Islam no longer dwelled on Husayn's temporal sacrifice in the interest of otherworldly gain, but recast him as a righteous hero willing to sacrifice his life in the struggle against corruption, tyranny, and oppression in this world.[10] It was the mobilization of myth par excellence.

Shi'ism segregates the archetypes of martyr and messiah. For the Shi'a, the Twelfth Imam is the anticipated *mahdi* (messiah) whose coming will herald God's kingdom on earth. The Imams, descendants of the Prophet through his son-in-law 'Ali, were the spiritual leaders of the early Shi'a community and venerated as temporal representatives of the continuation of God's covenant with Mohammed.[11] The death of the eleventh Imam in 874 CE without a known heir precipitated the rise of messianic Shi'ism.[12] The legend arose of the Twelfth Imam—the secret issue of his predecessor—who (like Moses and Jesus before him) became the target of a jealous ruler in his infancy. For his protection, and to maintain his purity from taint by the corrupt caliphate, the Twelfth Imam was spirited into celestial hiding, returning only occasionally for forty-day visitations and otherwise ruling through his designated representatives. This "lesser occultation" ended in 941 CE, when direct contact with the Twelfth Imam was lost, ushering in the "greater occultation," which the Shi'a believe will end at the judgment day. The greater occultation is to be an era of challenge and suffering for the faithful, the reward for which will be the triumphant return of the Twelfth Imam as the mahdi to guide the faithful in the establishment of God's kingdom on earth. Although he never made the claim himself, Khomeini was venerated by many Shi'a as the incarnation of the mahdi. Khomeini and his fellow ayatollahs have claimed status as the mahdi's designated representatives to this generation of Shi'a; their pronouncements are considered divinely inspired and, hence, infallible.

JIHAD AND THE GREAT SATAN

The element of Islam that precipitates the greatest misunderstanding between the Muslim and Christian worlds is the concept of *jihad*. In the West, it is the most universally recognized and widely misunderstood of all the tenets of Islam, in part because it has taken on a political prominence all out of proportion to its proper place in the Islamic faith. Jihad is a characterization of the struggle that the faithful face in an imperfect

---

10. Ibid., p. 59.

11. Armstrong, *A History of God*, pp. 162–163.

12. Lapidus, *History of Islamic Societies*, pp. 118–119.

world. Muslim theology defines two levels of jihad. The lesser jihad is the external struggle to defend and expand the territory and community of Islam. The greater jihad—the primary responsibility of the pious Muslim—is the spiritual warfare that takes place within the individual against the powers of perversity that tempt him from the path of righteousness. The greater jihad makes strict and constant demands on the faithful: that they so fix their eyes on God that worldly enterprises lose their importance; and that they be willing and eager to make earnest and sincere sacrifice of body, property, and if need be, life in God's service.[13] As the taxonomy implies, the individual act is more important than collective action, and peaceful jihad—writing, preaching, charity, or setting an example by living according to Islam—is far more meritorious in God's eyes than participation in brutality and war.[14] The Muslim concept of jihad, in this sense, differs little from similar notions of spiritual struggle in Zoroastrian, Judaic, and Christian traditions.

The Iranian brand of Shi'a Islam, however, downplays the internal spiritual struggle of the greater jihad and the existence of evil within its community; it denies the existence of an intrinsic "dark side" within its society. Instead, the Iranian-Shi'a national myth projects all the misfortune, suffering, and evil in Iranian society—past, present, and future—onto others and gives its rhetoric of jihad a distinctly intimidating air. To the Iranian Shi'a the jihad is less a spiritual struggle than a defensive stand against relentless barrages from their enemies, human and supernatural.[15] In earlier times, the projection of evil was aimed at the Arabs

---

13. Here again, the parallels with New England Puritanism are strong. Like the Shi'a, the Puritans saw themselves as members of the family of God first, and the family of man second. They viewed earthly life as a way-station: a period of preparation for the afterlife. Family life and social structures were a mechanism to become "better suited to God's service, and bring them nearer to God." Death was a "final rite of passage that frees believers from all earthly bonds, thus allowing them to return to the Father in heaven." By this calculation, material comforts and vain individuality only served to create sources of division and envy and distract both the individual and the community from their focus on preparation for God's kingdom. See John R. Gillis, *A World of Their Own Making: Myth, Ritual, and the Quest for Family Values* (New York: Basic Books, 1996), pp. 31–32.

14. Ali, *The Meaning of the Holy Qur'an*, n. 1270 to Surah 9: 20, p. 444.

15. The need to externalize evil is a common element in puritan movements. In New England Puritanism, the externalization of evil was most disturbingly manifested in the witch hunts that peaked in Salem, Massachusetts, in 1692. Historians differ as to the specific causes of the witch scares. Feminist historians have argued that the accused were independent-minded individuals, usually women, who threatened their neighbors' sense of spiritual well-being by their refusal or inability to conform to strict social norms. Another school argues that the witch hunts were the product of rising economic communal tensions that, likewise, threatened the spiritual unity and well-being

and the Central Asian invaders. Today, it is aimed squarely and unblink-ingly at the West in general and the United States—the Great Satan—in particular.

The Iranians' mythic image of the United States as the Great Satan is not without historical foundation. Iran's history over the past five dec-ades is peppered with instances of U.S. intervention in its internal affairs. The Central Intelligence Agency (CIA) masterminded the coup that over-threw the brief revolution led by Mossadegh and restored the shah to the Peacock throne in 1953. Through the mid-1970s, the United States sup-ported the repressive Pahlavi regime, and the CIA worked closely with the Shah's brutal internal security force, SAVAK. The Iranians believe the United States exploited their country—including the promotion of a costly Iranian military buildup—solely for the benefit of its Cold War strategy against the Soviet Union. The United States and its Western allies supported Iraq in its war against Iran, remained silent concerning Iraqi use of chemical weapons against Iran, and imposed a one-sided embargo on Iran in an attempt to pressure it into ending a war that the Iranians did not initiate. U.S. rhetoric regarding Iran remains consistently hostile, and in the Iranian view, unjust.

Iranian demonization of the West is born, in part, from a deep sense of insecurity. Western ideas about pluralism, democracy, and social equal-ity appear to conservative Iranians to threaten the traditional social order. Western consumerism bears the onus for undermining traditional values by emphasizing material and worldly concerns over spiritual ones. Ira-nian religious leaders see the United States as a particular threat as it aggressively exports its popular culture, secularism, consumerism, and individualism and seeks to create an "American Islam" among its secu-larized Muslim client regimes. In the Islamist view, Iranian society can avoid dilution of its faith only by a strict cultural quarantine.[16] The Iranian Islamists are convinced that the Western cultural threat to Islam

---

of the community—tensions between the rural agrarian elements and the increasingly capitalistic towns, between cities and rural landowners, and within and among con-gregations and ministers. Again, clear parallels with Shi'ism emerge: strict gender roles as the basis for community solidarity, the challenge of harmonizing economic growth with spiritual purity, conflict between the needs and ambitions of the individual and the pursuit of communal goals, tensions between rural and urban values, and conflicts between various power centers within the ranks of clergy. For the feminist interpreta-tion, see Carol F. Karlsen, *The Devil in the Shape of a Woman: Witchcraft in Colonial New England* (New York: Vintage Books, 1987); for the socio-economic interpretation, see Paul Boyer and Stephen Nissenbaum, *Salem Possessed: The Social Origins of Witchcraft* (Cambridge, Mass.: Harvard University Press, 1974).

16. Graham E. Fuller, *The "Center of the Universe": The Geopolitics of Iran* (Boulder, Colo.: Westview Press, 1991), pp. 252–253.

is not passive but part of an aggressive and intentional program of manipulation aimed at undermining Iran's spiritual and political cohesion. Western rhetoric about the "Islamic threat" and the "clash of civilizations" fuels this sense that the Judeo-Christian world fears the strength of Islam and is determined to destroy it.[17]

Iran's anti-Westernism is about controlling its own fate; yet a recurring pattern of Iranian history has been its inability to do so for more than short intervals. This paradox leads to an Iranian preoccupation with "saving face" that has a dual effect on Iranian national myth. First, it helps explain Iran's endless weaving of elaborate, and often implausible, conspiracy theories. Iran's extreme sensitivity to shame amplifies its disinclination to internalize responsibility for its weaknesses and failures: misfortune always has an external agent. While maintaining the conspiracy paranoia, the Islamic Revolution at least empowered the Iranian masses by giving them a voice in their economic and political future. But Iran's steady economic deterioration, and the spreading disillusion with a clerical hierarchy whose degree of corruption approaches that of the shah's elite, is leading to a renewed sense of powerlessness, especially among young people who increasingly see even the most basic goals—college, careers, marriage, or family—drifting beyond their reach.

Second, because of their keen sensitivity to "face," Iranians put great store in being inscrutable. In the past, they hid a determination never to accept a subordinate role behind a facade of elaborate courtesy and rhetorical hyperbole. Today, the extravagance of anti-Western revolutionary rhetoric serves a similarly defensive purpose. The political and religious leaders in Iran make frequent pronouncements that reveal almost nothing about the substance of Iranian policies and ambitions. In both internal and external affairs, dissembling has become a fundamental survival technique in Iran, since information is power in the hands of one's enemies.[18] The rigid social and political mores of postrevolutionary Iran have done nothing to reverse the tradition of Iranian diffidence; just as in the past, today's young Iranians have become cynical and quietly subversive, presenting a falsely pious front in public while privately embracing fashions and mores in direct conflict with the ideals of Islamism.[19]

---

17. Bernard Lewis, "The Roots of Muslim Rage," *Atlantic Monthly* (September 1990).

18. Fuller, *The "Center of the Universe,"* pp. 11, 15, 19.

19. Geraldine Brooks, "Teenage Infidels Hanging Out: In High Tops and Jeans, Iranian Youths are Quietly Subverting their Parents' Revolution," *New York Times Magazine*, April 30, 1995, p. 46.

THE GENESIS OF THE IRANIAN REVOLUTIONARY MYTH

The Islamic revolution itself is an exercise in mythogenesis. The Islamic Republic embodies the reconciliation of traditional Islamic behavioral norms with the demands of the modern world. The architects of Iranian Islamism do not seek to abandon modernism or return to the social order of the seventh century. Nothing about Islamic theology is inherently hostile to technology or modernization. But to Muslim societies experiencing the resultant social dislocation, the trappings of tradition—the reorganization of society based on the *shar'ia* (the Islamic legal code), the restoration of social virtue through dress and comportment, and the condemnation (*fatwa*) of those who challenge traditional norms—can provide a sense of continuity and collective security. Even modern Iranian "Islamic" dress codes are the stuff of national myth: with the exception of the *chador* (the head to toe black veil that never completely disappeared from rural and traditional Iran), *hijab* in Iran is a thoroughly modern expression of Islamic standards of modesty, although symbolically reminiscent of more traditional dress.[20]

Everyday life in postrevolutionary Iran mirrors the duality between the desirable, as reflected in revolutionary myth, and the actual revolutionary reality. Consumerism in Iran has continued more or less unabated despite the country's deep economic troubles. Regular mosque attendance has dropped dramatically since the revolution, and the calls to prayer bring barely a hiccup in urban activity. There are periodic conservative backlashes in Iran, particularly against the education system that clerics charge is not sufficiently Islamic, but the movement to revive strict Islamic revolutionary principles has garnered little popular support. The Islamic piety of the revolution was itself a mythic ritual of sorts. Revolutionary puritanism also has deep Zoroastrian roots: the spiritual battle between good—exemplified by the outwardly pious and inwardly spiritual life—and evil—the giving in to the baser, impure physical instincts and worldly temptations.

While steeped in Islamic myth and norms, the events of 1979 represented first and foremost a political revolution. Khomeini's revolutionary role models were secular and, for the most part, Western. During the revolution and since, revolutionary political goals have always taken precedence over religious goals. The centrality of the shar'ia in Iranian law is established in the constitution; thus, it is the secular law that establishes the legitimacy of religious law, and all new legislation must conform to both Islamic and constitutional standards. Moreover, Iranian law contains many non-Islamic concepts: legal (if not yet actual) equality

---

20. Roy, *The Failure of Political Islam*, pp. 56, 59, 181–182.

between the sexes concerning property, employment, and family rights; the lack of legal religious discrimination; citizenship requirements that are in no way tied to religious affiliation and give no preference to Muslim over non-Muslim aliens; and the retention of the Persian solar, rather than the Muslim lunar, calendar.[21]

THE STRATEGIC PERSONALITY OF IRAN

Iran's national myth dwells on two factors that lie at the core of its strategic personality: its cultural and religious singularity, and its long history of invasion and manipulation by outsiders it sees as inferior. This has rendered Iran a society that is introverted to the point of paranoia, one that holds fast to its cultural traditions, even as it longs to enjoy the benefits of modern technology. Iran is also a culture that has a long memory and holds grudges: the span of history that directly informs its decisions and behaviors ranges as far back as the ancient Persian Empire, and as far forward as Judgment Day. In short, Iran's model of time is much closer to the holistic Eastern cyclical concepts than to the compartmented, linear time of the West. Social and organizational structures in Iran are nebulous, generally consisting of complex webs of primary groups and conflicting centers of power (clergy vs. technocrats, *bazaaris* vs. industrialists, professional military vs. revolutionary guards, and numerous clerical and political subgroups). Communication among Iranians and between Iran and the outside world is heavily laced with metaphor, symbolism, and hyperbole; there are meanings behind meanings. Iran is a value-oriented culture, and sees its value structure as uniquely meritorious and constantly under siege from the outside. The Iranians are idealists, even visionaries, and believe their society is perfectible. In fact, the Iranians are so confident of their moral superiority that they feel compelled to attribute any shortfalls to outside interference. As a society, Iranians are only secondarily interested in their material success; Iran's cultural artifacts reflect this antimaterialism: its cultural expression tends toward the ethereal—poetry, literature, theology—rather than the concrete visual arts, architecture, infrastructure, and technology.

Iranian Foreign Policy and National Security Strategy through the Prism of National Myth

The central theme of Iran's mythic self-image and the key to its strategic personality is its sense of cultural and moral superiority—a superiority that the outside world has most often failed to honor. Since the earliest

---

21. Ibid., p. 178.

invasions of Persia by Huns, Greeks, and Arabs through the nineteenth-
and twentieth-century pressures from Ottoman and then European
imperialism, Persian sensibilities have been outraged by the abhorrent
spectacle of infidels manipulating the fortunes of true believers and
undermining the interests of Persia. It is a humiliation that still grieves
the Iranian psyche and drives its participation in international relations.
Persia has always labored under a certain nationalist narcissism,
imagining that it is the focus of international intrigue, jealousy, and
calculation.[22] Zoroastrianism and Shi'ism, both fairly insular faiths,
reinforced Iran's cultural arrogance with a sense of spiritual and
moral ascendancy over its Arab neighbors. Cultural arrogance toward
its Arab and Turkish neighbors was initially defensive, preventing
political and military domination from translating into foreign cultural
assimilation. After World War II, Iran pursued a foreign policy
that seemed designed to alienate its neighbors and cultivate its iconoclastic image. The shah's regime was openly hostile toward Arab nationalism, established cordial relations with Israel, and forged a close military and economic relationship with the United States. Even the
Islamic revolution put a premium on the distinctions between Iran
and the Arab world: Shi'ism versus Sunnism, Islamist Iran versus the
secular Arabs, champions of the victims of imperialism versus the
puppets of the West. The stark dualism of the Zoroastrian and Shi'a
myths are thus clearly reflected in foreign policy, giving Iranian diplomacy its stark zero-sum approach: We are good, you are bad. We win,
you lose.

Persia's once great empire and culture has, over the course of modern
history, been gradually ennervated and humiliated by foreign encroach-
ment and domination, and cultural imperialism. As a result, one of the
most important aims of Iranian foreign policy since the reign of Reza
Shah prior to World War II has been to restore Iran to its rightful place
as a great and feared power in the Persian Gulf and the world. All
developing nations want the West to treat them with respect, even if they
are realistic enough not to expect to be treated as equals. To Iran, however,
which sees itself as the heir to a once glorious and feared empire, to be
treated as a second- or even third-tier power is grievous. The current
mythic role of the United States and the Western powers as the "Great
Satan" follows from the perception that it was the arrival of the Europeans, their culture, and their ideas that undermined and eventually destroyed Iran's autonomy and dignity.

---

22. Fuller, The "Center of the Universe," p. 255.

IRAN'S REGIONAL STRATEGY

The Persians throughout history have believed themselves surrounded by expansionist aggressors. That historical paranoia has been enhanced by recent history: Western dependence on access to oil, Iran's strategic importance in U.S. strategy during the Cold War, its ability to influence traffic in and out of the Persian Gulf at the Straits of Hormuz, and the success with which the Islamic revolution disoriented Western policies have all served to intensify Western, and particularly U.S., concerns over Iran's hegemonic ambitions. Iran's present quest for hegemony is driven by the desire to force Western power out of the region and establish a more "natural" regional balance that extends Iranian influence but not Iranian territory. The fundamental difference in the security paradigms of Iran and its Gulf Cooperation Council (GCC) neighbors is this: Iran does not believe regional stability and security are possible so long as the United States retains a military presence in the Persian Gulf; the GCC is founded on the principle that regional security and stability are impossible without a sustained U.S. presence.[23]

There are few points of contention between Iran and its neighbors that are likely to result in open conflict in the near term. Although the territorial dispute with the United Arab Emirates (UAE) periodically takes on a more confrontational air, none of the parties to the conflict— Iran, the GCC, or the UAE—seem likely to press the issue to the point of a military conflict. Iran's pursuit of a blue-water navy (an ambition inherited from the Pahlavi regime) might also prove destabilizing. While Iran clearly seeks the ability to block ships entering the Persian Gulf, its interest in keeping the sea lines-of-communication open there is at least as great as that of the United States, and recently both sides have taken care to avoid potential naval confrontations. The Persian Gulf War eliminated Iraq as a potential aggressor for the time being; but legitimate concern lingers in Iran about the long-term possibility of a resurrected Iraqi military machine. There are still outstanding differences between the two states, and as a large, potentially wealthy, and populous state, Iraq is the most likely challenger to Iran's security in the Gulf Region. Since the 1970s, Iran and Saudi Arabia have engaged in an on-again, off-again Islamic Cold War over the future configuration of Persian Gulf security. In the past few years, the relationship has warmed as the Iranian regime has backed away from its declared intention to export its revolu-

---

23. Sharam Chubin and Charles Tripp, *Iran–Saudi Arabia Relations and Regional Order*, Adelphi Paper 304 (London: International Institute for Strategic Studies, November 1996), p. 28.

tion. Iran probably also senses that if the Arab-Israeli peace process bears fruit, it may suddenly find itself the new nemesis for Arabs in search of someone to replace the Zionists as the source of all misfortune in the Arab world. The peace process may contribute to a further sense of insecurity in Iran in another way. The West has made arms sales part of the currency that moves the peace process forward, building the military capabilities of states potentially hostile to Iran to spur their cooperation. The strategy may indeed contribute to Arab-Israeli stability by closing some of the capabilities gaps, but the United States should recognize that it is creating a source of anxiety for Iran.

In light of its dual aspiration for regional leadership and territorial and cultural security, Iran seems to be pursuing a two-track regional foreign policy. The Iranian regime has been making some conciliatory gestures toward mainstream Arab states—especially Saudi Arabia and Egypt—in the hopes of drawing them away from their relatively close security ties to the United States. Iran has also hinted at the possibility that it might forge a "revisionist axis" of Iran, Syria, and Iraq to balance the increasingly daunting U.S.-Saudi alliance. Syria has been Iran's only real ally in the region; the two regimes have in the past shared a bitter enmity with Iraq and, more recently, have moved closer in their position concerning the Arab-Israeli peace process. More surprising has been Iran's improved relationship with Iraq, once a mortal enemy, and the apparent rise of a mutual sense of shared Iranian-Iraqi interests. Rhetorical darts still fly, and Iran continues to support Shi'a rebellion inside Iraq; but Iran has also been outspoken in its criticism of continued UN sanctions and has been an important source of black market trade for Iraq. Talk of a future alliance with Iraq is likely little more than Iranian posturing, though. Iran has the most to lose from a full revitalization of Iraq's economic and military power and has no powerful allies to shore up its defensive position; moreover, its sole regional ally, Syria, is unlikely to share Iran's interest in rapprochement with Saddam Hussein's Iraq.

The tenor of Iran's national myth and the pull of history seem likely to continue to undermine Iran's relations with its Arab neighbors. The long history of Arab-Iranian enmity carries much weight in cultures with long memories, and the Arab states will remain suspicious of Iran's intentions in its recent attempts to improve relations. Recent Iranian behavior in the Persian Gulf—a military build-up coupled with an increased tempo of naval exercises in the Gulf, the recent dispute with the UAE over the mid-Gulf islands, and alleged Iranian connections to the anti-U.S. terrorist attacks in Saudi Arabia—belies Iran's statements of peaceful and conciliatory intentions. Far from moving the Arab states toward warmer relations, Iran's actions have led to increases in defense spending among the GCC states and increased reliance on Western security guarantees, and they have virtually insured that few of Iran's neighbors will join it in resisting the emerging U.S.-Saudi military dominance of the Persian Gulf region. Iran's relative military position has deteriorated since 1979, even given the crippling defeat Iraq suffered in the 1991 Persian Gulf War.[24]

Iran's mobilization of its myth of victimization and Western conspiracy holds little water with its Arab neighbors. Most Arab states agree (at least privately) that Iran and Iraq—not the Gulf Arabs or the United States—are responsible for the increased Western military presence through their long streak of reckless behavior: Iran's initiation of the 1988 tanker war in the Gulf; Iraq's invasion of Kuwait and its subsequent saber-rattling in late 1994; and Iran's naval arms build-up, militarization of the Tumbs islands and Abu-Musa, and acceleration of its nuclear program. Before it can win even cautious concessions from the rest of the Gulf states, Iran will have to revise one of the central themes of the Persian national myth: its natural claim to hegemony in South Asia and the Persian Gulf based on its moral and cultural superiority. Even then, the Arabs are likely to proceed only with great caution. They will not repeat their 1990 mistake of waiting for open aggression to appeal for outside help against an aggressively expansionist regime (in that case, Iraq). They are willing to "ally themselves with the devil" if necessary, to ensure security until Iran demonstrates a genuine desire for peaceful relations with its neighbors.[25]

One issue, more than any other, stands in the way of improved relations between Iran and its Arab neighbors: the declared intention of Iran's Islamist regime to export its revolution throughout the Muslim world, and its particular targeting of conservative, traditional Sunni regimes in the Persian Gulf region. The export of Islamic revolution is enshrined in the Constitution of Iran's Islamic Republic, but since at least 1985, the regime has gradually disengaged from its once robust program of subversion and violence in favor of building a positive image of Iran as a role model of the prosperous Islamic state.[26] Iran's efforts to export Islamist revolution provides two key elements of its strategy to force the

---

24. Dov Waxman, *The Islamic Republic of Iran: Between Revolution and Realpolitik,* Conflict Studies 308 (Leamington Spa, U.K.: Research Institute for the Study of Conflict and Terrorism, April 1998), pp. 17–18.

25. "Iran Forcing the Arabs into the West's Lap," *Mideast Mirror,* March 17, 1995, and April 1, 1995.

26. Sandra Mackey, *The Iranians: Persia, Islam, and the Soul of a Nation* (New York: Plume, 1998), p. 345.

West to fear and respect Iran, and by implication, the rest of the Muslim world. First, Iran paints itself both domestically and internationally as the modern-day heir of the great Muslim (albeit Kurdish) general Saladin who will lead Islam in wresting control of the rightful territory of Islam from the domination of today's crusaders: the Zionists, the secularists. Second, Iran stands to gain what it sees as important moral and psychological victories over the United States and its regional enemies by fostering domestic unrest and potential political upheaval in Arab states like Egypt and Saudi Arabia, whose regimes have aligned themselves with the West. In so doing, it hopes also to force other Gulf states to think twice before aligning themselves with the U.S.-Saudi security axis.

Iran's support of Islamist movements is the centerpiece of its effort to restore something of its ancient glory; it also has roots in another aspect of Iran's national myth—its concept of and quest for Islamic justice. Iran's Islamic interventionism is part of a more general revisionist foreign policy that seeks to overthrow the Western-dominated world system, at least as it functions in the Persian Gulf and Middle East, and replace it with a system that is more egalitarian—at least for Iran. The establishment of national systems based on Islamic principles is essential to Iran's view of a desirable future world order, because Muslim societies can only be treated with the dignity and justice they deserve within an Islamic context.

Despite the obvious mythic status that came with being the first successful Islamist revolutionary state, Iran has met with limited success in extending its influence beyond what one analyst called its "self-imposed Shi'a ghetto."[27] Even within Shi'a Islam, Iran's quest for paramountcy has met with frustration. The Shi'a outside Iran, it seems, are a difficult lot to manipulate: they are either loyal citizens of their home countries; or they are members of Islamist resistance movements that are fiercely independent. Shi'a clergy outside Iran have not undergone the radicalization of Shi'a myth that, in Iran, transformed the Imam Hussein into a prototypical Islamic revolutionary.[28] Of late, non-Iranian Shi'a—who number at least 150 million compared to Iran's 60 million—have come to resent Iranian efforts to dominate the clerical elite; local Shi'a populations in India, Pakistan, and Central Asia have challenged the right of the Iranian clergy to choose the next *marja* (the spiritual leader of the world's Shi'a faithful) and have chosen their own indigenous leaders.[29]

---

27. Roy, *The Failure of Political Islam*, p. 184.

28. Ram and Sabar-Friedman, "The Political Significance of Myth," pp. 60–63.

29. "Iran after Araki: 'Things will never be the same'," *Mideast Mirror*, December 1, 1994, p. 21.

Iran has recently attempted to expand its influence by offering support to Sunni-dominated Islamist movements in the Arab world and Farsi-speaking minorities in the Central Asian republics of the former Soviet Union. These efforts have been only moderately successful because the Shi'a influence is limited with Sunni Arabs, and ethnic and religious ties with the more secular Central Asian Muslims are tenuous.

Still, while Iran's influence may be limited, it is concentrated in areas of significant strategic interest in the West as well as the Arab world. As of 1988, 57 percent of the population in Iraq, 60 percent in Bahrain, 32 percent in Lebanon, 15 percent in Kuwait, and 12 percent in Syria were Shi'a.[30] Iran has made some inroads in areas where the Shi'a minorities have been radicalized and where politicized clergy lead resistance movements, particularly in the strategically vital but politically vulnerable Gulf monarchies like Bahrain and Saudi Arabia.[31] Nor is Iran's revolutionary Islamist myth without appeal outside Shi'a Islam. Iran established ties with Islamic resistance movements in Afghanistan and Pakistan, provided important moral and material support to Palestinian Islamists opposing the Palestinian-Israeli peace process, has provided financial support for radical Islamists in Saudi Arabia, and has been active in supporting and training Bosnian Muslim forces in the former Yugoslavia. The Iranian presence in Bosnia proved a major concern for NATO Implementation Force (IFOR) peacekeepers implementing the 1995 Dayton Accords. Iran has also succeeded in exercising a major influence on the course of events in Lebanon through its support of the Shi'a Hezbollah (Army of God) guerrilla forces and may also be playing a role in the course of the Syrian-Israeli bilateral talks.

Despite Iran's status as a role model and its short-term, episodic success in influencing Islamist movements, at least three factors argue against Iran's long-term success in its campaign for unchallenged dominance of global Islamist politics. First, while most Islamist opposition movements are not nationalist in the strict sense, the character of each follows directly from its own unique national culture and myth, and each pursues an agenda based on indigenous frustrations and aspirations.

---

30. Sharam Chubin, "Iran and Regional Security in the Persian Gulf," *Survival*, Vol. 34, No. 3 (Autumn 1992), p. 68.

31. The Saudi government does not release public data on the percentage of Shi'a within the Kingdom, as Sunni Islam is the official religion of all Saudi citizens. However, the Shi'a, while not a significant percentage of the Saudi population as a whole, are known to be either a majority or sizable plurality in small pockets along the Persian Gulf and Red Sea coasts. In addition, some Shi'a live in Saudi Arabia as guest workers (including some from Pakistan), although the number has likely declined since the 1991 Persian Gulf War.

Second, most Islamist movements are Sunni. They look for their revolutionary example to the Sunni Islamic Brotherhood rather than the Iranian revolution, and are unlikely to abandon the traditional Sunni denigration of Shi'ism to subordinate themselves to Iran. Islamists will continue to welcome Iran's material support but are likely to balk at attempts at ideological and political manipulation by Tehran. It is unlikely, in short, that Iran will become the great puppet master of the Islamist movement. Finally, the fundamental issues that create tension between Iran and its neighbors are essentially strategic rather than ideological in origin; these differences, for the most part, predate the Iranian revolution and would likely continue even if more "Islamically correct" regimes replaced the Gulf monarchies.[32]

The fact that Iran's myth lies especially close to the surface in its external relations does not rule out a certain expediency: the Iranians are nationalists first, coreligionists second. Iran has already demonstrated its willingness to set aside its ideological agenda given the right incentives and constrains. In fact, it has used its support of Islamist uprisings as a bargaining chip in international relations. Iran has carefully avoided the appearance of supporting Islamists in former Soviet Central Asia and Chechnya, and has played the role of peace-broker in the Nagorno-Karabakh conflict, all in the interest of expanding Russian-Iranian relations, which both seek for economic and strategic reasons. Iran may eventually have to back further away from its Islamist interventionism for three other reasons. First, if Iran's attempts to improve relations with its regional neighbors are part of a serious effort at rapprochement, then abandoning its support of disruptive extremists in the Gulf Arab states seems a prerequisite. Second, Iran's conciliatory overtures to the Mubarak regime indicate a concession that Islamist revolution is unlikely to overthrow Egypt's secular regime. Islamist movements in Bahrain and Saudi Arabia face similar frustration, although they are likely to remain active and may force some response from the existing regimes.

The third and ultimate motive for curtailing its Islamist offensive may be internal. Iran is a multiethnic nation, tied together by religion rather than language or homogeneous culture. It has been a remarkably stable entity over the centuries with comparatively little ethnic unrest, but the presence of large ethnic minorities (Azeris, Kurds, Armenians, Baluchis, Turkmens, Arabs, to name a few) along borders with potentially hostile or at least unstable neighbors could prove a source of future problems, in combination with Iran's deteriorating economy and declining standard of living. Domestic unrest is likely to spread as the Iranian population

---

32. Chubin, "Iran and Regional Security," p. 53.

grows impatient with the slow pace of economic and political reform and as the nascent conflict between Islamic conservatives and more progressive elements heats up. The frustrations of a highly educated but underemployed middle class and the disillusioned, religiously zealous veterans; the pressures of the fast (although decelerating) rate of population growth; and the destabilization of urban populations by an influx of undereducated rural peasants all add fuel to an already incendiary social situation. The Iranian regime may be beginning to suspect that meddling in the internal affairs of other states—at least those with the means to retaliate in kind—is not a regional norm that Iran would like to see spread. A substantial expatriate resistance movement has made no secret of its ambition to return to Iran, and its government in exile, while unpopular inside Iran, has gained allies in Washington, raising the specter of an Iranian "Bay of Pigs." In short, Iran may find itself the occupant of a glass house: interference in the domestic affairs of its neighbors could boomerang on the Islamist regime with disastrous effect.[33]

Persian tradition leaves Iran with a very different calculus of the balance of ends and means in international security than that of the Western world. Iranians have never been pragmatic realists in their approach to international relations. Iran, like the United States, is partial to crusades: both tend to imbue foreign policy and conflicts with the moral baggage of the conflict between good and evil. Long before the revolution, the shah enthusiastically signed on in a partnership with the United States and Israel in the Holy War against communism. Iran's current Islamist leadership puts great store in moral and psychological victories and can view what might, to Western eyes, seem a setback as a net victory. Iran takes great pride in what it sees as two great victories over the United States: the seizure of the U.S. Embassy in Tehran in 1979 and the subsequent defeat of President Jimmy Carter in 1980, and the Iran-Contra affair, which seriously undermined a major Reagan administration foreign policy initiative. In both cases, the Iranians believe (not without cause) that their actions triggered major political crises in the United States and undermined both its self-confidence and its international image. Unlike Saddam Hussein, Iranian leaders understand that they cannot hope to challenge U.S. military power directly, and have thus opted for more indirect challenges to U.S. power: psychological and moral offen-

---

33. Roughly 60 percent of the Iranian population is too young to remember the 1979 revolution; hence, they have little memory of the "bad old days" under the Shah, or of the hard times of the early revolutionary years. They compare their lot to the outside world, and, whether radical Islamists or moderates, find the present theocracy sclerotic and wrong-headed. See "The Mullah's Balance Sheet," *Economist*, January 18, 1997, "Survey Iran" insert, p. 3.

sives rather than head-on military confrontations. Iran can be expected to continue its efforts to seize any opportunity to make the United States look weak, to undermine U.S. credibility as a friend of the Arabs, and to exploit rifts between the United States and its allies (as it has over economic embargoes and, more recently, punitive air strikes against Iraq).

IRAN'S MILITARY MODERNIZATION

A key theme in Iranian foreign policy is its ambition to restore Iran to a position of regional, if not global, respect; it is logical to assume that its principle objective in national security strategy should be the construction of a military capability consonant with such respect. In this regard, the Islamic Republic has remained consistent with its predecessors. Like the shah, who wanted to build Iran into a self-reliant power that did not depend on anyone else for its security and which could, if need be, strike fear in the hearts of potential aggressors or challengers, the Islamic Republic seeks to establish a sort of prickly autarchy that would inspire outsiders to leave it alone but would allow Iran to venture out and engage with the outside world at times and by means of its choosing. The shah exploited Iran's immense oil wealth and close ties to the West— especially the United States—to build, among other things, a modern, professionalized army and air force that could impose Iran's (and the West's) will throughout the Persian Gulf region. Revolutionary Iran will likewise strive to build a military capability that strikes fear and respect into the hearts of its potential enemies, but in a way suited to its goals and social and political capabilities. The desire to regain something of its former prestige and to do so by becoming an alternative to rather than a clone of the West provides much of the motivation for the modernization of Iran's military establishment.

Iran's anti-Western, revolutionary national myth provides two important hints as to the nature of future military modernization. First, Iranian Islamists have been highly critical of the extent to which Middle Eastern regimes—first the shah, now the Gulf Arabs—mortgaged the economic futures of their societies to acquire advanced Western military technologies. This, the Iranians believe, keeps them dependent on the West for training and maintenance (as Iran learned the hard way when its 1970-era air force deteriorated for lack of spare parts and know-how). Second, the Iranian Islamists have an iconoclastic view of how best to withstand the economic and military prowess of the West. Iran's view of itself and its relationships to the outside world points to a strategy aimed at deterrence; like the porcupine, Iran may not need to "win" but merely to inflict sufficient pain to cause its adversaries to back away and keep their distance. Iran will modernize its military capabilities according to its own unique vision of its strategic environment. The challenge for the United States and Iran's neighbors will be to anticipate what that model will be, how it might shape the direction of future Iranian military strategies and capabilities, and what might be done to modify or respond to Iran's strategic vision. The Iranians will have learned from the Persian Gulf Crisis of 1990–91 and are likely to strive to avoid Saddam Hussein's missteps. The most important of these lessons might be the risk inherent in constructing a large-scale Western-style military establishment. Most likely, Iran will attempt to build capabilities that can respond to the kind of conventional aggression that might come from a resurgent Iraq and defend Iran's interests in the Persian Gulf without triggering a massive Western military intervention. At the same time, the perception that the West is unable to eliminate Iraq as a nuclear or chemical threat might lead Iran to step up its own efforts to develop defensive capabilities in those areas.

An understanding of Iranian myth also suggests that the evolution of a formal, hierarchical military establishment on the European model is highly unlikely. The character of Persian social and political interaction—diffidence, dissembling, and conspiracy-mindedness—have perpetually undermined its political and military effectiveness.[34] Iran's decision-making structure reflects its decentralized social and political structure: there is no dominant voice in international security policy, competing power centers often work at cross-purposes, and decisions usually reflect unsatisfying and inefficient compromises.[35] Moreover, the mullahs who dominate the current regime would be unlikely to welcome the competition for power that would follow the founding of a robust military establishment. Political purges since the 1979 revolution have further undermined the professionalism of the Iranian military establishment. Iran's President Muhammed Khatami, elected in 1997, is likely to continue reforms initiated under Hashemi Rafsanjani's pragmatist regime to moderate the emphasis on Islamic correctness over technical expertise in the Iranian military, as well as in industry, transportation, and energy production. The pragmatists' increased efforts to liberalize society, professionalize the military, and mend fences with Iranian expatriates stems, in part, from the realization that Iran has paid a steep price for the politically driven depletion of its technical and professional classes.[36] Nonetheless, the creation of a large professional military—even along the lines of the pseudo-professionalism of the Iraqi Army—is unlikely.

---

34. Fuller, *The "Center of the Universe,"* pp. 11, 15–19.

35. Chubin and Tripp, *Iran-Saudi Arabia Relations and Regional Order,* p. 51.

36. "Iran Seeks to Deter Israeli Threat to Nuclear Facilities," *Mideast Mirror,* January 10, 1995, pp. 10–11.

The political and cultural limits on Iran's conventional military capabilities provide an important motive to acquire new technologies. Iran began acquiring ballistic missiles in 1985, when it became obvious that it was incapable of maintaining and operating an effective air force.[37] Iran's "human wave" tactics, which relied on mass armies of untrained but religiously zealous soldiers, were effective in defending against Iraq's more advanced military technology. Such mass armies are much less effective offensive forces, but they fit well into Iran's porcupine strategy as a disincentive to a conventional attack. An arsenal of nuclear, biological, and chemical weapons would complement Iran's conventional strategy, providing an additional layer of deterrence in the event that the human-wave defense failed or Iran's human and material resources were depleted. The nature of Iran's foreign policy and national security objectives and its perception of its role in the international community also leave it highly motivated to acquire both nuclear and biological or chemical weapons and their delivery systems.

Iran's military modernization and nuclear programs serve an internal political purpose as well. While the general trend toward moderation is likely to continue, Khatami and the pragmatists still face powerful opposition from more conservative quarters: hard-line Islamists are angry over what they regard as the secularization of Iranian society and the abandonment of revolutionary principles; moderates are frustrated and disillusioned by the snail's pace of democratization and modernization; and everyone is alarmed at the deterioration of the Iranian economy and the government's apparent inability to turn things around. Military modernization, especially the acquisition of a nuclear capability, continues to enjoy great political popularity and is widely perceived as a source of international respect and prestige. In fact, the military modernization is the greatest political success of a regime that has seen precious few in recent years. Iran's leadership also remembers the psychological impact that the mere threat of Iraqi chemical attacks against civilian populations had on Iranian morale in the final stages of the Iran-Iraq War, contributing to the Khomeini regime's decision to accept a negotiated peace in 1988.

However, the ultimate success of Iran's quest for a credible nuclear capability is uncertain, as is its commitment to its nuclear weapons program. Iran is clearly determined to develop or purchase nuclear capabilities. Iran currently spends roughly $2 billion a year on nuclear technology from Russia, China, Pakistan, North Korea, and the former

---

37. Seth Carus and Janne Nolan, "Arms Control and the Proliferation of Ballistic Missiles," in Alan Platt, ed., *Arms Control and Confidence-Building in the Middle East* (Washington, D.C.: United States Institute of Peace Press, 1992), p. 68.

Soviet republics. Given Iran's economic crisis, however, it is far from certain that it can maintain this pace of spending, especially given the resources it has recently devoted to conventional modernization. Iran's nuclear program is also vulnerable to political setbacks, should domestic economic realities force the diversion of resources away from military spending—though U.S. and European expressions of alarm and attempts to put pressure on Iran have breathed new political life into the program in the short term and will be a strong incentive to continue however high the economic price. A significant proportion of Iranian decision-makers, including some conservative hard-liners, oppose nuclear weapons on Islamic grounds. (Khomeini abandoned the shah's nuclear program, denouncing nuclear weapons as inhumane and un-Islamic.)[38] This is not to say that the Iranians do not intend to keep their options open. But any nuclear program seems likely to remain a low priority for the time being. Still, U.S. counterproliferation efforts, even given their limited formal success, have slowed the progress of the Iranian nuclear program, largely by discouraging Iran's most likely nuclear suppliers from fully supporting its proliferation efforts.[39]

Ultimately, Iran's Islamist leadership most likely believes it need not build a force capable of challenging the West, just one capable of deterring its Arab neighbors—particularly Iraq—and keeping the Western powers and their regional allies uncertain about Iran's intentions. (Inscrutability is, after all, a Persian high art.) Both the Persian tradition of conflict (whether interpersonal or international) and Islamist revolutionary myth put great stock in psychological maneuvering and moral victories. Hence, Iran will be perfectly capable of declaring victory in defeat, much as Saddam Hussein did, simply because it survived the onslaught from the West. Iran and Iraq share a determination to resist what they see as a Western and Israeli-dominated status quo in Middle East regional security. So far, Iran has sent a clear message that it intends to challenge the status quo, but has conducted its revisionist campaign below the level of a clear casus belli. If the Iranians have found the effort to rebuild and modernize their conventional capabilities and acquire nuclear weapons frustrating, those efforts have nonetheless netted Iran some political capital. The purchase of Russian-built submarines, the militarization of the disputed mid-Gulf islands, an increasingly robust tempo of naval exercises in the Gulf, and an aggressive program to procure nuclear weapons has brought Iran once again to the forefront of Western security concerns.

---

38. Gary Sick, "Rethinking Dual Containment," *Survival*, Vol. 40, No. 1 (Spring 1998), p. 17.

39. "Israel Digs In: No Signing of the NPT," *Mideast Mirror*, January 11, 1995, p. 8.

After the Iran-Iraq War, the West wrote Iran off for a time as a defanged, marginal threat. To the extent that at least part of Iran's motivation is to regain status, the Western reaction is a clear message that the strategy is working.

Iran's security objectives, including its pursuit of nuclear weapons, follow from its strategic personality and its deep sense of insecurity. Iranian security policy is based on three assumptions: that some form of U.S-led military aggression against Iran is inevitable in the near future; that Iraq will eventually rebuild its military capabilities and pose an equally certain threat; and that Iran could not stand up to either of its potential adversaries with its projected conventional military capabilities. Iran also knows it desperately needs to modernize and revitalize its economy but that it cannot do so without considerable help from Western Europe and Asia. Iran's technological and military backwardness and its lack of regional influence conflict with its arrogant self-image. The resulting dissonance could lead to two policy responses: shame and appeasement, or the stepped-up, compensatory bravado of the schoolyard bully. For the time being, at least, Iran has opted for the latter.

Iranian Nuclear Strategy and the Counterproliferation Agenda

The conventional wisdom says that Iran seeks to acquire nuclear capabilities because they are the "great equalizer" that would allow it to challenge Western military dominance directly, and perhaps offensively. This conventional wisdom overstates the similarities between Iran and Iraq. While Iran's military build-up and hegemonic ambitions do pose a threat to regional stability, Iran does not share Iraq's aggressive, territorially expansionist ambitions or its personalized and, hence, unpredictable security objectives. Moreover, Iran is ruled by a regime that, while certainly not a liberal democracy, is nonetheless accountable to its people and more democratic than the United States' Arab allies in the region—and certainly more than Iraq.[40] That accountability makes it

---

40. One school of analysis draws conclusions concerning Iran's likely nuclear strategies based upon its casualty-intensive "human wave" tactics during the Iran-Iraq War. (See, for example, Paula A. DeSutter, *Denial and Jeopardy: Deterring Iranian Use of NBC Weapons* [Washington, D.C.: National Defense University Press, 1997], p. 10.) Citing a so-called "cult of death," these analysts assume that Shi'a religious fervor and the myth of martyrdom leave Iran less risk-averse and more willing to accept casualty levels, even among civilians, that most other states would deem unacceptable. Apart from the cultural arrogance that such a conclusion implies, the facts do not bear it out for at least two reasons. First, the Iranian "human waves" were the product of desperation, not ideology (although Islamic fervor was clearly mobilized to gather volunteers for what amounted to a suicide mission)—the Iranian army had exhausted

much less likely to risk its survival recklessly for self-serving ends, as the Iraqi regime has done. Iran seems to be constrained by the same sorts of norms that govern Western concepts of the rules of war, although those norms are clearly colored by Iran's Shi'a myth. For example, Iran initially conceived of the Iran-Iraq War as a defense of righteous Islam against an onslaught from the blasphemous Saddam-Yazid (a mythic reference to the Umayyad caliph, Yazid, responsible for the murder of Imam Husayn). Therefore, Iran pursued a military strategy that sought to limit the impact of the war on the Iraqi people and avoid collateral damage, especially in the predominantly Shi'a provinces. The underlying assumption of this strategy was that the Iraqi people, given the opportunity, would rise in rebellion against the apostate Ba'athist regime. Only after several years, as it became clear that the Iraqi population had no inclination to overthrow Saddam Hussein, did the Iranian leadership lift restraints on the conduct of the war. The Iranians stopped viewing the Iraqi population as innocents and started treating them—the Iraqi Shi'a included—as willing followers of "Yazid." The redefinition of the Iraqi civilian population paved the way for Iran's decision to begin shelling Iraqi cities in 1984.[41] So, in this case, Iran's particular interpretation of its war with Iraq, at least initially, inspired even greater restraint than standard norms might have dictated.

Iran can produce chemical weapons and will soon be able to produce ballistic missiles to complement its North Korean *Nodong* missiles. To date, however, Iran has treated its chemical weapons as a deterrent capability and has demonstrated misgivings about their use. In fact, one of Iran's most bitter grievances against the West is that the United Nations and the Western powers issued no condemnation of or sanctions against Iraq for its frequent use of chemicals against Iranian forces after 1985.[42]

---

all other options. Public and clerical opinion quickly turned against the human cost of a war that, once Iran was on the offensive in Iraqi territory, was no longer seen as defense of the Iranian nation but an ill-considered and unpopular Islamic crusade. Second, the bombing of Tehran during the "War of the Cities," which is a better indicator of the Iranian response to nuclear attacks, led the Iranian public, the clerics, and eventually even the Ayatollah Khomeini to conclude that the survival of the Iranian nation outweighed even the loftiest Islamic objectives. (See Mackey, "The Iranians," pp. 324–333.)

41. John Kelsay, *Islam and War: A Study in Comparative Ethics* (Louisville, Ky.: Westminster/John Knox Press, 1993), p. 75.

42. Iran probably employed some captured Iraqi chemical weapons shells during the Iran-Iraq War, but their use was episodic, opportunistic, and on a very small scale. Anthony Cordesman, *Weapons of Mass Destruction in the Middle East* (London: Brassey's, 1991), p. 92.

Iran principally seems to see its chemical arsenal as a deterrent to Israeli nuclear weapons, as do some of the Arab powers. In the name of fairness and justice, Iran proclaims its right to have chemical weapons so long as the Western powers and Israel maintain their nuclear capabilities.

What if the Iranian porcupine grows nuclear quills? Three elements of the Iranian myth should figure prominently in any attempt to counter an Iranian nuclear strategy. First, Iran believes it is the center of the universe and the eventual seat of paradise; hence, it is culturally and morally stronger than any of its adversaries, especially the corrupt West. Moreover, its Zoroastrian and Shi'a traditions instill in Iran the confidence that it is destined, sooner or later, to defeat the forces of evil through the power of its righteousness and the favor of God. Therefore, it is not necessary or even desirable to pursue extremely risky strategies, especially ones in which the stakes are high (survival) and the chances of prevailing nearly nonexistent. Second, Iran will assume (as did Saddam Hussein) that the Great Satan does not have the mettle to stand up to pain and suffering—that the United States is unlikely to risk significant casualties in any conflict with Iran. Third, Iran's concept of "victory" is driven by its sense of shame over past foreign domination and the determination to defend its territorial, cultural, and religious integrity. It is not necessary that Iran defeat its adversaries, merely that it prevent their violating Iran's frontiers.

Iran's national myth will constrain its use of nuclear weapons. Because it sees the United States as the Great Satan that operates without moral constraints and with the aim of destroying the Islamic way of life, Iran has to assume that if it uses its nuclear weapons, the United States will not hesitate to retaliate in kind. The Iranians also contend that Iranian lives are expendable in the U.S. view, as demonstrated in its failure to condemn Iraqi gas attacks against Iran. Given these assumptions, Iran almost certainly will assume that U.S. retaliation would be far greater than the degree of damage Iran could inflict on the United States, Saudi Arabia, or Israel. Similarly, Iran (like its Arab neighbors) is acutely aware of Israel's vast military superiority, and its ability and willingness to punish far in excess of any pain Iran could inflict on Israel. Iran is also aware of Israel's national myth: that it will fight to the last Israeli to defend its right to exist and will be little constrained by international sanction or criticism.

Iranian nuclear weapons will almost surely be weapons of last resort aimed at Iraqi cities and military facilities, U.S. military forces in the Persian Gulf region, Israeli cities, and those Arab states that might cooperate with a U.S. military action against Iran. While it is clearly important to Iran to acquire nuclear weapons, Iran is no more likely than any other nuclear power has been to consider the employment of nuclear weapons with anything but the utmost gravity. Iran is not a risk-taker in its efforts to challenge the West; rather, it fusses, fumes, and tests the waters, while maintaining enough plausible deniability to back off without losing face when it meets real resistance.

Will Iran attempt to capitalize on its nuclear status in future saber-rattling and regional pressure tactics? Almost certainly. Is it likely to initiate what it certainly knows would be a suicidal first strike before all other options had been exhausted and Iran's survival was in imminent danger? There is no reason to think so, any more than for any other rational power. The message to the West would be "if you attempt to destroy us, we may go down, but we will take as many Americans, Europeans, and Israelis with us as we possibly can." The message to potential Arab members of an anti-Iran coalition would be similar: "If the Great Satan invades Iran, and threatens our sovereignty and Islamic way of life, then we will visit the wrath of God on those who invited them in and gave them haven." The underlying assumption of an Iranian nuclear strategy would be that no matter how much the United States, the West, and the Arabs might vilify Iran, they would not regard its defeat as worth that kind of risk.

IRAN'S INCENTIVES TO A PURSUE NUCLEAR WEAPONS CAPABILITY
Three factors drive Iran's quest to develop a nuclear weapons capability. Only one, the development of peaceful nuclear power, is openly declared. The official line is that Iran seeks to build a nuclear power infrastructure that would eventually provide 20 percent of Iran's electrical power.[43] Both Presidents Rafsanjani and Khatami have been vehement in their denial that Iran seeks nuclear weapons, and the insistence that Islam forbids it, that Iran hates nuclear weapons, and that "it is not a weapon that can be used in human confrontation."[44] Perhaps this is Iranian politics at its dissembling best. Few doubt that Iran's ultimate if not immediate goal is to build a nuclear weapons production capability, but peaceful nuclear power provides a fig leaf, both for the Iranians and for their potential suppliers (especially Russia and China) that do not wish to directly challenge the U.S. counterproliferation agenda. However, the economic argument also affords something of a lever for U.S. counterproliferation policy. The construction of a nuclear power infrastructure makes little

---

43. "Friends and Foes: How Iran Sees Itself in the World," *Economist*, January 18, 1997, "Survey Iran" insert, p. 10.

44. Hashemi Rafsanjani interview with Christiane Amanpour, CNN (Cable News Network) Specials, Transcript #534, July 2, 1995.

economic sense considering that Iran sits on one of the world's largest reserves of oil and natural gas. The Iranian argument that it needs nuclear power to modernize its economy could be deflated if European and Asian companies were encouraged to help Iran modernize its conventional electric power infrastructure. Such an approach could only minimally affect Iran's ambitions, but it could complicate Iranian efforts to find suppliers among states with an economic interest in maintaining reasonably good relations with the United States.

Iran's second objective in seeking a nuclear arsenal is defensive. Iraq continues to be a major security concern for Iran, despite the latter's recent threats to forge a revisionist alliance. Iran was traumatized by its military's inability either to defend against or retaliate for Iraqi missile attacks in 1988; its leadership is determined not to be caught short again. Iran feels doubly threatened by the unprecedented level of U.S. military presence in the Persian Gulf (which even the pro-U.S. shah would likely have found unacceptable). Quelling Iran's defensive concerns would be a complicated task, but there are steps that can be taken to reduce their destabilizing potential. One approach would be to recast the rhetoric of U.S. policy to minimize (or eliminate) its ideological aspects in favor of stressing international norms and concrete national interests. The United States and Iran share a number of common security interests in the region—starting with the continued containment of Iraqi military power. They also share a penchant for ideological crusades; but, in this case, the realpolitik and ideological agendas are irresolvable and work at cross-purposes. Iranian rhetoric radicalizes when its leadership feels cornered, but has abandoned its ideological agenda given the proper incentives and constraints (as in its relations with former Soviet Central Asia). It will likely be a very long time before the United States and Iran once again see themselves as strategic allies, but as Iran's neutrality in the 1991 Persian Gulf War demonstrated, the Islamic Republic can recognize (although not acknowledge) certain common interests with the United States. For that to happen, however, the United States will have to moderate its "dual containment" rhetoric and follow the lead of its European, Asian, and Arab neighbors that have resigned themselves to the legitimacy and likely longevity of the Iranian Islamist regime. The challenge will be to find a way to do so that avoids the appearance of backing down, which Iranians will characterize as weakness and exploit in its internal and regional propaganda. There are ways for the United States to decouple its approaches toward Iran from the containment of Iraq without easing legitimate pressure on Iran to improve its record concerning human rights, terrorism, interference in the internal political affairs of its neighbors, threatening military activity, and nuclear proliferation.

A second approach to defusing the destabilizing potential of Iran's strategic insecurity might be to revise the current U.S. policy concerning the embargo on conventional arms transfers to Iran. The current approach hands Iran a moral victory every time it circumvents U.S. efforts (which it does with reasonable frequency), and the difficulty Iran has had acquiring conventional weapons and spare parts increases the incentive to acquire weapons of mass destruction, especially nuclear weapons. While the denial of all Iranian military modernization might be the U.S. ideal, it is an unrealistic goal. If some modernization is inevitable, better that it be in conventional forces than in nuclear capabilities, especially since resources devoted to conventional forces may be resources taken out of the nuclear weapons program. A conventionally stronger Iran will complicate the regional security challenge, but not nearly so much as would a nuclear-armed Iran.

Iran's final incentive for becoming a member of the nuclear club is its desire to be a regional hegemon, a major regional power, or at least not marginal. Leadership of the Islamic world is a popular objective in Iran and one of the regime's few reliable sources of political consensus. Iran, in addition, has always seen itself as the strategic center of gravity in the Persian Gulf. The Iranians would prefer to be admired as a positive role model of a righteous Islamic society; but failing that, they will settle for being feared and pretend that is what they wanted all along. Ironically, while this is the aspect of Iranian policy that looks most ominous to the United States and the Gulf Arabs, it may be the least realizable for a number of reasons. First, Iran's ideological excesses consistently work against its ambitions. The intensely practical Saudis and most other GCC states have good cause to be highly suspicious of Iran and very skeptical of Iranian gestures of conciliation; it is a skepticism born of long experience and likely to be of long duration. Even those Western European states that have pursued a policy of "critical dialogue"—including Germany, until recently Iran's best European friend—have lost patience with the intemperance of Iran's ideologically driven misconduct. Second, Iran is beginning to show signs that it recognizes that it will eventually need to moderate its ideological stance to have any hope of becoming economically or militarily competitive. Internally, pragmatists and technocrats within the Iranian regime recognize the need to balance Islamic correctness with technical competence. Externally, Iran has acted surprisingly responsibly in Central Asia, largely in order not to alienate Russia, an important economic partner and military supplier. Likewise, for purely expedient reasons, Iran has made common cause with China, which treats its Muslims notoriously badly. In short, Iran is likely to continue to moderate its external behavior—albeit slowly and erratically—as the incentives to do so increase. Finally, even if Iran moderates its external behavior, regional dominance is likely to remain beyond its grasp. The pragmatic strategic differences that divide Iran and its regional competitors—especially Saudi Arabia—are wider and more intractable than those that separate Iran and the United States.[45] In fact, it is not beyond the realm of possibility that Iran and Saudi Arabia will at some point become competitors for U.S. support in the region, as they were before 1979.

U.S. AND IRANIAN BARGAINING STRENGTHS

Iran has a few advantages in resisting the United States' counterproliferation agenda. First, Iran is not alone in its suspicion of the major powers' motives in pressing the counterproliferation agenda. The fairness issue that Iran consistently raises, especially regarding Israeli nuclear capabilities, resonates throughout the Middle East and the developing world. Moreover, Iran's status as an uncompensated victim of Iraqi chemical weapons attacks gives its protestations a moral viability among the weapons of mass destruction (WMD) have-nots. Iran's rhetorical flourish and penchant for playing the victim have served it well in this regard. Fairness is an issue for which the antiproliferation major powers must eventually formulate an answer. Second, the Iranian nuclear and conventional military modernization programs are politically very popular in Iran, despite their cost. They seem to cause their Saudi and U.S. rivals heartburn, providing Iran with a symbolic victory and prestige. In fact, the revitalization of Iranian military might is a rare political success for a regime that has had precious few. The declared U.S. dual containment strategy tends to reinforce the popularity of the modernization program by lending credence to Iran's national paranoia. Finally, experience has demonstrated the difficulty of preventing the acquisition of nuclear technologies by a state that is determined to have them. U.S. efforts to cut off Iran's access to nuclear technology at the source have been somewhat successful; but U.S. relations with Iran's major supplier, China, chilled at the same time that Iran and China have begun to see common cause, especially in their mutual sense of grievance at perceived U.S. attempts to coerce them into internal reforms.

The United States is not without a few advantages of its own. First and foremost, Iran desperately needs an economic turnaround, which will depend on foreign investment and technical assistance. European and Asian investors are not likely to make significant capital commitments in Iran until they are sure they do not stand to lose their investments, as some did during the 1990–91 Persian Gulf Crisis. To reassure

---

45. Chubin and Tripp, *Iran-Saudi Arabia Relations and Regional Order*, p. 74.

them, Iran will have to demonstrate convincingly its ability to function as a responsible international agent. It cannot do so unless it achieves at least minimal rapprochement with the United States. The Islamist leadership will continue to protest loudly that it is unconcerned by U.S. opinion of its regime; at the same time its public ritual—daily flag-burning (a practice long staged by the Iranian regime, but which Khatami denounced in a 1998 address to the U.S. people on Cable News Network) and "Death to America" displays before the old U.S. embassy—betray their obsession. And the Iranian population will continue in its apparently unquenchable fascination with the United States and its culture, will continue to learn English, and will keep alive rumors of secret talks to improve U.S.-Iranian relations.[46]

The second U.S. advantage lies in pragmatic strategic and economic calculations: no state would suffer more than Iran were there to be a major disruption of traffic in the Persian Gulf, because Iran currently has no overland outlets for its oil. The United States is the only guarantor of security and reasonable stability in the region and the only naval power capable of keeping the Gulf open. Iran's mythic image of the United States (amplified by the 1991 Persian Gulf War) leads it to assume that Iranian actions will not be met with moderate or even proportional responses from the United States. Iran harbors little doubt that the United States will retaliate for any WMD first strike with catastrophic effect against which Iran could not hope to defend itself. Iranian leaders also know that, in that event, world opinion would be unlikely to take their side; even if it did, the United States would not be deterred.

Countering Nuclear Proliferation in Iran

The real challenge for U.S. counterproliferation policy toward Iran is to find the middle way between looking weak, vulnerable, or desperate and looking ominous and belligerent. Iran's national myth points to some approaches. U.S. policy should avoid actions that trigger Iranian national myth in ways that run counter to its objectives: arguments that focus on ideological issues rather than concrete national interest; rhetoric, such as the "Islamic threat"; policies or actions that hand Iranian hard-liners moral victories; or policies that overtly aim to overthrow or destabilize the current political order, and hence feed Iran's xenophobic paranoia. At the same time, it is possible to craft approaches that mobilize or respond to the Iranian myth in a way that advances the U.S. agenda. The United States must continue aggressively to occupy the moral high ground

---

46. "Friends and Foes," *Economist*, p. 10.

concerning weapons of mass destruction: point out the immorality of wasting scarce resources on expensive weapons of questionable defensive utility; turn Iran's self-righteousness and its antinuclear rhetoric against it by emphasizing the unjust, inhumane, immoral, and fundamentally un-Islamic nature of such weapons; and emphasize that Iran, a past victim of chemical weapons, bears a particular responsibility to lead the Islamic world in resisting the spread of such weapons. Nor can it hurt to prick the Iranian ego by championing new norms of international behavior that denigrate WMD as the last resort of failed, corrupt leaders trying desperately to acquire the empty trappings of power at the expense of building truly robust and successful societies. The "first tier" powers of the post–Cold War era—the United States and its European allies, Russia, and even South Africa—are trying to find ways to denuclearize. In the process, they are creating a new norm: strong, internally secure, and externally respected powers do not need the crutch of nuclear dominance and would certainly not bankrupt their societies to get it. At the same time, a new regime of international norms should hold up as paragons those developing countries (like the "Asian tigers") that have achieved international respect and influence, often without natural sources of wealth like oil, and that have avoided "Westernization" by investing in their economies and keeping their investment in military infrastructure proportionately low.

Finally, the United States can best avoid the appearance of either weakness or belligerency by dropping its side of the ideological battle and tailoring its rhetoric to a measured, realist worldview that focuses on Iran's behavior rather than its ideology and defines benchmarks for improved relations. In so doing, the United States need not abandon but only recast the ideological principles that define its own strategic personality. Instead of insisting that Iran is the new evil empire (which feeds both its ego and its paranoia) and demanding that our allies conform to that view, the United States can state that it finds Iran's behavior unacceptable—its support for international terrorism, its interference in the internal political stability of its neighbors, and its nuclear and chemical weapons proliferation. A unilateral stand that withholds U.S. economic and political approval but also states clear conditions under which the United States would be willing to relax its pressure on Iran shifts the burden for improving relations back to the Iranian regime. European advocates of "critical dialogue" are coming to the realization that they have been no more successful than the United States in moderating Iranian behavior. The Iranian Islamist regime, no matter how important its economic ties with states like Germany and France, seems unable to restrain its Jacobean tendencies—as it demonstrated in sending hit squads (under cover of diplomatic passports) to assassinate exiled dissidents in Germany. Those states, especially in Europe, that are now inclined either to buck U.S. policy as a sign of policy independence, and to attribute Iran's more intemperate behavior to U.S. provocation and cultural arrogance, may eventually run out of rationalizations and find themselves compelled to put greater emphasis on the "critical" dimension of their dialogue.

Upon encountering difficulties, those who deal with Middle East policy often throw their hands up and mutter about "inscrutable Orientals" and the unpredictability of non-Western cultures. The objective of this brief exercise has been to offer some insights into the mythic foundations of Iranian culture and strategic personality in an effort to make its rationality less opaque and its behavior less mystifying. Iran is not unpredictable; we just have not been very good at reading it. Iran is not irrational; we just have not fully deciphered the cognitive models that inform its behavior. The temptation to classify Iran as an irrational rogue that operates outside the limits of accepted norms is counterproductive. It leaves us with few positive options, since it is virtually impossible to devise rational approaches to modifying irrational behavior. The national myth–strategic personality approach offers a better option: it predicts the possible range of future Iranian behavior and likely reactions to U.S. strategic initiatives by seeking insight into how the Iranians think about their security and relationship to the outside world. To say the Iranian mind is inscrutable tells us nothing useful; to attempt to put ourselves inside the Iranian mind, however, can point to some constructive approaches if not immediate solutions.

Part II
Potential Evolution and Consequences of a
Nuclear Crisis with the United States

Chapter 5

Nuclear Proliferation and Alliance Relations

Stephen Peter Rosen

The recent multiple nuclear tests conducted by India and Pakistan in 1998 showed that countries that have a latent nuclear capability could move quickly and reliably from "bombs in the basement," that is, untested and unassembled weapons, to usable nuclear or thermonuclear weapons. The possibility is now more obvious that a latent or clandestine nuclear weapons state might move quickly to the status of a demonstrated nuclear power after it had initiated a military crisis. The opponents of that state might then face a state after it had committed an act of aggression, only to find out that it was capable of using nuclear weapons.

How will further nuclear proliferation affect the pattern of alliances between the United States and regional powers? During the Cold War, the creation of new nuclear powers tended to draw the superpowers into closer relations with their regional allies. The possibility of regional nuclear weapons use in the context of the U.S.-Soviet competition created the possibility that U.S. allies and Soviet clients might either be able to achieve decisive gains by nuclear threats or use, gains that would affect the U.S.-Soviet balance of power, or that U.S. allies and Soviet clients might become engaged in a nuclear conflict that could escalate and involve the superpowers in a war in which nuclear weapons were being used. As a result, the United States and Soviet Union tended to intervene to stabilize regional conflicts and prevent regional nuclear weapons use. The perception of this dynamic provided the foundations for the de facto nuclear strategies of several smaller nuclear powers. With the demise of the Soviet Union, regional nuclear weapons use may still occur, but one incentive for the United States to become involved has been greatly decreased. The costs and risks of intervention in a regional nuclear conflict were always high. Now that the need to balance against another superpower is reduced or gone, U.S. military intervention after nuclear threats have been made in the context of a regional war may be less likely. The regional allies of the United States may well have less confidence that the United States will intervene militarily in the face of regional nuclear threats. The belief that the United States would project military power to ensure the survival of its regional allies provides the basis of our asymmetrical alliances with regional powers, alliances in which the United States extends guarantees to them in return for access to military bases on their soil. These alliances are unlikely to survive regional nuclear threats to which the U.S. does not militarily respond. The result will be a sharp decrease in the military access granted by our former allies to bases on their soil or to the maritime areas and air space over which they claim jurisdiction. In the next chapter, Barry Posen argues that the large and negative long-term consequences of nonintervention might lead the United States to intervene, despite the near-term risks; this chapter explores the reasons why and circumstances under which the United States might not intervene.

To support this argument, I first review the information that has emerged about the new nuclear states' plans for nuclear weapons use developed during the Cold War. Second, I try to assess the impact of regional nuclear proliferation on the incentives of the United States to intervene militarily in regional conflicts by mentally "replaying" the 1991 war against Iraq with the assumption that Iraq had a small nuclear arsenal. Based on the air war against Iraq, what can we say about the United States' nuclear strategy in a hostile encounter with a regional nuclear power? The effectiveness of the U.S. bombing of Iraqi nuclear facilities can now be evaluated on the basis of some of the reports made after on-site inspections of those facilities conducted under United Nations auspices, interviews with members of inspection teams, and the report submitted to the U.S. Secretary of the Air Force, the Gulf War Air Power Survey.[1] Based on the experiences of 1990–91, what might have been the characteristics of an encounter between the United States and a nuclear-armed Iraq? What might Iraq's nuclear strategy have been, given what we know about its actual regional political-military strategy? This chapter examines what the events of August 1990 through August 1991 might have looked like if Iraq had completed a nuclear weapons development program *before* it invaded Kuwait, but did not *reveal* its capabilities until *after* it invaded Kuwait. I then develop a partial answer to the second question by examining in detail what we now know about the actual effects of the U.S. bombing campaign directed against Iraqi nuclear

---

1. Thomas A. Keany and Eliot A. Cohen, *Gulf War Air Power Survey: Summary Report* (Washington, D.C.: Government Printing Office, 1993).

weapons–related facilities in 1991. Together, the analysis of the actual bombing and the review of the hypothetical Iraqi use of nuclear weapons suggest the nature of the problems the United States might face in encounters with regional nuclear powers over the next five to ten years. This look at the future is supplemented by surveys of how members of the U.S. policy community actually responded during 1993 when they were asked to develop policy options for the United States in simulated encounters with regional nuclear powers.

The result of this examination can only be tentative, but it is striking. There appear to be no good non-nuclear military options open to the United States when facing a regional nuclear power. There is little detectable sentiment in the U.S. policy community today to use U.S. nuclear weapons against aggressive regional nuclear powers. Without the global strategic need to prevent military gains by clients of the Soviet Union or to prevent regional nuclear wars from escalating to U.S.-Soviet nuclear war, there is less incentive for U.S. military intervention. Hostile regional nuclear powers are, therefore, in a good position to deter U.S. military intervention against them, and so may be in a better position to threaten the allies of the United States that do not have nuclear weapons and are within reach of the non-nuclear forces of the aggressive power. The main purpose of regional nuclear weapons use is likely not to be against cities or military targets in the region. Nuclear weapons use is most likely to follow an act of aggression, and to be purely demonstrative, designed to deter U.S. military intervention and to convince the other powers in the region that they cannot rely on U.S. security guarantees, so that the aggressor can obtain its foreign policy objectives without going to war with the United States. The asymmetrical U.S. relations with its regional allies in Europe, the Middle East, and the Pacific are unlikely to survive if the patterns of behavior forecast in this essay take place. The chapter concludes with a discussion of what measures the United States might take to increase its ability to preserve its alliances by increasing its ability to intervene militarily against small nuclear powers.

Nuclear Weapons Doctrine in New Nuclear States

What do we now know of the ideas concerning nuclear weapons use in India, Pakistan, South Africa, and Israel?

In 1990 and 1993, I interviewed the retired chief of staff of the Indian Army, General K. Sundarji, and the head of Secretary of the Defense Research and Development Organization of the Indian Ministry of Defense, Dr. V.S. Arunachalam. The timing of these interviews, which were given on the initiative of Arunachalam, may have been related to the decline and then the demise of the Soviet Union, and the perceived need in India to develop better relations with the United States. One element in the effort to improve U.S.-Indian relations would be some explanation of the nuclear weapons custody arrangements in India that would reassure the United States that weapons would not be used inadvertently or in thoughtless ways. Arunachalam has subsequently denied in print the substance of these interviews. In this paper, I present what he told me in December 1990, and what he confirmed as accurate in the spring of 1993, and what I reported to the United States government in December 1990.

In 1990, Arunachalam, though formally denying that India had nuclear weapons, adopted the following position. It would be reasonable, he said, for non-Indians to assume that since the 1974 Indian nuclear weapons test, the government of India had worked very hard at resolving the command and control problems for nuclear weapons. In fact, the civilian leadership had fought a long and difficult struggle with the Indian military to decide who would control nuclear weapons and how they would be used. This struggle was resolved in favor of the civilians.

The outcome was that the Indian military would not be and has not been told how many nuclear weapons India could have, nor will it be told in peacetime how nuclear weapons would be used in war. The inescapable implication was that the Indian military does not have custody of nuclear weapons components in peacetime. Since the military cannot plan for nuclear war, a set of detailed instructions on how to get access to nuclear weapons and how to employ them has been prepared by civilians. These sealed instructions have been given to a military officer with instructions to open them in the event that nuclear weapons have been used against India and have destroyed the Indian national command authority at New Delhi. To quote Arunachalam, "If New Delhi goes up in a mushroom cloud, a certain theater commander will go to a safe, open his book, and begin reading at page one, paragraph one, and will act step by step on the basis of what he reads." The technical means for command and control are rudimentary, but adequate for Indian needs. No provision has been made to harden Indian communications against the effects of electromagnetic pulse (EMP), since it is thought that the number of nuclear weapons that would be used against India would be sufficiently small and the campaign sufficiently short that EMP blackout would not be substantial enough to interfere with the execution of Indian plans.

Much of what Arunachalam said was placed in broader context by the remarks of General Sundarji in interviews in 1990 and 1993. He said that India has adopted a pure minimum deterrence posture, for two reasons. First, India cannot afford to develop a nuclear force to destroy

Pakistani nuclear weapons because the Pakistanis have dispersed and hidden their nuclear weapons and India does not have enough resources to find and strike them. Second, India has sufficient non-nuclear forces to handle Pakistani non-nuclear attacks and thus does not need to threaten Pakistan with a nuclear first strike in order to deter war. When asked what India would do if Pakistan achieved some limited initial territorial gains in an invasion of Kashmir, halted, and then threatened to use nuclear weapons if India tried to push it back, Sundarji said India would simply proceed with its non-nuclear counterattack.

After the 1998 nuclear weapons tests, the only public statement about Indian nuclear doctrine was given by Prime Minister Atal Behari Vajpayee, who publicly stated that India needed only a credible deterrent, and would not need a large infrastructure to control or safeguard Indian nuclear weapons.[2] While consistent with the Arunachalam-Sundarji statements, it does not confirm them.

The alleged Indian posture has several advantages. It would make the Indian problems of obtaining warning and of executing its nuclear strike much simpler. There would be no need for India to retaliate quickly. Although India's retaliation after a Pakistani use of nuclear weapons would not be delayed indefinitely, it would be acceptable for India to retaliate in something like twelve to forty-eight hours. Capabilities to execute a slow-motion nuclear war would be much cheaper and easier for India to build. Most importantly, the alleged strategy would prevent military custody of nuclear weapons or any control over nuclear weapons in peacetime.[3]

On the Pakistani side, the retired Chief of Staff of the Pakistani Army, Mirza Azlam Beg, wrote in April 1994 that Pakistan had acquired a nuclear weapons capability but had capped it "because we thought that Pakistan had acquired the maximum deterrence level that is needed to avert the threat that we perceived and . . . could also quantify, for example, the correlation between India and Pakistan . . . conventional forces," forces which he wrote favored India by "three to one in ground forces and five to one in air forces." As a result, he quoted approvingly the statement by Sundarji that "between India and Pakistan a substantial degree of nuclear stability exists," and Beg went on to write that the existence of Indian and Pakistani nuclear weapons "certainly has reduced the possibility of war." The Pakistanis, he wrote, had created a nuclear

2. Kenneth Cooper, "Leader Says India Has A 'Credible' Deterrent," *Washington Post*, June 17, 1998, p. 21.

3. Interviews conducted by author in Pune, India, December 1990, and Cambridge, Massachusetts, March 1993.

chain of command that was run by the head of state operating through a National Nuclear Command Authority that issued orders to a military Joint Operations Center in Rawalpindi.[4]

The Indian-Pakistani nuclear relationship, thus, appears to have emerged as a stable relationship in which non-nuclear military imbalances are neutralized by the existence of nuclear weapons. At the same time, there are persistent reports that the possibility of nuclear weapons use in South Asia has been manipulated by the Pakistanis in order to increase the level of U.S. diplomatic intervention.

Some evidence suggests that other smaller states have explicitly considered the use of nuclear weapons in ways that would lead to intervention by the United States. For example, the 1993 statement by the president of South Africa explicitly asserted that while the purpose of South Africa's seven nuclear weapons was related to deterrence, the South African strategy for nuclear weapons "use" had nothing to do with counter-city or counterforce targeting, nor even directly with the regional adversaries of South Africa. Rather, "the strategy was that if the situation in Southern Africa were to deteriorate seriously, a confidential indication of the deterrent capability would be given to one or more of the major powers, for example the United States, in an attempt to persuade them to intervene."[5] This plan for nuclear weapons use resembles that imputed to Israel. It has long been hinted that the strategy that Israel employed in the early hours of the 1973 war was similar to the South African nuclear strategy, and that Israel initiated "preparations to turn the nuclear option into usable nuclear weapons," and then communicated to the United States Israel's "immediate need for arms," as well as "the desperate alternative it contemplated if it did not get help."[6]

The concepts concerning nuclear weapons use in the four cases of India, Pakistan, South Africa, and Israel appear to fall into two categories: minimum deterrence nuclear strategies and strategies to draw the intervention of the superpowers. They do not appear to consider the use of nuclear weapons to deter superpower military intervention, but this is a logical possibility in Northeast Asia or Southwest Asia.

---

4. Mirza Alam Beg, "Who Will Press the Button?" *The News* (Islamabad, Pakistan) April 23, 1994, p. 6, reprinted in Foreign Broadcast Information Service, FBIS-NES-94-086, May 4, 1994, pp. 69–71. See also the remarks by Beg in *Pakistan Observer*, December 2, 1993, p. 1, reprinted in FBIS-NES-93–230, December 2, 1993.

5. Speech by F.W. De Klerk, March 24, 1993, released by the embassy of South Africa to the United States, March 26, 1993.

6. Nadav Safran, *Israel: The Embattled Ally* (Cambridge, Mass.: Harvard University Press, 1978), p. 488.

There is no hard information that could provide us with certain estimates about how nuclear weapons might be used by regional powers to deter intervention against them. Could small and medium-sized nuclear powers be confident that their small arsenals could survive a non-nuclear attack by the United States? If so, would they then be confident of their ability to deter U.S. military intervention? Given the data available from the 1991 Gulf War, what can be said about the ability of the United States to destroy the nuclear weapons of small states by non-nuclear means?

U.S. Targets in Iraq in the 1991 War

DISARMING STRIKES: THE AIR WAR AGAINST IRAQI BALLISTIC MISSILES

Two sets of Iraqi targets that were attacked by the aircraft and missiles of the United States in the 1991 war were relevant to Iraqi nuclear war-making capabilities. The first was the Scud ballistic missile, one of the possible delivery systems for Iraqi nuclear weapons. The Gulf War Air Power Survey (GWAPS) analysis of the air campaign against the Iraqi Scud missile launchers is thorough and candid. It notes that possible Scud hiding places, suspected fixed launch sites, production facilities, and mobile Scud launchers were attacked. One thousand five hundred aircraft strikes of all kinds were launched against Scuds, out of a total of some 42,000 strikes by all Coalition fixed-wing (as opposed to helicopter) aircraft during the war. One thousand aircraft were sent on Scud "patrols." Only 215 strikes were directed at what the pilots thought were mobile Scud launchers. The Iraqis appear to have relied totally on mobile launchers, and did not employ any of the suspected fixed launch sites. The effectiveness of the air attacks against mobile Scud launchers was extremely limited, in three senses. First, because many Scuds were launched at night and the launchers were small and hard to see from the air, even in the forty-two cases when the flare of light from a Scud launch was observed by a U.S. aircraft pilot in a strike aircraft circling in the area, in only eight cases was the pilot then able to *see* a target well enough to launch a weapon against it. Second, while the number of Scuds the Iraqis were able to launch does appear to have been reduced because they were being hunted by U.S. aircraft, this reduction was limited. Although the number of Scud launches did decrease from the first to second weeks of the war, and then dropped further in the third week, the number of launches then began to increase, and in the last eight days of the war there were as many Scud launches as in the second week of the war. Third, in the words of the report:

The fundamental sensor limitations of the Coalition aircraft, coupled with the effectiveness of Iraqi employment tactics (including use of decoys), suggest that few mobile Scud launchers were actually destroyed by Coalition aircraft or special forces during the war. . . . Once again, there is no indisputable proof that Scud mobile launchers—as opposed to high-fidelity decoys, trucks, or other objects with Scud-like signatures—were destroyed by fixed-wing aircraft.[7]

One cannot infer directly from the 1991 campaign against Iraqi Scuds armed with non-nuclear warheads that a similar campaign waged against Iraqi missiles armed with nuclear warheads would have looked the same. (This issue is discussed under the section that considers hypothetical Iraqi nuclear weapons use.) However, one issue that would have been common to both the actual non-nuclear ballistic missile force and a hypothetical nuclear force would have been the physical aspects of the command and control system used to give orders to missile commanders to launch their weapons. Though the Iraqi command and control system was subjected to 840 precision and nonprecision attacks, "by the end of the second week of the war, Coalition air planners and intelligence analysts became increasingly convinced that Iraq's national-level telecommunications system had not collapsed as a result of attacks on central switching and microwave relays. . . . Fiber-optic networks and computerized switching systems proved particularly tough to put out of action. . . . Moreover, the Iraqi government had been able to continue launching Scuds during the final days of the campaign."[8]

THE AIR WAR AGAINST THE IRAQI NUCLEAR WEAPONS INFRASTRUCTURE

The second target set was composed of facilities that produced the materials necessary for the production of Iraqi nuclear weapons. A military campaign to destroy Iraqi nuclear weapons would necessarily have attacked this infrastructure as well in order to prevent Iraqi nuclear rearmament. How effective was the bombing of the Iraqi nuclear infrastructure?

To begin with, how should "effectiveness" be defined? It is routine for the U.S. Air Force to identify a set of targets, attack them, and then to assess the effectiveness of the campaign by counting the number of targets on the list against which attack missions were flown, and then, using aerial photography, to count the number of targets hit and damaged. The percent of targets on the original list damaged constitutes one measure of effectiveness. A second, more appropriate measure of effec-

---

7. Keany and Cohen, *Gulf War Air Power Survey,* pp. 83–90.

8. Ibid., pp. 69–70.

tiveness is the one adopted by the Economic Objectives Unit of the Office of Strategic Services in its work to support the U.S. Eighth Air Force during World War II. In that assessment, the measure of effectiveness adopted was how the industrial systems—composed of a network of factories, warehouses and so on—were affected by bombing. Since the industrial systems that produced aircraft, engines for armored vehicles, and petroleum could be rebuilt, and since the consumers of the output of those systems could draw supplies from stockpiles of produced end items, the relevant question was how many months' worth of industrial production had been destroyed by bombing. This could then be compared to rates of consumption to achieve a meaningful measure of the effectiveness of the bombing.[9] A modified version of this measure of effectiveness appears to be appropriate for the Iraqi nuclear weapons infrastructure. There, too, facilities that were damaged or destroyed by bombing could be repaired or replaced by stockpiled output of those facilities or by stockpiled, redundant production capabilities. Since there was and is no observable rate of consumption for Iraqi nuclear weapons, the final measure of effectiveness is simply how many months of delay would be introduced into the Iraqi nuclear weapons production system. The air campaign, according to Iraqis interviewed by the International Atomic Energy Authority/United Nations Commission (IAEA/UN-SCOM) teams, halted the Iraqi nuclear program in mid-January 1991.[10] But what was the capacity of the Iraqi program to recover and begin work again once the bombing stopped? On the last day of the air war against Iraq, how long would it have taken the Iraqis to restore their nuclear weapons production capacity to where it had been when the air campaign began?

Unfortunately, this straightforward question has no straightforward answer. Events only indirectly related to the bombing added to the actual delays the Iraqis faced at war's end. The victors insisted that IAEA inspection teams be given unhindered access to all known and suspected Iraqi nuclear weapons–related facilities. This led Iraq to dismantle, hide, or destroy certain portions of its nuclear infrastructure. In addition, the IAEA teams were themselves able to order the destruction of equipment related to the production of nuclear weapons, and to monitor Iraqi sites after the war. The U.S. victory in the war increased the ability of the

---

9. Walt W. Rostow, "EOU and its Doctrine," *Pre-invasion Bombing Strategy* (Austin: University of Texas Press, 1981), pp. 15–23.

10. Report by the Director General, IAEA, "The Implementation of United Nations Security Council Resolutions 687, 707, and 715 (1991): Addendum to GC(39)/10," September 4, 1995, p. 2.

United States to restrict the flow of oil revenue and foreign technology into Iraq, and raised global sensitivities to selling any kind of industrial and scientific equipment to Iraq. In part, these were indirect results of the bombing of Iraq, which increased the difficulty Iraq faces in trying to reconstruct its nuclear weapons program. As a result, the answer to the question of practical policy interest, "how many months *will* it take Iraq to rebuild its nuclear program to where it was in January 1991?" is quite different from the question, "by how many months did the bombing itself set back the Iraqi nuclear weapons program?" It is the answer to the second question in which we are interested, so that we can understand the value of bombing isolated from the value of militarily occupying a hostile regional nuclear power and from the value of alternative postwar control regimes.

The most important source of information to answer this question comes from the IAEA inspection reports, which are unclassified.[11] The members of the IAEA teams were on loan to the United Nations from various U.S. and European government agencies or from jobs that gave the team members technical expertise that allowed them to judge the development of and damage to a nuclear weapons production infrastructure. Twelve inspections were carried out between the end of the war and June 1992. There is some disagreement among the IAEA analysts as to exactly how close the Iraqis were to a bomb before the war, and about which directions the Iraqi program was likely to take in the years after the war; however, when asked to provide their best judgment on how much time Iraq would have needed to rebuild its nuclear weapons program on the day the bombing stopped, given where they thought it was when the war started and assuming the same access to international resources that was available before the war, the analysts expressed a high degree of agreement on seven points.

First, the component of the Iraqi nuclear program that was furthest from completion in January 1991 was the production of the fissile material for an atomic bomb. There are debates about how long it would have taken Iraq to have produced enough fissile material for a bomb, but the analysts agreed that all other problems of bomb manufacture were solved or were solvable within the shortest period of time that would have been needed to manufacture the necessary amount of fissile material. Iraq had in 1991 workable detonators, high-explosive lenses, neutron initiator technology, and a valid design for an implosion bomb incorporating

---

11. For a brief summary of the state of the Iraqi nuclear infrastructure in January 1994, after the war and after twenty-two IAEA inspection visits, see *IAEA Action Team for Iraq: Fact Sheet* (Vienna: IAEA, January 18, 1994).

design features that the United States introduced into its fission weapons in the late 1940s to reduce their size and weight.[12] Making enough weapons-grade fissile material was the task that determined the pace of the overall Iraqi program. It was the long pole in the tent.

Second, after the bombing of the Osirak reactor by the Israelis in 1981, the Iraqis switched from work on a plutonium bomb to an enriched uranium bomb that did not require a highly visible, vulnerable reactor to produce weapons-grade fissile material. Large amounts of uranium ore were obtained from a superphosphate mine in Iraq at Al Qain.[13] Of the many routes for producing highly enriched uranium in which Iraq had invested time and money (electromagnetic isotope separation, centrifuge enrichment, ion-exchange technologies, gaseous diffusion, jet nozzle), electromagnetic isotope separation—known variously as EMIS and Calutrons and Baghdadtrons—was the route most likely to have produced enough fissile material for an Iraqi bomb first. The other methods may have proven useful for providing partially enriched uranium for further enrichment in the EMIS process. The Iraqis manufactured EMIS devices in Iraq. Some necessary items of electrical equipment to support the EMIS devices were purchased abroad. Iraq had at least thirty EMIS units running or being assembled and plans to build forty more. There is uncertainty about when the forty devices would have been finished. One assessment is that seventy EMIS devices could have been operating by January 1992; seventy EMIS devices could have produced enough enriched uranium for a bomb that was not particularly efficient in its use of uranium in three years, that is, by January 1995. If the other separation technologies had been used to provide partially enriched feeds to the EMIS systems, that time might have been cut to one year, that is, to January 1993, but there is debate about how far along the other separation technologies were.

Third, the mine at Al Qain that supplied Iraqi uranium oxide was

---

12. See the report of the fourth IAEA visit on its trip to Iraq, July 27–August 10, 1991, which found exploding bridge wire detonators, and polonium and beryllium for neutron initiators at Al Qa Qaa. See also the report of the fifth IAEA visit to Iraq of September 22–30, 1991, which found high explosive lenses for an implosion device, and documents at Al Atheer covering Iraqi activities up to May 31, 1990, which include an Iraqi design for a weapon incorporating a flying plate core to increase the efficiency with which energy is transferred to the core of fissile material. See also the independently prepared briefing charts of IAEA team member David Kay, "Iraq and Beyond: Challenges in Controlling Nuclear Proliferation," February 1992.

13. Uranium oxide (UO_4) can be produced as a byproduct of superphosphate extraction. Iraqis reported to the eleventh IAEA team in Iraq in April 7–15, 1992, that they had produced 168 tons of uranium oxide powder, and had also purchased natural uranium from Brazil.

heavily bombed during the 1991 war. The first IAEA team to visit it, in July 1991, estimated that it would take the Iraqis two to three years to put it back into operation. In fact, the mine was back in operation producing superphosphates by September 1991, within six months after the end of the war. IAEA interviews noted that in any case, enough uranium oxide was available before the war began and remained after the war to sustain the enrichment program. Overall, the delay created in the Iraqi nuclear program by the bombing has to be judged at between zero and six months. However, uranium oxide must be turned into a gas before it can be enriched by the EMIS process. The production of that gaseous form of uranium was performed at Al Jazira at a turn-key plant built from the ground up by a foreign firm. If the Iraqis had been free to go back to that firm, and if that firm, or another one like it, had been willing to do the work, rebuilding the plant would have taken about two years. It is not known if Iraq had learned something about how to build the plant after having operated it for a number of years. If Iraq had learned a lot, and had been free to devote unlimited resources to its reconstruction, a crash program might have rebuilt it in eighteen months.

Fourth, the EMIS separation process consumes enormous quantities of electrical power, as do other enrichment processes. The electrical power generating and distribution system in Iraq was heavily bombed during the war, and some 88 percent of its installed generating capacity was hit and damaged. But the impact of this destruction was far less. Iraq had installed almost twice the generating capacity it needed, and so, in effect, had redundant electrical power generators. IAEA inspectors also found that Iraq had numerous stockpiled transformers that could be used to replace electrical power substations that were damaged by bombing. Sheds containing over 100 spare substation transformers were observed at nuclear facilities during initial inspections. They disappeared by the time of follow-up inspections, and the assumption was that they had been used to help repair the national power grid. The net effect was that adequate electrical power production relative to consumption was quickly restored by May 1991. The summer of 1991 was particularly hot in Baghdad, and all available air conditioners in that city ran flat out, with no observed electrical brownouts or blackouts. Air Force planners had thought that it would take two years to restore full power to Baghdad.[14]

Fifth, the impact of U.S. bombing on EMIS devices and processes was limited by the redundancy that Iraq had built into the program, by its efforts to protect the EMIS devices from bombing, and by the ease with

14. Keeney and Cohen, *Gulf War Air Power Survey*, pp. 71–74.

which the devices could be put back into operation after disruption. The Iraqis went to considerable lengths to create redundancy in their EMIS program. The first EMIS enrichment facility was constructed at Tuwaitha. Duplicates of that plant were built at Tarmiya and Ash Sharqat.[15] There may have been additional unidentified sites. The first IAEA team to visit known EMIS installations found that some EMIS machines themselves (the disks, vacuum chambers, and pole magnets) had survived the bombing, either inside damaged buildings or underground, where they had been buried to protect them from bombing or from IAEA inspection. Some EMIS equipment was found to have been destroyed, but inspectors estimated that 95 percent of the damage observed had been done by Iraq itself in the process of moving the EMIS machinery in order to conceal it from IAEA inspectors, not by bombing. IAEA teams reported Iraqi efforts to recover hidden EMIS equipment in July 1991. Estimates of how long it would take Iraq in the absence of United Nations monitoring and controls to dig up and restart hidden EMIS devices vary. There is disagreement about how well the EMIS device survives burial, and disagreement about how long it takes to repair the fragile electronic and hydraulic equipment that supports the EMIS device. That supporting infrastructure was more heavily damaged than the EMIS devices themselves. The EMIS devices themselves are constructed from heavy pieces of steel, are hard to damage, and are not technologically difficult to build with appropriate machine tools. Totally rebuilding that infrastructure at all known sites would take about two years. However, considerable amounts of equipment survived the bombing, and repairs to the damaged EMIS devices themselves were estimated to take no more than twelve months.[16]

Repairs to the EMIS devices or to other components of the Iraqi nuclear program would have depended on the availability of machine tools. EMIS devices and centrifuges for uranium isotope separation can be built only with modern machine tools such as vertical and horizontal flow forming machines and multiple axis, computer-controlled milling and boring machines. IAEA inspectors who were experts in machine tools surveyed the sites where such machine tools were known to have been in operation. Many of these machine tools, by virtue of the heavy work they perform, are themselves constructed out of heavy pieces of steel, and are inherently hardened against bombing. IAEA inspectors visited sites with bombed buildings that housed machine tools still in operable condition. They also saw sites from which heavy machine tools had obviously been removed, and large stockpiles of spare machine tools. In April

---

15. Report of the fourth IAEA inspection trip in Iraq, July 27–August 10, 1991.

16. Ibid.

1992, IAEA inspectors identified 122 machine tools "which have technical characteristics required for producing key components needed in a nuclear program." All ten of the computer-controlled machine tools that were known to have been delivered to Iraq before the war and that would be needed to make key parts of centrifuges for isotope separation survived the bombing. IAEA inspectors found fifty-four boring mills that would be used to make EMIS disks, and estimated that there were an estimated one hundred such mills in Iraq in 1992. In Al Atheer, the special presses needed to produce the high explosive lenses for implosion-type atomic bombs survived the war. The extent to which Iraq protected itself against bombing by buying machine tools and storing them away from existing factories is suggested by an IAEA report that an Iraqi factory that had been leveled by bombing was by June 1992 fully rebuilt and equipped with 150 theretofore unknown milling machines; it was busy making artillery shells.[17]

Sixth, nuclear reactors still played a role in the Iraqi program, and were inherently hard to hide, protect, or make redundant. Although the Iraqis were not relying on reactors for the production of plutonium for use as the fissile material in a bomb, their reactors were still important for the production of short-lived isotopes of hydrogen, polonium, and plutonium, all useful as neutron initiators to begin the fissioning of the core of an atomic bomb at the proper moment. The Iraqi research reactor used for the production of those isotopes was destroyed by bombing. Rebuilding that reactor would take, under favorable conditions, an estimated five years. The impact of the destruction of that reactor is harder to estimate. The necessary isotopes could have been produced and stockpiled before the war, but would decay in a few years or less. However, devices are commercially available that can produce neutrons for initiation without the use of isotopes. The use of such devices imposes some design problems on weapons builders. The Iraqis showed IAEA inspectors neutron initiators on which they were working, but it is unlikely that they were completely candid about the true extent of their competence in this area. As a result, the full impact of the destruction of the Iraqi nuclear reactor cannot be confidently assessed.

Finally, in many ways, the most important component of the Iraqi nuclear weapons program remains the highly skilled people working in it who had solved many problems and gained the expertise that would aid them in solving others.[18] Iraq invested at least as much time and effort

---

17. Report of the eleventh IAEA inspection in Iraq, April 7–15, 1992.

18. For a discussion of the importance of the availability of highly competent scientists and engineers to regional powers, like Iraq, which are working on an atomic

in creating redundancy in this cadre of technical workers as it did in creating redundancy in its physical plant. For example, teams of Iraqis were brought in to operate the original EMIS facility at Tuwaitha and then were moved out to operate another EMIS facility at Tarmiya, while a new team was put in to train at Tuwaitha.[19] Approximately 20,000 people were estimated to have been employed in the clandestine Iraqi nuclear weapons program.[20] It is not clear that all key members of that program have been identified. There is, for example, the possibility that Iraq had a large program concentrated on the use of centrifuges for isotope separation that was staffed by a group of people separate from the major program and facilities. Given the lack of information about the identities of these people and the ease with which people can be dispersed and protected against bombing, the impact of the bombing on the Iraqi nuclear weapons program personnel must be assumed to have been minimal.

HYPOTHETICAL IRAQI NUCLEAR WEAPONS IN 1990
If the available data suggest that the United States would have had considerable difficulty in destroying or otherwise neutralizing Iraqi nuclear weapons capabilities, they do not tell us how Iraq might have employed any nuclear weapons that it might have had but could not actually construct before the war began. For the purposes of this exercise, some arbitrary assumptions must be made about what the Iraqi nuclear weapons program might have achieved in 1990 and 1991. To avoid debate about what the Iraqis really might have achieved and when, questions that are important but not central to this exercise, this chapter simply stipulates a notional Iraqi nuclear force that represents the kind of regional nuclear threat that the United States might have to face at some time in the future. The Iraqi nuclear capability will be assumed to have been between five and ten fission weapons that could have been launched by ballistic or cruise missiles. This particular capability was certainly not available to Iraq in 1991. (The exact time when Iraq could have achieved this capability is, once again, an important question but not central to this paper.) Iraq would have had to have developed a nuclear warhead that could have been moved around with a mobile missile and then mated with it when the time for launch came. It is worthwhile noting that the United States had developed such a weapon, the Mk-7/W-7 weapon,

bomb, see Russell Seitz, "Interesting Times," Olin Institute for Strategic Studies, Monograph No. 2.

19.   Report of the fourth IAEA inspection team in Iraq, July 27–August 10, 1991.

20.   David Kay, "Iraq and Beyond."

based on highly enriched uranium, the same fissile material Iraq was trying to produce. The missile warhead version of this warhead was available to the United States in 1954 for delivery by submarine-launched cruise missiles and land-based ballistic missiles.[21] The Iraqis would have to have solved to their satisfaction the civil- military problems of command associated with giving Iraqi military commanders live nuclear weapons and the means to deliver them over long distances, and then giving them the freedom to move the weapons about. Ensuring that these commanders, who may have been harboring grievances toward their own regime or others, or who might have entrepreneurial ambitions, would fire their weapons only when told to and only at the target intended by the government, is not a trivial problem. We know from the Soviet case that despotic regimes have been able to reconcile the problems of ensuring the survival of nuclear weapons with their particular civil-military problems of nuclear command and control. Since those methods usually involved the uses of secret police and military forces to keep the nuclear forces in check, rather than sophisticated technology, there is no reason to assume that there would be insuperable technical barriers to Iraq's doing likewise.[22]

What might the deterrent value of nuclear weapons have been to Iraq at the time of the 1990 invasion of Kuwait and in the 1991 war? The U.S. Army and Air Force were able to deploy massive forces and the logistics to support them to a relatively small number of bases in the Persian Gulf area. The Navy was able to send multiple carrier battle groups into the Persian Gulf. The existence of Iraqi nuclear weapons could have deterred or interfered with this deployment in two ways. First, Iraqi nuclear weapons could have deterred countries like Saudi Arabia from allowing the United States access to its bases. How could the United States have reacted if the Saudis had decided not to allow U.S. forces on their soil because they feared Iraqi nuclear attack? The United States could have threatened nuclear retaliation against Iraq if it attacked Saudi Arabia with nuclear weapons, and could have threatened to carry out that threat with long-range forces not in the theater. Would such a threat have been

---

21. See the discussion of the 1,700-pound W-7 warhead, which went into production in August 1954 and was the warhead for the U.S. Army's "Corporal" and "Honest John" ballistic missiles, in Chuck Hansen, *U.S. Nuclear Weapons: The Secret History* (New York: Orion Books, 1988), pp. 133–136, 180.

22. See Stephen M. Meyer, "Soviet Nuclear Operations," in Ashton Carter et al., eds., *Managing Nuclear Operations* (Washington, D.C.: Brookings, 1987), pp. 490–493, for a discussion of the KGB role in controlling nuclear weapons that were under the control of Soviet military commanders.

credible to the Saudis and to the Iraqis? Such a question cannot be answered definitively, but some observations can be made.

First, despite cultural and political ties to Western Europe that were stronger than those to Saudi Arabia, the United States was never able entirely to reassure the governments of Western Europe that the Strategic Air Command, by itself, was sufficient to deter Soviet nuclear attack on Western Europe, once the Soviet Union acquired even very modest and uncertain means of attacking U.S. cities with nuclear weapons. A combination of forward deployed non-nuclear and theater nuclear forces were consistently deployed over a period of almost forty years in order to neutralize Soviet nuclear threats and to reassure U.S. allies in Europe that Soviet nuclear attack was deterred; however, the government of Japan never felt the need for U.S. nuclear weapons on Japanese soil. Allies of the United States facing a nuclear Iraq, Iran, or North Korea might be harder to reassure than Great Britain, France, and Germany, particularly if nuclear threats against U.S. allies were coupled with implicit fears of nuclear attacks on U.S. cities if the United States used nuclear weapons. Even if the enemy had only marginal capabilities to attack the United States, this threat would create considerable uncertainties within its alliances. The United States would be in the difficult position of trying to prove to its allies either that the common enemy had no means of attacking the United States—for example, that it had never been and would never be able to introduce a clandestine nuclear weapon into the United States—or that the United States would be willing to use nuclear weapons if Riyadh were attacked, even if it meant losing New York.

Second, even if the United States were able to convince its regional allies that it was safe for them to allow it to use bases on their soil, the leadership of the U.S. military would itself have considerable doubts about the wisdom of concentrating hundreds of thousands of U.S. soldiers and airmen in a few bases within range of enemy nuclear weapons. Even though the United States could have retaliated with nuclear weapons if the Soviets had conducted nuclear attacks against U.S. forces in Western Europe, the U.S. Army consistently worried about the military problems created by the presence of enemy nuclear weapons that could strike at troop concentrations. While the army never developed a totally satisfactory solution to this problem, it invested considerable time and money in partial solutions ranging from the Pentomic division to helicopter air mobility to Air-Land Battle.[23] Similarly, the U.S. Navy spent

---

23. Kevin Patrick Sheehan, "Preparing for an Imaginary War" (Ph.D. dissertation, Harvard University, 1988).

billions of dollars and structured its fleet tactics to limit the ability of the Soviet Union to strike successfully at concentrated naval formations with nuclear weapons. Despite the deployment of the F-14/PHOENIX system, and Aegis air defenses, the reluctance of the carrier battle group commanders to operate in closed waters when within range of enemy nuclear weapons suggests that these costly programs were at best partial solutions to the problem of potential nuclear attack. Even if the U.S. government had adopted a declaratory policy of striking at Iraq with nuclear weapons if Iraq used nuclear weapons against bases on land or against U.S. Navy ships at sea, prudent military planners would very likely have altered their deployment and operational plans to take into account the existence of Iraqi nuclear weapons. These alterations would have reduced the ability of the United States to concentrate its forces, increased the amount of resources devoted to defense, rather than offensive strike capabilities, and imposed significant virtual attrition on the expeditionary force.

Would the United States have been able to circumvent the difficulties associated with nuclear threats against our regional allies and U.S. military personnel within range of a hostile nuclear-armed power by executing a successful preemptive strike against Iraqi nuclear weapons *before* initiating Desert Shield, that is, before the deployment of U.S. tactical aviation to bases in the region? A preemptive attack conducted by U.S. military units stationed outside the Persian Gulf area would have faced considerable difficulties. Only B-52 and F-111 bombers had the necessary range for such missions, and it is not clear that they had the sensors for the kind of reconnaissance-strike missions later conducted by F-15s and supporting aircraft during Desert Storm. No F-111s remained in the U.S. Air Force inventory as of 1998. And even the reconnaissance-strike team of F-15s, as we have noted above, was unable to suppress Iraqi Scud ballistic missiles on mobile launchers. As noted above, the lesson of the 1991 war was that if Iraq had been willing to mate its nuclear weapons with its ballistic missile launchers, mix those launchers with non-nuclear launchers with similar signatures, and then disperse the ballistic missile force and keep it mobile, it is difficult to see how the United States could have had confidence in its ability to destroy even a high proportion of the deployed Iraqi nuclear weapons.

It is entirely conceivable that Saddam Hussein would have been reluctant to put operational, nuclear-armed ballistic missiles into the hands of several Iraqi military commanders left to their own devices in the Iraqi desert. With such power in their hands, and under the command of a despot who was not well liked, Iraqi military commanders with nuclear weapons might not have proved completely reliable. It might well have been difficult for Iraq to implement command and control arrangements to prevent unauthorized use of nuclear-armed, mobile ballistic missiles that would also have ensured prompt execution of plans while under attack, if necessary. Such considerations might well have led Iraq to keep its nuclear weapons in concealed, very hard storage sites. Such sites would be very difficult for the United States to find. In fact, in January 1991, the U.S. list of nuclear-related targets in Iraq consisted of two sites. At the end of the war, UN inspectors operating in Iraq identified over twenty nuclear-related facilities in Iraq, sixteen of which were designated as "main facilities."[24] A separate question would be the ability of the United States to have successfully attacked superhardened sites with non-nuclear weapons. Available evidence suggests that by the end of the war, the United States had constructed special kinetic energy penetrators that could destroy deep underground bunkers. Tunnels into mountains would require a kinetic energy penetrator either to penetrate hundreds, perhaps thousands of feet of rock, or to execute difficult maneuvers in order to fall vertically and then turn and fly horizontally into the tunnel entrance. Cruise missiles might fly into tunnel entrances and seal the known entrances into a nuclear weapons storage site. This might be adequate to neutralize the nuclear weapons for a period, but with enough time, the weapon could be recovered. Fully neutralizing the weapon would require occupying the site and taking the weapon out.

A tentative conclusion is that the United States would have faced considerable difficulties in executing a preemptive attack with weapons based outside the region that would have given the U.S. president high confidence that all or most of the Iraqi nuclear weapons had been destroyed. The inability of the United States to conduct such a preemptive disarming attack could have deterred U.S. military deployments to the region, and certainly would have forced U.S. military planners to take precautions that would have reduced the offensive military capabilities of the expeditionary force. It also seems fair to conclude that such a disarming attack would still have had uncertain prospects even after the U.S. Desert Shield build-up had been completed.

Thus, the overall lessons learned from the 1991 war by regional powers contemplating military action threatening U.S. interests might be that developing nuclear weapons creates only a small risk of U.S. military action before a nuclear weapon is complete, particularly if adroit diplomacy is employed; that much nuclear weapons–related activity is concealable, even from intrusive inspections, that nuclear weapons can create considerable uncertainties within U.S. alliance relations that might pre-

24. Keeney and Cohen, *Gulf War Air Power Survey*, p. 79.

vent the U.S. use of regional bases, and that a substantial proportion of the nuclear weapons of Iraq might not have been easily destroyed by a preemptive U.S. attack employing non-nuclear weapons. As a result, a U.S. president might decide not to deploy forces to the region, or might deploy a force with reduced offensive military power that would require more time and face less favorable force ratios when engaging a regional nuclear power. All of these tentative conclusions could and should be tested by focused studies employing the best available political and technical intelligence and commentary from U.S. military commanders with operational responsibilities.

The 1991 War and Hypothetical Nuclear Weapons Use

The previous section addressed what might have happened if Iraq had simply possessed nuclear weapons when it invaded Kuwait in August 1990. This section explores the problems and issues the United States would have faced if Iraq had actually used nuclear weapons. In the absence of any concrete information about Iraqi nuclear weapons employment doctrine, the discussion proceeds on the assumption that the government of Iraq would have had the following set of strategic objectives. These objectives would apply as well to other expansionist regional powers with nuclear weapons. First, Iraq would have sought to deter or prevent the U.S. military build-up in the region, which gave the United States effective military options against it. Second, if Iraq had failed to prevent the U.S. build-up, it would have sought to deter U.S. employment of its military forces in the region. Third, it would have sought to minimize the likelihood that a U.S. president would use nuclear weapons against Iraq. Fourth, while pursuing all of these objectives, the government of Iraq would have tried maintain central control over Iraqi nuclear weapons, out of fear that Iraqi commanders might not be entirely reliable custodians of the few nuclear weapons in the Iraqi arsenal, and to make sure that if Saddam Hussein issued an order to use the nuclear weapons, his order would be implemented quickly and faithfully.

To be sure, this set of objectives represents an idealized set of Iraqi objectives that are not entirely consistent with Iraq's behavior during the crisis. For example, Iraq did not mount armored attacks on Saudi Arabia immediately after invading Kuwait, though by doing so Iraq might have been able to deny the United States the areas in which it later built up its forces. Saddam Hussein did not fully exploit the confused diplomacy of the period from August 1990 through January 1991 to deny the United States a solid political basis for military action. The set of objectives, therefore, represent a "worst case" in which a hypothetical nuclear-armed

Iraq acts as a completely informed, rational actor that does everything within its power to frustrate U.S. strategies, and tests the capacity of U.S. military forces and strategy to the limit.

How might Iraq have translated the broad objectives listed above into military plans? Four operational goals for Iraq are consistent with the objectives listed above.

First, the timing of the revelation of Iraqi nuclear weapons capabilities would have been important. If Iraq had announced its nuclear weapons capabilities *before* invading Kuwait, the United States presumably would have become more alert and more focused on the possibility of Iraqi aggression. U.S. diplomacy and military preparations would have been affected by the announcement, and Iraq might never have been presented with what it perceived as a favorable opportunity to invade Kuwait. But if Iraq invaded Kuwait and then credibly revealed its nuclear weapons capabilities, the consequences might have been serious. Iraq might have used nuclear weapons to create the impression within Turkey, Saudi Arabia, and other countries in the region that it was dangerous to those countries to allow the United States the use of their military bases in response to the Iraqi invasion of Kuwait. There would be a natural tension between the United States and its regional allies as a result of the fact that the United States would be out of range of nuclear-armed Iraqi missiles and aircraft, while the allies of the United States were in range and vulnerable to nuclear attack. Before the Desert Shield buildup was complete, there were tensions between the United States and Saudi Arabia because of the differential in non-nuclear Iraqi threats to Saudi Arabia as compared to the United States. Iraqi nuclear weapons capabilities would have exacerbated that tension.

Second, Iraq might have used nuclear weapons to maximize the perception in the United States that military action against Iraq was extremely dangerous and carried the chance of very high U.S. casualties and ultimate political failure. While U.S. leaders would have been confident that U.S. nuclear superiority would make possible a nuclear response to Iraqi nuclear weapons use, such a response might not have prevented hundreds of thousands of U.S. casualties and a politically disastrous outcome for the United States in the Middle East. Iraq might be able to create a perception of such risks without mounting nuclear attacks on U.S. forces. An Iraqi detonation of a nuclear weapon plus the dispersal of nuclear-capable delivery systems within Iraq, for example, could have created the impression that Iraq could inflict massive damage on U.S. forces and that a U.S. nuclear counterforce strike in retaliation would also have to be massive. This perception could have been maximized by deceptive Iraqi practices designed to encourage U.S. assessments that Iraqi nuclear forces were larger than they were or larger than they had previously been estimated to be. Multiplying units with the same signatures as nuclear weapons units and revealing the existence of previously clandestine nuclear material production sites are two obvious, hypothetical deceptive options. Imagine the impact that an Iraqi nuclear test in August 1991 would have had along with the revelation of the Calutron facilities for uranium enrichment, if outside observers could not go and assess the production capacity of the facilities.

The fact that U.S. perceptions might have been manipulated by nuclear explosions without an attack on U.S. forces is important because the third Iraqi operational goal would likely have been to avoid the use of nuclear weapons against U.S. military personnel, the military personnel of close U.S. allies such as Israel, and, perhaps, to avoid the use of nuclear weapons against civilians. This operational goal would be adopted if Iraq chose to minimize the likelihood that the United States would use nuclear weapons against Iraq. Nuclear attacks against politically salient targets such as U.S. or Israeli troops would practically guarantee U.S. military retaliation.

Fourth, and finally, Iraqi operational plans for nuclear weapons use would likely have been simple, tightly scripted, and preprogrammed in order to cope with the command and control problems faced by a despot with potentially unreliable commanders, few nuclear weapons, and a communications system that is vulnerable to military attack. Nuclear-armed ballistic missile units operating from fixed, concealed locations, with the coordinates of fixed targets locked into the guidance system, under the command of an elite military unit commanded by a relative of the ruler, guarded by secret policemen reporting back to the ruler through separate communications channels, prepared to execute a set of simple shoot/don't shoot instructions communicated by secure land lines were more likely to have been characteristic of Iraqi military plans than nuclear-armed aircraft on airborne alert with orders to search for and destroy U.S. carrier battle groups at sea or maneuvering ground units.

Given these objectives and constraints, what kinds of nuclear attacks might Iraq have conducted? The nuclear option with the fewest material effects might have been the option with the greatest net political benefits to Iraq. Iraq could have detonated a nuclear weapon over the territory of Saudi Arabia in the period August through October 1990 that inflicted no major damage on military or population targets, but which would have dramatically demonstrated Iraqi nuclear capabilities before the United States had been able to build up operationally significant military capabilities in the region. What would have been the advantages and disadvantages to Iraq from such an initial use of nuclear weapons? Such an attack would certainly have delayed U.S. military intervention, and would have complicated U.S. military deployments if they had proceeded. Compared to other conceivable Iraqi uses of nuclear weapons, it would have given the United States fewer reasons to use nuclear weapons of its own. It would have been well within the assumed limits of the Iraqi command and control system. The demonstration shot would, at a minimum, have brought about a pause in the deployment of U.S. and allied forces to Saudi Arabia while the situation was being clarified and reconsidered by the governments of the United States and Saudi Arabia. The Saudi government would certainly have asked the U.S. government for guarantees and protection against subsequent nuclear attacks if the Saudis continued to allow the military buildup on their soil. In the United States, as in past military crises, the Congress and the U.S. people would have rallied around the commander-in-chief and, at least initially, would have given him their support in whatever policy he chose to pursue. But if the crisis was not resolved within a week or so, debate within the United States military, the executive branch, and in the larger political system would have become intense.

On the negative side for Iraq, the attack would have alerted the world to its nuclear capabilities and willingness to use them, and this would very likely have solidified universal diplomatic hostility to Iraq. This, in turn, might have facilitated the development of the anti-Iraqi coalition. U.S., Israeli, and perhaps even Soviet inhibitions on the use of nuclear weapons against Iraq would have dropped.

Once the initial shock had passed, how might the United States have reacted to such a demonstration shot? One option would have been simply to continue the buildup of non-nuclear U.S. military forces in the region, and to declare that any Iraqi nuclear attack on U.S. or allied civilian or military personnel would be met with a full-scale U.S. nuclear retaliation. In effect, this strategy would declare that Saddam Hussein's nuclear threat was a bluff, and would call the bluff. This might or might not have been accompanied by a tit-for-tat demonstration of U.S. nuclear capabilities. If the Saudi government hesitated or refused to issue an invitation to the U.S. forces to deploy to Saudi territory, the United States could have moved in without permission. Deployments would have continued, but with greater care to disperse and defend U.S. forces against ballistic or cruise missile attack. At the operational level, the deployment would have proceeded in a more orderly fashion if U.S. military planners had already explored the problems of deploying an expeditionary force against a regional nuclear enemy. If Saddam Hussein declared in response that he had already managed to introduce a clandestine nuclear weapon into the United States, this would also be treated as a bluff, since any nuclear destruction of a U.S. city would certainly result in the nuclear destruction of Iraq.

This option would call for a high tolerance for the risk of nuclear destruction, perhaps more tolerance than major U.S. political actors or our regional allies might possess. If the Saudis were not willing to go along with the United States, this would add the costs of acting against the wishes of a friendly Arab state. The ability of the United States government to reduce the risk of nuclear attack by military preemption would be problematic. In this scenario, the United States would not yet have deployed sufficient non-nuclear air power in the region to execute any effective air strikes against Iraqi nuclear weapons. A U.S. nuclear strike against Saddam's nuclear weapons, while within U.S. capabilities, might have been politically difficult to execute, particularly if uncertainties about the location and hardness of the Iraqi weapons storage sites forced the United States to conduct a fairly massive nuclear attack to ensure itself that most of the Iraqi weapons were destroyed. For example, let us say the United States delivered 100 nuclear weapons by cruise missile with yields in the 200-kiloton range to destroy a maximum of ten estimated actual weapons in individual storage sites, with five decoy storage sites for every real site, and two U.S. weapons per target. If the nuclear weapons storage sites were collocated with Iraqi population centers, the United States would face a greater dilemma. Other means of reducing the perceived risk of nuclear attack on U.S. allies might include the deployment of nuclear offensive systems to Saudi Arabia to "couple" any Iraqi nuclear attack on Saudi Arabia to a U.S. nuclear response by forces in theater, or the deployment of U.S. systems to defend against Iraqi nuclear attack. In the aftermath of an Iraqi nuclear weapons demonstration, the use of nuclear warheads on U.S. defensive systems, including Patriot, Phoenix, and Aegis/Standard 2 systems, might have been both politically acceptable and necessary to enhance the real and perceived effectiveness of U.S. defenses against Iraqi nuclear attack. If such warheads were available, if plans had been made to deploy them, and if exercises had familiarized crews with the modes of employment for nuclear-armed defensive systems, this option would have been more readily available to U.S. leaders to help them reduce perceived and actual risks of Iraqi nuclear attack.

Another option might have been a U.S. ultimatum to Iraq to withdraw from Kuwait or face U.S. nuclear attack. The kind of nuclear attack needed to carry out the threat inherent in this response would have been well within the capabilities of U.S. nuclear weapons outside the region, but even more politically difficult to execute than continued military deployments to the region.

Moving one rung up the escalation ladder, once Iraq decided to take the risk of employing nuclear weapons at all, it might have decided that its objective of denying the United States the ability and desire to intervene militarily in the region would best be served by nuclear attacks on the few ports and airfields in the region through which U.S. forces would have to approach the theater. Given the limited number of harbors in the region that can unload heavy equipment, such an attack might have halted the deployment of U.S. armored forces until the ports were decontaminated and repaired. Coupled with nuclear attacks on Saudi Arabia's airfields with runways and logistics support facilities to handle C-5 class aircraft and heavily laden F-15s, this attack could have created physical barriers to U.S. military intervention that could have lasted for many months, if not years. Preplanned strikes against fixed targets would, again, have been within the assumed constraints on Iraqi command and control.

On the downside for Iraq, such an attack would presumably have killed thousands of Saudi Arabians. While Saddam Hussein displayed little reluctance to kill his "fellow Arabs" in Kuwait, and might even have wished to demonstrate his ruthlessness against his enemies, a nuclear attack on centers of population would have reduced the political constraints on U.S. use of nuclear weapons against Iraq.

What might the U.S. option have been in this scenario? Since the attack would severely constrain the deployment of conventional forces to the region, Iraqi forces would still be in control of Kuwait. Nuclear strikes against Iraq by U.S. forces based in or out of the region might be threatened, and those threats would have had somewhat more credibility than in the first scenario because of the civilian casualties in Saudi Arabia. But what targets might the United States have credibly attacked? Iraqi occupation forces in Kuwait would be inappropriate targets. Iraqi ports or urban-industrial centers could have been destroyed in tit-for-tat reprisal/punishment attacks, though this might well have been politically difficult for a U.S. president to order, particularly if the initial Iraqi attack had killed only a few U.S. military personnel in Saudi Arabia. In the first days after the Iraqi attack, U.S. blood would be boiling for nuclear retaliation, but after that, the United States would debate the morality and strategic utility of killing tens of thousands of Iraqi civilians, and of weakening the psychological restraints on the use of nuclear weapons even further, the future of U.S. relations in the region after a nuclear attack on Arab cities, and so on. There would, therefore, very likely have been a search for Iraqi military targets that were not near Iraqi centers of population: nuclear weapons storage sites, ballistic missile deployment areas, elite military units, lines of communication from Iraq to Kuwait.

Once again, there might have been hesitation to use nuclear weapons against these targets because of the risks outlined in the section above, particularly if Saddam had chosen to place his nuclear weapons production and storage sites near population centers. But even if these hesitations had been overcome and the attacks had been executed, Iraq would still have been in possession of Kuwait and Baghdad; Iraqi oil production centers would still have been intact; Saddam Hussein would still have remained in power; and the United States would still have faced physical obstacles to near-term non-nuclear military action against Iraq.

What might the United States have done then? There would appear to be few options other than to conduct whatever nuclear reprisal/punishment attacks seemed appropriate, use nuclear weapons against Iraqi nuclear weapons where possible, proceed slowly either to reconstruct the air and beachheads into Saudi Arabia or build new ones, build up U.S. military forces in the theater, and then liberate Kuwait. Faced with such a task stretching over months in a post–nuclear war world, it is unclear what the national interest or the actions of the United States would have been.

In an unlikely variant of the scenario just reviewed, Iraq could have attacked not only Saudi ports and airfields, but also the Saudi National Command Authority. This might even further reduce political constraints on the use of U.S. nuclear weapons against Baghdad. It would also create a disturbing problem for the United States: The Saudi government would have disappeared, and it would be unclear who controlled the country and who the U.S. "allies" were. Very few U.S. troops would have been killed. If the United States had decided to intervene militarily with conventional forces, it would have had to intervene unilaterally, as much to restore order in Saudi Arabia and to prevent the emergence of a hostile regime as to liberate Kuwait. In this scenario, a nuclear attack on Iraqi military forces would not have liberated Kuwait and would have left the Iraqi government in power. If, despite the absence of U.S. casualties, the United States had destroyed Baghdad with nuclear weapons in retaliation, it would not only have killed tens of thousands of civilians, but would also have faced the need to intervene in Iraq to restore order there as well. An Iraqi leader willing to think all this through and to take his chances might have concluded that the United States would be self-deterred from taking any military action in response to the nuclear destruction of the Saudi government. Indeed, it is difficult to construct any interesting U.S. policy options for this scenario.

The next level of Iraqi nuclear attack might have involved attacks on U.S. forces after they had arrived in the region. This is one of the least likely scenarios because it would virtually have guaranteed U.S. nuclear attacks on all Iraqi military units and political command posts, for emotional as much as for strategic reasons.

An Iraqi detonation of a man-portable "suitcase" bomb or bomb in a merchant ship or light civilian aircraft in a major U.S. city, as opposed to threats to execute such attacks, represents the least likely of all scenarios. Nuclear weapons, by logic and common agreement, have severely limited strategic utility, but they are very good at deterring nuclear attack on cities. This is true whether the weapons are delivered by ballistic missile or by suitcase. For Iraq to extract any strategic benefit from the use of nuclear weapons smuggled into the United States, Iraq would have had to identify itself as the party responsible for the destruction of a U.S. city. It is impossible to imagine any circumstances in which the U.S. government would not have executed a massive retaliatory strike in response to such an attack, and impossible to imagine circumstances in which the certain knowledge of retaliation would not have deterred this kind of attack by Saddam Hussein. *Threats* of such an attack to frighten the United States, to complicate the tasks of the U.S. government, and to force it to divert resources into searches for such a weapon are, however, completely plausible.

This extended hypothetical consideration of regional nuclear weapons use utilizes a Middle Eastern enemy and allies, but its logic is not confined to such allies and enemies. Middle Eastern nuclear enemies may well try to intimidate U.S. allies in Europe who provide the military bases on their soil that make U.S. military intervention in the Middle East feasible. Northeast Asian nuclear enemies of the United States may try to intimidate Japan to prevent the U.S. use of military bases in that country. The essential logic of the problem is that U.S. regional military alliances have been based on credible U.S. military guarantees of the survival of the regional ally, in return for basing rights. Regional nuclear proliferation undermines that asymmetrical alliance relation wherever it exists.

Nuclear Weapons Use and U.S. Policy

How is the U.S. policy community responding to the problem of possible nuclear weapons use? A survey conducted by David Andre reviewed political military simulations conducted by or for twenty U.S. government agencies involving the use of nuclear weapons by hostile regional powers. A second survey, conducted by RAND, was based on a series of simulations conducted by RAND specifically to elicit the views of the policy community toward the problem of regional nuclear weapons use. Twelve named analytical institutions and several unnamed organizations sent a total of 239 of their members to participate in the RAND simula-

tions. Both surveys found an overwhelming reluctance by the participants in the simulations to recommend the use of U.S. nuclear weapons
in response to a wide variety of simulated nuclear weapons use. There
was a strong preference to seek nonmilitary solutions to nuclear crises,
to avoid engaging U.S. military forces in the crisis at all, and to use
long-range, accurate, non-nuclear strike forces if the United States had to
act militarily. There was a notable lack of consensus that it was in the
interest of the United States to punish nuclear proliferation or regional
nuclear weapons use.[25]

The survey data appears to reveal a disinclination by the U. S. defense
policy community to come to grips with the difficulties of handling
hostile regional nuclear powers. This disinclination is reflected in U.S.
military doctrinal publications in which the problem of hostile regional
nuclear weapons use is generally handled with broad assumptions and
wishful thinking. The most recent version of the basic U.S. Army field
manual for operations, FM-100-5, published in 1993, simply states that
the problem of nuclear weapons use by a hostile regional power may best
be handled by rapidly bringing the war to a conclusion favorable to the
United States before the enemy's nuclear weapons can be brought to
bear.[26] A 1994 U.S. Army Training and Doctrines Command (TRADOC)
publication that looked forward at military operations in the twenty-first
century, TRADOC pamphlet 525-5, notes the phenomenon of the proliferation of nuclear weapons. Though it deals in some detail with the
changes in military operations that will be the result of improvements in
non-nuclear warfare, and particularly in information technologies, the
pamphlet is silent on how the U.S. Army might conduct military operations against a regional nuclear power.[27]

IMPLICATIONS FOR U.S. POLICYMAKERS

If the United States is not likely to be able to disarm a hostile nuclear
power by force, if it is reluctant to use military force at all against regional

---

25. David Andre, "The Third World Few Nuclear Weapons Problem: Policy Issues
and Implications for Military Planning Guidance as Derived from Gaming" (McLean,
Va.: SAIC, April 1993); and Marc Dean Wilmot, Roger Molander, and Peter Wilson,
"The Day After . . . " Study: Nuclear Proliferation in the Post–Cold War World (Santa
Monica, Calif.: RAND, 1993), vols. I and II.

26. U.S. Army, Field Manual 100-5 Operations (Washington, D.C.: Headquarters, Department of the Army, 1993) pp. 6–10.

27. U.S. Army Future Battlefield Directorate, TRADOC Pamphlet 525-5: Force XXI
Operations (Fort Monroe, Va.: Headquarters U.S. Army Training and Doctrine Command, 1994), pp. 2–7, 3–19.

nuclear powers and particularly unwilling to use nuclear weapons, what is the result?

If regional states adopt policies of minimum deterrence to each other, the results appear to be minimal. The increased security and stability of the military relations between two nuclear-armed states with minimum deterrence doctrines seem not to affect the interests of the United States in any significant way. In addition, U.S. military guarantees, assistance, and intervention will be much less in demand from states that have established stable nuclear deterrence. There will be little that the United States will be able to do militarily to help or hurt nations locked in a relationship of stable nuclear deterrence, short of offering to provide credible first strike capabilities or credible antinuclear defenses—and practical and political constraints on the United States will limit its ability to do either. The influence that the United States can exert over states in this kind of relationship will be likely to decline.

States will be less able to draw the military intervention of the United States by means of the demonstrative use of nuclear weapons when there is no competitive superpower that might profit from the failure of the United States to intervene in the aftermath of nuclear weapons use. However, regional wars that threaten the existence of a U.S. ally remain possible, and the use of nuclear weapons as a distress signal to the United States remains possible.

If regional allies of the United States are presented with regional nuclear threats, the United States will face the familiar problem of extended deterrence in a new context. Will the threat of U.S. nuclear retaliation protect a third party from nuclear or conventional attacks? In the past, there were doubts, in the United States and Western Europe, about whether the United States would actually use nuclear weapons on the battlefield or against the Soviet Union if Western Europe were invaded by non-nuclear Warsaw Pact forces or attacked by Soviet nuclear weapons. Yet during the Cold War, the security of Western Europe was understood to be vital to the security of the United States; as a result, the United States was willing to station troops and nuclear weapons in Western Europe and to adopt declaratory policies linking nuclear weapons use in Europe to intercontinental nuclear war. After the Cold War, many of the areas that are affected by regional nuclear weapons threats are likely to appear much less vital to the United States, and less worth the risks of nuclear war. Measures to extend U.S. nuclear deterrence to those regions are likely to be much less vigorous and credible.

In the cases of both stable regional nuclear deterrence and hostile regional nuclear threats, the result is likely to be a diminution in the influence that the United States can exercise over a region's politics by military means. Over the next decade it is possible that there will be additional nuclear proliferation, and little in the way of U.S. responses that effectively reassure our allies that we remain credible and useful allies. In consequence there could follow a weakening of the military alliances of the United States with countries in regions where there are regional nuclear threats, specifically, South Asia, Northeast Asia, and Southwest Asia. This consequence is consistent with the abstract logic of nuclear deterrence. If nuclear weapons create absolute threats to national survival, and there are no external protectors, countries will tend to respond either by capitulation or by developing their own nuclear weapons. This may restore stability, but the process weakens any existing military alliances. Nuclear proliferation reduces the willingness of allies to run risks for others, and it reduces the need for allies by states with nuclear weapons. As a somewhat unexpected consequence of nuclear proliferation, we should expect not a world full of nuclear war, but a world in which alliances are weak or nonexistent. The consequences for U.S. foreign policy may be significant: U.S. citizens are habituated to fighting overseas assisted by and to protect our allies. They are increasingly reluctant to pay the costs of unilateral military action, and eager to find justification for their military actions in the support of their allies. Nuclear proliferation and a world without allies may make U.S. military intervention abroad, for good or ill, increasingly difficult.

IMPLICATIONS FOR U.S. MILITARY PLANNING

One possible reaction to the arguments developed in this chapter is that a fundamental change in the pattern of U.S. military alliances after the demise of the Soviet Union is natural and acceptable, though possibly uncomfortable or even fatal for some former allies. Hence, no major changes in U.S. military policy are called for. Another possible reaction is that the level of international conflict and chaos that would accompany this change in alliance relations is unacceptable because the direct human costs to the regional powers would be appalling and the indirect costs to the United States would be large, though hard to predict. How robust is the U.S. military's ability to disarm small nuclear powers?

The ability of the United States to respond effectively to small nuclear powers with offensive and defensive military measures has been affected by the virtual denuclearization of U.S. power projection forces. Given the hardness and position location uncertainties that are currently associated with anticipated regional nuclear forces, and given the difficulty of intercepting regional ballistic and cruise missile capabilities, the U.S. use of smaller nuclear weapons for military purposes, offensive and defensive, appears to have some value from an operational military perspective. The

U.S. military, for a variety of reasons, has reduced and is reducing its deployment of such weapons. A proper assessment of the impact of this denuclearization on the ability of the United States to respond to regional nuclear powers is beyond the scope of this paper, but appears to be negative.

The ability of the United States to respond militarily to smaller nuclear powers is also being affected by military research and development. Over the longer term, U.S. capabilities may be changed by what has been called a "revolution in military affairs." In the near term, the ability of the United States to defend against medium- and theater-range ballistic missiles may be improved by the successful completion of programs such as the Theater High-Altitude Area Defense system, improved versions of the Aegis system, and airborne laser systems for use against cruise and ballistic missile targets. Over the longer term, the integration of offensive and defensive systems into a theater anti-missile system appears to be of some promise, and is being investigated by both the United States and Israel.[28] The United States may be able to use new technologies to collect, process, and distribute information for the purposes of neutralizing a small hostile nuclear power. Sensors on aerial platforms, both manned and unmanned, unmanned ground sensors, and other inputs can now be integrated with greater rapidity and coherence to provide, in the absence of enemy countermeasures, a radically improved picture of the targets the United States may wish to strike. Linked with long-range precision strike capabilities, deep penetration light ground forces, and other systems, the combination of improved information and coordinated offensive and defensive antimissile forces may yield greatly improved results; such systems are being explored by the United States. Radically improved integration, management, distribution, and utilization of target-tracking data for fleet air defenses are being explored by the United States Navy. Operational concepts for moving an expeditionary force with dramatically fewer soldiers close to a target while under the defensive umbrella of fleet air defenses, then establishing ground-based antimissile defenses, and then maneuvering in a fashion designed to force the enemy to deploy and expose its nuclear weapons are being explored by the U.S. Marine Corps. These experimental technical and operational capabilities are far from operational status, and may

---

28. See, for example, General John M. Shalikashvili, *Joint Vision 2010* (Washington, D.C.: Office of the Joint Chiefs of Staff, n.d.). pp. 22–24. For insights into Israeli preparation for war against states with nuclear-armed missiles, see Eliot A. Cohen, Michael J. Eisenstadt, and Andrew Bacevich, *Knives, Tanks, and Missiles: Israel's Security Revolution* (Washington, D.C.: Washington Institute for Near East Policy, 1998), pp. 93–94, 125, 127.

never reach that level. If, however, the United States judges that it is worthwhile to develop military measures that would increase its ability to act against small nuclear powers and so increase the ability of the United States to maintain its alliance relations, these efforts should be encouraged. Of course, the United States will not be the only country able to profit from time and improved technology. Hostile countries will be able to take advantage of technology to improve their ability to hide from or otherwise thwart U.S. forces. How this interaction between hiders and finders plays out is very much in question.

What will be the political utility of improved technology? Will improved U.S. antinuclear capabilities, many of which will be secret and undemonstrated, be credible to our allies in the face of demonstrated nuclear threats? Will U.S. technology be substitutable for the lives of U.S. citizens put at risk when the question is the credibility of U.S. nuclear guarantees? And will the United States be inclined to put itself in harm's way if worst does come to worst, to help bring about a more stable and secure world? Or will it accept a world in which the United States cannot use its military power to resolve regional conflicts in its favor?

Chapter 6

U.S. Security Policy in a Nuclear-Armed World, or What If Iraq Had Had Nuclear Weapons?

Barry R. Posen

Presuming that the United States remains a global power, it may well one day confront a situation where a nuclear-armed regional power threatens to trespass upon, or actually has trespassed upon, a very important U.S. national interest. The United States may still lack an array of offensive and defensive weaponry that would permit a senior military commander to assure the president of the United States that the adversary cannot successfully explode a nuclear weapon on the territory of a U.S. ally, or over a U.S. military force, or even on the United States itself. As far as one can tell, the consideration of how the United States would handle this situation has not proceeded very far. The national security policy community, in and out of government, is beguiled by the possibility of a happy ending.[1] It hopes that nonproliferation policy will prevent most proliferation and that a combination of weapons and tactics will negate the capabilities of those who slip through the policy net.[2]

Special thanks to my research assistant, David Burbach, and to the numerous colleagues who have offered comments on the chapter and on presentations based on the chapter.

---

1. For a review of United States post–Cold War nuclear weapons and nonproliferation policy, and policy debates, see Stephen Cambone and Patrick J. Garrity, "The Future of United States Nuclear Policy," *Survival*, Vol. 36, No. 4 (Winter 1994–95), pp. 73–95. They divide policy preferences on nuclear weapons into two rough schools, traditionalists and "Marginalisers." The former is self-explanatory, the latter "argue that the United States has a unique opportunity to place nuclear weapons on the road to ultimate extinction," pp. 75–76.

2. Marc Dean Millot, Roger Molander, and Peter A. Wilson, *"The Day After . . . " Study: Nuclear Proliferation in the Post–Cold War World* (Santa Monica, Calif., RAND, 1993), vols. I, II, and III. This document is a must-read for students of current U.S. nonproliferation policy. It reports the results of a series of small-group policy simulations conducted by a representative cross-section of mid-level, mainly U.S., civilian

This chapter assesses why the United States might, in fact, act force-fully in a confrontation with an expansionist state in possession of a modest nuclear retaliatory capability, and how it might proceed. The vehicle for this assessment is a counterfactual historical analysis of the U.S. reaction to Iraq's invasion of Kuwait in 1990. For heuristic purposes, I introduce a small nuclear force into the Iraqi arsenal, a half-dozen weapons and the means to deliver them regionally.[3] These weapons cannot reliably be located for conventional or nuclear preemption, and their delivery systems are sufficiently capable that extant defensive weap-onry cannot intercept them reliably. I postulate that Saudi leaders invite the United States to send forces to help deter an Iraqi attack on the kingdom immediately following the invasion of Kuwait.[4] I then ask two questions. Should the United States consider the military liberation of Kuwait?[5] If so, what strategy should follow?

---

and military policymakers and practitioners and policy analysts. The study conducted multiple iterations of four simulations involving nuclear weapons employment in four different regions, the Persian Gulf, Korea, the former Soviet Union, and South Asia. I have reviewed the results of the first two with some care, and the overview report. It is pretty clear from the simulation that regional nuclear crises are so scary in prospect that people seem inclined to focus the bulk of their energy on the "happy ending." While many participants recognize the improbability of the happy ending, the docu-ment shows a relative paucity of recommendations about how the United States ought to analyze and prepare for the most likely and dangerous contingency, a nuclear-armed regional adversary against whose nuclear forces we will not have reliable offensive and defensive options. "In the exercises, this kind of threat introduced enormous conservatism into the calculations of participants contemplating U.S. mili-tary intervention." Vol. I, p. 15. I do not believe that this "conservatism" is the fault of the simulation designers, but rather a fundamental insight gleaned from the exercise about the mindset of the U.S. policy community. For the authors' brief summary of the insights to be derived from these simulations, see Roger C. Molander and Peter A. Wilson, "On Dealing with the Prospect of Nuclear Chaos," *Washington Quarterly*, Vol. 17, No. 3 (Summer 1994), p. 32.

3. Al Venter, "How Saddam Almost Built His Bomb," *Jane's Intelligence Review*, Vol. 9, No. 12 (December 1997), pp. 559–566, suggests that Iraq might have completed a bomb by perhaps the end of 1992, had the Kuwait crisis not occurred.

4. This assumption is arguable, but it is sufficiently plausible that it does not vitiate the value of my analysis. See the essay in this volume by Steven Walt.

5. Many seem to think it self-evident that the United States would be unwilling to act. For example, Lawrence Freedman writes, "One of the advantages of a nuclear arsenal may be its role in discouraging Western involvement in local conflicts, thereby hastening Western disengagement from the security arrangements in many parts of the world. One only needs to contemplate the impact of a completed Iraqi nuclear program on Western calculations during the Gulf crisis to appreciate the importance of such a step." Freedman, "Great Powers, Vital Interests and Nuclear Weapons," *Survival*, Vol. 36, No. 4 (Winter 1994–95), p. 47. See also Michael M. May and Roger D. Speed, "The Role of U.S. Nuclear Weapons in Regional Conflicts" (Center for Interna-

Inaction too costly . . .

I develop three recommendations for the U.S. government about the hypothesized scenario. First, the United States should have tried to liberate Kuwait because of the general strategic consequences of inaction, not because of the intrinsic strategic value of Kuwait. Indeed, the whole definition of the crisis should have changed to "the first post–Cold War nuclear crisis." The likely consequences of inaction are developed at length, and provide the core of the argument for action.

Second, to support a liberation campaign, the United States ought to have pursued a strategy of "intrawar deterrence." It would have needed to make explicit and ferocious threats that nuclear retaliation would occur if Saddam Hussein used nuclear weapons on any member of the coalition. It should have explained to Saddam Hussein through a very systematic diplomatic campaign why the United States was compelled to liberate Kuwait, and would be forced to retaliate with nuclear weapons if Iraq employed them.

Third, the coalition should have pursued a military strategy of "limited war" in this operation: both ends and means ought to have been restrained. This would have been the best way to control the risks. An intensive discussion on the nature and scope of these military operations among and between civilians and soldiers would have been essential.

The first three sections of the chapter develop these recommendations. In each section, I first offer the policy advice that I think I would have given immediately after Iraq invaded Kuwait. Then I speculate on the answers to two additional questions: To what extent would my advice *to act* have found a sympathetic hearing among some of the actual participants in the crisis? To what extent would my advice on *how to act* have found a sympathetic ear? I show that the participants in the crisis confronted issues analogous to those I raise. I reason that the decision-makers' actual behavior suggests that the three recommendations I develop would have found a sympathetic hearing.

To complete the counterfactual analysis I briefly discuss how successful the conventional war might have been under the tactical constraints I recommend. I also discuss the strategic benefits of such a war. The chapter concludes with a critical assessment of current U.S. nonproliferation policies and some advice for possible changes in those policies. My

---

tional Security and Arms Control, Stanford University, June 1993). Keith Payne, *Deterrence in the Second Nuclear Age* (Lexington: University of Kentucky Press, 1996); Barry Schneider, "Radical Responses to Radical Regimes: Evaluating Preemptive Counter-proliferation," McNair Paper No. 41 (Washington, D.C.: Institute for National Strategic Studies, National Defense University, 1995); and Martin Van Creveld, *Nuclear Proliferation and the Future of Conflict* (New York: Free Press, 1993) all broadly concur with the view that regional nuclear powers can easily deter the United States.

tools throughout are offense-defense theory, deterrence theory, and limited war theory.

This chapter does not argue that the policies advocated here would have been followed. There were too many players with too many diverse interests involved in the crisis to argue comprehensively that this advice would have become the basis for action. For example, the analytic premises and tools that I employ do not permit me to speak to the reaction of the U.S. public, or other publics. Neither can I predict the behavior of the U.S. Congress, or the legislatures of allies.

Some pieces of the analytic puzzle that might profitably be pursued with my premises and tools are also omitted, largely for the sake of brevity. For example, if the United States had known that Iraq possessed nuclear weapons, would the preinvasion crisis have developed as it did? How should the United States have acted during such a period? Presuming that the invasion of Kuwait did happen, would the leaders of Saudi Arabia have invited U.S. forces into their country?[6]

The Risks of Inaction

THE VIEW FROM 1990–1991

Iraq's conquest of Kuwait immediately confronts U.S. decision-makers with one very important fact: This is the first post–Cold War nuclear crisis. It is a defining moment. The actual stakes—Kuwait's oil and Iraq's power—are probably secondary to the more fundamental question: What will nuclear weapons mean? The "consensus" at the end of the Cold War is that nuclear weapons deter nuclear attacks on oneself or one's allies, and arguably deter conventional invasion of one's own territory, and to a lesser and more debatable extent, one's allies' territory.[7]

---

6. Molander and Wilson argue, "A regional predator will find a small nuclear arsenal a powerful tool for collapsing regional military coalitions that the United States might craft to oppose such a future opponent." "On Dealing with the Prospect of Nuclear Chaos," p. 32.

7. Robert F. Jervis, *The Meaning of the Nuclear Revolution* (Ithaca, N.Y.: Cornell University Press, 1989), especially chap. 1, "The Theory of the Nuclear Revolution," pp. 1–45. He believes that nuclear weapons do all of these things; his essay essentially summarizes the consensus view among nongovernmental deterrence theorists, and what I suspect was the "median" view among most analysts, about the impact of nuclear weapons on international politics in the 1970s and 1980s. There were, of course, many who held opposite views. Since the end of the Cold War, however, one has heard fewer challenges to the "Jervis" view. It is the basis for current U.S. nuclear declaratory policy, strategic nuclear weapons acquisition and deployment policy, strategic nuclear arms control policy, and nuclear nonproliferation initiatives. Indeed, current policy seems to aim to sell the argument that nuclear weapons are good for almost nothing

Will the United States and its allies, through inaction, allow other countries' nuclear forces to become potent instruments of aggression against non-nuclear powers?[8] If the Iraqi conquest of Kuwait is permitted to stand, nuclear weapons will come to be viewed as a shield that protects conventional conquests from *any* challenger, including a great power heavily armed with its own nuclear weapons. The new nuclear truism will be that conventional forces take; nuclear forces hold! The context will matter as well. Very important interests of the world's sole surviving superpower will have been successfully trampled by a state of modest conventional and tiny nuclear capabilities.

The United States has asserted a vital interest in the independence of the various Gulf oil states, and in the free flow of oil out of the region, since the fall of the shah of Iran. The United States has programmed and deployed military forces and assets for the explicit purpose of protecting this interest. The United States has employed military force in operation Earnest Will to defend directly the free flow of oil. Since 1973, the United States has devoted considerable diplomatic effort and financial resources to achieve a settlement of the Arab-Israeli conflict. It is fair to say that there is no place else in the world where the United States has been so active diplomatically and militarily without formal treaty relationships. Capitulation to Saddam Hussein's coup de main will therefore enhance the image of nuclear weapons as great equalizers. Indeed, their successful exploitation to deter a conventional *counterattack* to liberate ill-gotten gains would make them seem even more valuable.

If Iraq were to deter a direct challenge to its aggression, how big a change would it be? The change would be radical because nuclear weapons have not been used by any state in this way since the dawn of the nuclear age. No state has engaged in large-scale conventional aggression, and then explicitly or implicitly tried to protect its gains from a conven-

---

but deterrence of *nuclear* attack on oneself and one's allies. Following India's nuclear tests in 1998, Indian policymakers expressed strikingly similar views about nuclear weapons, but, of course, they may be telling us what we want to hear. John Burns, "India Defense Chief Calls U.S. Hypocritical," *New York Times*, June 18, 1998, p. A6.

8. Some do assume that this would occur. For example, May and Speed write, "Regional tactical nuclear deterrence could be difficult for the U.S. to counter, depending on the actual military and political situation. Thus, if the U.S. wanted to roll back an aggression, as it did in the Gulf War, it would have to initiate the use of force against at least a tactical nuclear threat, and potentially a strategic one as well. The feasibility of doing this successfully and the feasibility of keeping the use of nuclear weapons local might be so questionable as to deter any crossing of the military lines, just as it did in Europe for forty years. In addition, any coalition the U.S. were to put together for this purpose would come under extreme stress." May and Speed, "The Role of U.S. Nuclear Weapons in Regional Conflicts."

tional counterattack through nuclear deterrent threats. This gambit is so alien to the ways both theorists and political leaders have understood nuclear weapons that there is almost nothing written about it even from a theoretical perspective. The situation would be analogous to a half-successful Warsaw Pact invasion of West Germany, which stalled after a 100-kilometer gain, going uncounterattacked by NATO's conventional forces because of a Soviet claim that its nuclear deterrent umbrella now covered its new real estate. An explicit consideration of this possibility probably exists somewhere in the classified or open literature, but I have neither found it nor heard of it.[9]

The closest the nuclear deterrence literature comes to an explicit treatment of the question of how conventional aggression could occur between nuclear-armed adversaries is the "stability-instability" paradox. This predicts that two states that can destroy each other may resort to conventional warfare with each other because they do not fear escalation. The strategic balance is so stable, and the consequences of nuclear war so obvious, that neither side would find it reasonable to employ nuclear weapons, and both would know it.[10] But the paradox logically cannot predict conventional attacks without also predicting conventional counterattacks.[11]

---

9. During the Cold War some military planners and analysts believed that nuclear weapons would make it difficult to mount an amphibious landing in Europe if initial enemy successes expelled the United States entirely from the Eurasian landmass. This was a largely tactical belief, i.e., that amphibious forces were so vulnerable to nuclear attack during the landing that the United States would never risk such an approach. General Omar Bradley averred in 1950 that "appraising the power of the atomic bomb, I am wondering whether we shall ever have another large-scale amphibious operation. Frankly, the atomic bomb, properly delivered, almost precludes such a possibility." See Roger Hilsman, "NATO: The Developing Strategic Context," in Klaus Knorr, ed., *NATO and American Security* (Princeton, N.J.: Princeton University Press, 1959), p. 19, n. 7. Similar views were expressed by Sir John Slessor, *Strategy for the West* (New York: William Morrow, 1954), p. 90; and Glenn Snyder, *Deterrence and Defense* (Princeton, N.J.: Princeton University Press, 1961), p. 139.

10. Robert Jervis, *The Illogic of American Nuclear Strategy* (Ithaca, N.Y.: Cornell University Press, 1984), pp. 29–34, 148–150. In *The Meaning of the Nuclear Revolution,* Jervis argues at greater length that the "paradox" seems not to operate very powerfully (pp. 19–23). Direct conventional clashes between the superpowers did not occur during the Cold War. This may suggest that the paradox is weak. Alternatively, perhaps neither superpower actually coveted much of what was in the other's sphere of interest. Or, given the nuclear warfighting elements of both superpowers' strategies, the nuclear balance was somehow just unstable enough to discourage risk-taking. Or, the relationship of NATO and Pact conventional forces was sufficiently balanced that neither side had high confidence of victory.

11. *The Illogic of American Nuclear Strategy*, p. 155. "Whereas the stability-instability paradox could lead to relentless probing in the calculation that the risks are low

Even states that achieved large-scale territorial gains through conventional *counterattacks* did not try to secure their gains with nuclear deterrence; nor did the mere possession of nuclear weapons deter challengers. For example, the United States took North Korea in a counteroffensive, but did not make nuclear threats to dissuade a Chinese challenge to that offensive, even after the appearance of Chinese troops in October 1950.[12] And the Chinese were obviously not deterred from entering the war by the known U.S. nuclear capability.[13] Israel never formally invoked nuclear

---

enough to be worth running, decision-makers have taken what Patrick Morgan calls a more 'sensible' approach and have generally been willing to forego the chance of gains in order to keep the risks of war as low as possible." This might suggest another alternative pattern to the crisis under consideration; as a nuclear weapons state, and hence a legitimate nuclear target, Saddam Hussein might have chosen not to invade Kuwait.

12. Nuclear-configured B-29s were dispatched to the Far East at the end of July 1950, when United States and Republic of Korea forces had retreated to the Pusan perimeter. This was publicly announced but was accompanied by no explicit statement of their purpose. Apparently the force had defensive and deterrent purposes both in Korea, and against a Chinese attack on Taiwan. Roger Dingman, "Atomic Diplomacy During the Korean War," *International Security*, Vol. 13, No. 3 (Winter 1988/89), pp. 60–65. The bombers were returned to the United States before the Chinese crossed the Yalu. Nothing that even looks like a nuclear deterrent threat is made again until President Truman's confusing allusions to nuclear weapons in a press conference on November 30, five days into the first major Chinese offensive of the war, and more than a month after the first strong Chinese appearance. This nuclear threat, which was widely reported in the press, appears to have been unintended, and if anything was driven by a concern to avert a tactical military disaster. In any case, it precipitated a wave of concern in the United States, but more particularly in Britain, and was quickly disavowed. Ibid., pp. 65–69. A third nuclear threat was effectively made in April, when aircraft and atomic weapons were dispatched to Guam, in part out of fear of a possible combined theater-wide Chinese and Soviet offensive, suggested by intelligence indicators. This deployment was accompanied by indirect and somewhat oblique warnings through contacts in Hong Kong that there were limits to United States restraint. Ibid., pp. 50–91. For an even more elaborate account of these events see Rosemary Foot, *The Wrong War* (Ithaca, N.Y.: Cornell University Press, 1985). Her account does not include any explicit U.S. warnings to China or the Soviet Union that they must accept the unification of Korea. Indeed, the United States issued no such warning even after it received significant warnings from China that the United Nations should not cross the 38th parallel.

13. Thomas J. Christensen, "Threats, Assurances, and the Last Chance for Peace: The Lessons of Mao's Korean War Telegrams," *International Security*, Vol. 17, No. 1 (Summer 1992), pp. 137–138, describes Mao's complex reasoning as revealed in the telegrams to Stalin and Zhou Enlai. While they mentioned Mao's fear of U.S. military forces camped on his border as a reason *for* military intervention in Korea, they do not mention any fear of nuclear bombing. Ejecting the United States from the peninsula was seen as a way to prevent a battlefield stalemate, which Mao feared would precipitate intensive bombing of China's cities. If nuclear bombing had been feared, this belief would have

weapons to secure its gains of 1967, and its rumored possession of nuclear weapons was not enough to deter an Egyptian and Syrian challenge.[14]

Effectively, the toleration of an Iraqi success in Kuwait would change nuclear weapons from "defenders of the status quo" to instruments of aggression. The consequences for world politics are likely to approximate those that theorists have associated with "offense-dominant worlds," at least until most states can acquire secure second-strike capabilities.[15] International politics will suddenly turn very competitive.

Ambitious states will be the quickest to learn lessons from the episode. Aggressors will strive harder to get nuclear weapons because their utility as "conquest protectors" will have been demonstrated starkly. States that get or already have nuclear weapons, and have claims against their non-nuclear neighbors, would feel substantially freer to enforce those claims. And because their neighbors understand this, and will surely try to get their own nuclear weapons to ensure themselves, the ambitious will rush to push their claims.

The failure to act to reverse Iraqi aggression would thus also increase the desire of status quo powers for their own nuclear weapons. Status quo states would have to assume that non-nuclear states with claims against them would soon embark on nuclear weapons programs. Whatever deterrent benefit non-nuclear states have derived from tacit or formal alliances with any of the five major declared nuclear powers will also have been eroded, once the strongest nuclear power is deterred by a few nuclear weapons from acting to secure a vital interest. Threatened status quo states will thus find it prudent to move quickly to try to acquire an independent nuclear capability.

---

been senseless, because the United States could have devastated the small number of important Chinese industrial areas with nuclear bombers from bases outside of Korea. Mao would likely have understood this.

14. Jonathan Shimshoni, *Israel and Conventional Deterrence* (Ithaca, N.Y.: Cornell University Press, 1988), pp. 31–33, notes that "declaratory policy on both sides between 1967 and 1973 was completely devoid of references to Israeli nuclear weapons."

15. Stephen Van Evera, "The Cult of the Offensive and the Origins of the First World War," in Steven Miller, Sean Lynn Jones, and Stephen Van Evera, eds., *Military Strategy and the Origins of the First World War* (Princeton, N.J.: Princeton University Press, 1991), pp. 64–67. He lists aggressive foreign policies; first strike and first mobilization advantages; wider windows of vulnerability and opportunity; competitive diplomacy to include brinkmanship and faits accomplis; and tighter secrecy as plausible consequences of widely perceived offense dominance. Because the nuclear offensive advantage discussed here is "strategic" rather than tactical, the "first-strike and first mobilization advantages" ought not to be a direct consequence of the United States' failure to act. They may arise as an indirect consequence, however, since many of the states trying to acquire nuclear weapons quickly will lack the resources to field a secure retaliatory force.

The utility of distant allies in a conventional war will have been reduced if not eliminated. It will have been demonstrated that the nuclear aggressor can deter the intervention of allies to take back lost territory. If a status quo power is conventionally inferior to a nuclear-armed challenger, it would have to act quickly to try to redress the conventional balance on its own. It might also seek the permanent presence of substantial allied conventional forces on its soil. But even if a status quo power can compete conventionally, it will never be certain that its conventional defenses will hold in war. Thus, it will want its own nuclear weapons to ensure that if a nuclear adversary achieves military gains, that adversary will not be able to deter a conventional counterattack. Against a nuclear-armed aggressor, only the country that stands to lose its sovereignty or real estate has the necessary will to risk a nuclear confrontation.

Ambitious states will see many incentives for preventive war. In particular, evidence of nascent nuclear weapons programs in states they see as prospective victims could spark a conventional attack, because this would be the last opportunity for aggrandizement. Aside from the inclination to take long-coveted land when a temporary military advantage arises, one can also imagine a wave of "Osirak"-type attacks, in which both status quo and aggressor states attempted to wipe out one another's nascent nuclear capabilities conventionally as Israel attempted on June 5, 1981, in its air raid on an Iraqi research reactor. To evade preventive attacks, both conventional and nuclear, even status quo states would have powerful incentives to pursue secret nuclear weapons programs, and to try to acquire black market bombs. (Other weapons of mass destruction that might be secretly developed, such as biological weapons, would also become attractive.) This would in turn create new opportunities for those with small nuclear weapons programs in one region to sell weapons to threatened states in other regions, and it would also create incentives for theft. Rumors of secret nuclear weapons programs would be rife, causing still more countries to contemplate nuclear acquisition. The Nuclear Non-Proliferation Treaty (NPT) would simply be swamped by violations and suspected violations.

This singular failure to act could thus usher in a dynamic pattern of international security competition in which even the "good guys" would have to play a rough game. It would certainly make for regional nuclear and conventional arms races. The incentives for preventive war would intensify. Nuclear weapons states would perceive aggression against non-nuclear states to be relatively easy whenever their conventional capabilities appeared sufficient to achieve success. After years of trouble, the world might settle down to many stable relationships of mutual deterrence, a series of "micro-cold wars." This process could take a long time, and be very exciting.

Taiwan and South Korea will face the starkest choices. These non-nuclear states face nuclear or near-nuclear rivals with strong claims against them. Both are competitive conventionally with their adversaries, but hardly dominant. The South Koreans have a direct U.S. military presence but they will likely clamor for even more U.S. troops—sufficient to block confidently any North Korean attack virtually at its initiation. Seoul's proximity to the border may prove too tempting a target for North Korea's large standing army, if the United States shows itself to be too intimidated by a few nuclear weapons to even try to take back lost territory. Taiwan is not even as well placed as South Korea, since U.S. forces are not present, and U.S. diplomacy has to some extent conceded the legitimacy of the Chinese claim to Taiwan. Both Taiwan and South Korea have the scientific and industrial capability, and the wealth, to build nuclear weapons; indeed, both once had nascent nuclear weapons programs that U.S. policy initiatives managed to stop.[16]

It is implausible that Japan can hold fast to its non-nuclear status. Though it is free of claims by any state against most of its territory, can it ride out a sudden surge of proliferation on its periphery? It seems unlikely that such a capable state would want to remain dependent on U.S. deterrent promises in such a competitive world. The United States will also likely find itself facing a new round of doubts in the Federal Republic of Germany about the credibility of the U.S. nuclear commitment.[17]

---

16. See Mitchell Reiss, *Without the Bomb* (New York: Columbia University Press, 1988), pp. 86–108, who suggests that the Republic of Korea did initiate a nuclear weapons program in the 1970s. See also Joseph Yager, "Taiwan," in Joseph Yager, ed., *Nonproliferation and United States Foreign Policy* (Washington, D.C.: Brookings Institution, 1980), pp. 66–81, suggesting that the evidence for a deliberate Taiwanese program at that time was weak. Nevertheless, the United States did insist that a particular reactor be shut down, which it was. This was not the end of the story, however, as a program continued that brought the Taiwanese very close to a bomb in 1987, when additional U.S. restraining pressure was exercised. David Albright and Cory Gay, "Taiwan: Nuclear Nightmare Averted," *Bulletin of the Atomic Scientists*, Vol. 54, No. 1 (January/February 1998), pp. 54–60.

17. At the time of Desert Shield and Desert Storm, the Soviet Union was still intact. Had it collapsed following a U.S. failure to act against Iraq, the subsequent consolidation of Soviet tactical nuclear weapons would probably not have proceeded so smoothly. It is implausible, for example, that Ukraine would have contemplated giving up its nuclear weapons. Given the probability that more than one nuclear power would have emerged from the wreckage of the Soviet Union, it seems likely that states on the periphery of the old empire, seeing these new nuclear states arise, would themselves begin to think more seriously about their own nuclear weapons capability.

The course of world politics can be changed dramatically for the worse if the United States fails to act to reverse Iraqi aggression. The relative U.S. power position will be undermined. The perceived utility of nuclear weapons will be enhanced. Thus, the incidence of war in general will probably increase, at least in the near term. The probability of nuclear accidents and ecological disasters associated with crash, low-budget nuclear weapons programs will also increase. Most seriously, there will be a greater risk of regional nuclear war. Can the United States find a plausible strategy to live comfortably in such a world? If not, then these dangers seem sufficiently compelling to warrant U.S. action against Iraq.

COULD THE UNITED STATES "LIVE" IN SUCH A WORLD? Three alternative U.S. strategies suggest themselves if a new era of intense strategic competition and stronger proliferation incentives is set in motion by a failure to act against Iraq: an intensified "counterproliferation" policy; a military build-up to allow the United States confidently to defend allies against conventional aggression; and a "fortress America" policy.

Failure to act against Iraqi aggression would show U.S. decision-makers that a few nuclear weapons in the hands of a prospective enemy can utterly paralyze U.S. policy. If the United States is to maintain any freedom of action in the world, then nuclear proliferation must not occur. The political aftermath of the conquest of Kuwait would make it difficult to place a new emphasis on the NPT. Too many states would have incentives to get nuclear weapons to be placated with great power promises. Instead, the United States would have to concentrate much more on active measures. States bent on getting nuclear weapons, with policies inimical to U.S. interests, would have to be denied the completion of their programs. Any means, fair or foul, would be employed to this end. U.S. intelligence agencies would be turned loose to sabotage nascent programs. If these measures failed, then preventive conventional war would be undertaken. These measures would only be acceptable before aspiring nuclear weapons states completed a working bomb. States with a working weapon would have to be left alone.

A second policy would create "little NATOs"—forward-deployed U.S. capabilities sufficient for a tenacious local defense, plus institutionalized nuclear guarantees—wherever the United States has important interests that might be threatened by a nuclear state. To convince pro-

---

The most likely candidate here would have been Iran, which would have perceived itself under "nuclear pressure" from every side—Iraq, India, Pakistan, and the Transcaucasian nuclear fragments of the former Soviet Union (Azerbaijan, Armenia, or Georgia). Turkey could not have remained quiescent in the face of these developments.

spective victims not to acquire nuclear weapons, the United States or another superpower would have to guarantee their security with sufficient force to stop a conventional aggressor. Programs of nuclear cooperation, similar to those arranged between the United States and its NATO allies, might also have to be considered. "Desert Shield," the little NATO of the Middle East, might last for decades. U.S. forces in Korea would have to be beefed up. The planned drawdown of U.S. troops stationed in Europe may need to be canceled. The fundamental decision about whether or not to fight for Taiwan could no longer be postponed.

The United States would now be called upon to offer nuclear hostages to any ally threatened by a nuclear aggressor—one important function that U.S. troops served in Germany during the Cold War. If Germany had been subjected to nuclear attack, U.S. troops (and their dependents) would also have suffered. This increased the probability that the United States would retaliate. The likely indignation of U.S. leaders alone raised the risk of a nuclear response. Failure to respond might also have eroded the credibility of the U.S. nuclear deterrent to discourage nuclear attacks on North America. If U.S. soldiers and citizens could be attacked with nuclear weapons with impunity abroad, U.S. decision-makers would have reason to fear that the next step would be an ambitious aggressor misperceiving the freedom to threaten U.S. citizens at home. This fear would have provided an additional incentive for a strong response.

"Extended nuclear deterrence" is unlikely to go even as smoothly as it did during the Cold War. The failure to act to counter aggression by a small nuclear weapons state like Iraq would surely raise new doubts about the value of forward-deployed U.S. troops as guarantors of U.S. nuclear retaliation. Even during the height of the Cold War, many asked whether the United States would trade "Boston for Bonn," if it was called upon to retaliate for a Soviet nuclear attack. Greater doubts would arise under these new conditions.

Though the previous two policies could be executed separately, they would probably be combined. Perhaps, after considerable assertive counterproliferation and a substantial U.S. military buildup, the proliferation incentives that would arise from a failure in Kuwait could be dampened. Aggressors would find it dangerous to initiate nuclear programs; status quo powers would find the United States ready to offer the coin of real, local conventional military power to forestall invasions, and incidentally provide hostages against coercive nuclear threats.

The human and material costs of this strategy are substantial. One cannot be sure that even these measures would prove adequate to deter all future proliferators, so costlier measures could prove necessary. Forcible occupation and disarmament of some countries is the next step.[18] This seems probable in the cases of North Korea and Iran, for example. These countries have large populations and conscript armies; they will be difficult to conquer and even more difficult to police.

The "little NATOs" aspect of the strategy could require a force structure equal to or greater than that sustained by the United States during the last years of the Cold War. U.S. forces in NATO's center region might shrink somewhat, but greater standing forces would be necessary in the southern region—particularly in Turkey. Forces stationed in South Korea might have to grow. A new permanent force in the Persian Gulf region will prove necessary. Ready forces for new commitments would prove useful to create a general expectation among states that the United States would come to their assistance if their neighbors turned nasty. If coupled with the "preventive war option," an additional "strategic reserve" offensive force would prove necessary.

"Fortress America" is a distinct, third alternative national strategy. The task of U.S. policy would be to divert nuclear competition and the risks of nuclear war away from its own territory. The United States would simply accept that a hellishly competitive world will emerge. The first step would be a change in U.S. foreign policy toward disengagement. By staying out of overseas political competitions, the United States would reduce the incentives for new nuclear powers to threaten the country or its armed forces. U.S. theater forces would not be jeopardized by regional nuclear forces because there would not be any U.S. theater forces.[19]

The United States would rely more forthrightly on its own nuclear deterrent power; policymakers would stress their intention to respond to any direct attack on the United States with a devastating retaliation. Renewed efforts in the realm of strategic defense would be justified; just in case others doubt U.S. will, they will also face formidable obstacles to

---

18.   For example, it is clear in retrospect that the Iraqi nuclear weapons program that existed prior to Desert Storm could not have been catalogued, much less destroyed, without considerable direct access to the entire country. This was only achieved after Iraq suffered a terrific military defeat and became convinced that it faced the threat of a potentially endless economic embargo.

19.   Freedman, "Great Powers, Vital Interests and Nuclear Weapons," p. 48, sees this as a general consequence of extensive nuclear proliferation. "The spread of nuclear weapons, in terms of political control as much as absolute numbers, encourages strategic disengagement and thus a loss of influence in regions where important, if not quite vital, interests are involved . . . proliferation feeds on and then reinforces an existing tendency to reduce the security links between the declared nuclear powers and those parts of their 'far abroads' that are not covered by a well-established alliance."

a successful attack. Moreover, in a world where many states have nuclear weapons and not all are competent, an errant weapon could come the way of the United States. U.S. intelligence services would focus intensively on reducing the possibility that nuclear weapons might be smuggled into the United States. More importantly, intelligence agencies would act to back U.S. deterrent threats by creating the strong expectation that the United States would determine the source of any weapon smuggled into the country and find some group or place against which a retaliation can be directed.

This policy would require a major change from the way the United States has conducted its foreign policy since World War II, but disengagement from world politics is not unprecedented for this country. The United States would require enormous self-control to stay out of political disputes abroad that may arouse its passions or idealism. And it would require steady nerves.

INFERENCES FROM THE ACTUAL CRISIS

The thinking and behavior of U.S. decision-makers during the Desert Shield crisis suggests that the foregoing analysis, or something similar, would have been conducted and would have found a ready audience. Though the nuclear issue apparently did not arise in early discussions, there is considerable evidence that key decision-makers worried about the global and regional consequences of letting the Iraqi success stand. It was feared that Iraqi power would grow, and that the credibility of U.S. threats to deter future Iraqi action would weaken. More broadly, U.S. credibility would suffer, perhaps enough to set off a wave of challenges.

Brent Scowcroft, the president's national security adviser, favored strong action against Iraq's seizure of Kuwait. In one of the earliest National Security Council meetings on the Iraqi invasion of Kuwait, on August 3, Scowcroft reportedly opened the meeting by saying, "We have got to examine what the long-term interests are for this country and for the Middle East if the invasion and taking of Kuwait become an accomplished fact. We have to begin our deliberations with the fact that this is unacceptable. Yes, it's hard to do much. There are lots of reasons why we can't do things but it's our job."[20] The conversation in this meeting

---

20. Bob Woodward, *The Commanders* (New York: Simon and Schuster, 1991), paperback edition, pp. 217–218. See also p. 211, on Scowcroft's critical reaction to the discussions in the preceding day's meeting: "Scowcroft indicated to the President that the meeting had seemed to miss the point about the larger foreign policy questions." Unfortunately, Woodward does not spell out precisely what Scowcroft meant. It seems to have been a combination of a failure to meet a longstanding U.S. declaration that Persian Gulf Oil could not be allowed to fall into the hands of a dangerous adversary

reportedly moved in the direction of a consensus that Iraq had created a major strategic problem for the United States.[21]

U.S. Secretary of Defense Richard Cheney reportedly had ambitious objectives from the outset of the crisis. He was apparently concerned about the future political and military threat Hussein could pose if he controlled Gulf oil. This must be interpreted as fear of what Iraq could and would do if it became even more powerful. Already on August 2, he saw the question as whether the U.S. goal should be just the liberation of Kuwait, or also the overthrow of Saddam Hussein.[22] Dennis Ross, the State Department's Director of Policy Planning, and a key aid to U.S. Secretary of State James Baker, saw Saddam Hussein's aggression as an indicator that Iraq had become a threat to the entire Middle East. He apparently concluded that Saddam Hussein's regime would need to be eliminated.[23] UN Ambassador Thomas Pickering believed that U.S. credibility in the Middle East would suffer if nothing was done about the invasion of Kuwait.[24]

Deputy Secretary of State Lawrence Eagleburger was also a strong early proponent of rollback. He worried about the precedent that successful Iraqi aggression would set for all the "Quaddafis and Kim Il Sungs of the world," indicating to them that the end of the Cold War had created exploitable power vacuums.[25] In the August 4 Camp David meeting, one unnamed participant made a mirror image argument. "I think if we succeed this time, the next such crisis might not take all this much agony. In other words, if potential aggressors believe in advance that the civilized world is going to behave in a certain way, they will tend to tailor their actions."[26]

President George Bush seems to have learned the 1930s lesson on appeasement. He was concerned that if he practiced "appeasement," it would have the same consequences it had in the 1930s. One unnamed

---

and a more general desire to show that the United States was not afraid to use military force to protect its interests. Both boil down to credibility.

21. Ibid., p. 218.

22. Michael Gordon and Bernard Trainor, *The Generals' War: The Inside Story of the Conflict in the Gulf* (Boston: Little Brown, 1995), pp. 32–33. Woodward, *The Commanders*, p. 208, asserts that Cheney only favored the defense of Saudi Arabia at this point.

23. Gordon and Trainor, *The Generals' War*, p. 35.

24. Ibid., p. 36.

25. Ibid., p. 37.

26. U.S. News and World Report, *Triumph Without Victory: The Unreported History of the Persian Gulf War* (New York: Random House, 1992), p. 72.

presidential adviser observed, "George Bush is deathly afraid of appeasement. His generation had to fight a war over it, and he feels that if he blinks today, he will be leaving a real mess for the next generation to clean up."[27] President Bush characterized his views this way in subsequent communications with the staff of *U.S. News and World Report*. "The bottom line was that aggression could not stand. If he was permitted to get away with that, heaven knows where the world would have gone and what forces would have been unleashed."[28]

This perspective was also expressed by British Prime Minister Margaret Thatcher in a meeting with President Bush in Aspen, Colorado, on August 2. She analogized explicitly to Nazi aggression in the 1930s, an analogy that Bush employed repeatedly.[29] Later in the crisis, in a December 24 meeting, Bush indicated that he would no longer settle for an unconditional Iraqi withdrawal; he wanted a military campaign and victory. Chairman of the U.S. Joint Chiefs of Staff Colin Powell reports President Bush's reasoning: "If the Iraqis withdrew now, it would be with impunity for their crimes. A pullback would also mean that Saddam would leave Kuwait intact with his huge army intact, ready to fight another day."[30] The president appears to have been concerned that a lesson needed to be taught; the aggressor must be punished, and deprived of capabilities for future mischief.

Had the United States confronted a nuclear-armed Iraq, the choices would have been difficult. A decision to launch a military operation to liberate Kuwait would have involved immediate and obvious nuclear risks. The consequences of inaction outlined above, and the relatively unpleasant range of possible U.S. alternative policies in the new world thus created, would surely have been enumerated by someone. But the costs and risks of inaction would have seemed theoretical and distant. The risks of confrontation with a nuclear Iraq would have seemed clear

---

27. Steven J. Wayne, "President Bush Goes to War," in Stanley Renshon, ed., *The Political Psychology of the Gulf War: Leaders, Publics, and the Process of Conflict* (Pittsburgh, Penn.: University of Pittsburgh Press, 1993), p. 39. The themes of "no appeasement" and "stopping aggression" both surface in the August 2 National Security Council meeting. Woodward, *The Commanders*, pp. 235–237.

28. U.S. News and World Report, *Triumph Without Victory*, p. 48. An unnamed adviser who was with Bush when the invasion occurred observed, "He thought that the aggression should not stand, especially in a place where we had staked out as our vital interests . . . . That was the President's strong view." Ibid., p. 48.

29. Gordon and Trainor, *The General's War*, p. 36; U.S. News and World Report, *Triumph Without Victory*, p. 72.

30. Colin Powell with Joseph Persico, *My American Journey* (New York: Random House, 1995), p. 499.

and imminent. One cannot say with confidence how the consensus among the key decision-makers would have actually evolved. But this analysis does suggest that the often implied proposition that the United States would easily have been deterred by Iraq should be subjected to much closer scrutiny. There were strong reasons for the United States to act. And there is much circumstantial evidence from the way the crisis was actually handled to suggest that these reasons would have resonated with the decision-makers. Yet a final decision to act would have required a credible plan for success. Someone would have had to propose a strategy that promised political and military success, with low risks of escalation. Below, I try to develop such a strategy.

Planning the Reversal of Iraqi Aggression

THE VIEW FROM 1990–1991

As we approach the end of autumn, Desert Shield is a great success. Sufficient Western military power has reached the Persian Gulf to defend successfully against any new Iraqi conventional attacks. Though the economic embargo on Iraq is taking hold, and ought to be given a fair chance to pressure Saddam Hussein into withdrawal from Kuwait, the United States cannot count on the success of this strategy. It may fail altogether, or simply "succeed" at such a slow pace that Iraq will have plenty of time to devise diplomatic strategies to erode the integrity of the U.S.-led coalition. Though Iraq is dependent on many imports, it has vast reserves of the cheapest and most industrially flexible energy source—oil. This alone must give the economy a substantial adaptive capacity. The economic embargo may weaken. Iraq's reserves of oil, to include its new Kuwaiti reserves, are a seductive prize for some prospective trading partner. On the diplomatic side, the Arab-Israeli conflict provides a lever that Saddam will try to exploit diplomatically to fissure the coalition. Saudi Arabia may feel squeezed by the trade-off between external security and internal stability. Elements within Saudi society may begin to ask how long the Americans intend to stay. "Decades" may not be an acceptable answer.

Sooner or later, the United States and its allies will have to consider military action to eject Iraq from Kuwait. There is no reliable algorithm that can tell the United States exactly when it should switch from embargo to war. Given that embargo is a slow and insidious weapon, if the economic embargo does not soon *persuade* Saddam Hussein to change his policy, it will take years for its cumulative effects to *force* him to do so (if ever.) It is therefore unreasonable to wait until the coalition begins to suffer erosion to go to war, and more useful to choose a cut-off point for giving the embargo a chance to work. For planning purposes, if Iraq is not out of Kuwait by autumn of 1991, the United States and its partners should place themselves in a military position to initiate offensive operations at that time.

INTRAWAR DETERRENCE

Perhaps the most important issue that the coalition faces is how to neutralize any attempt by Iraq to exploit its nuclear weapons to deter a coalition attack of any kind, or to thwart a coalition victory. Two methods suggest themselves, a "splendid first strike," or intrawar deterrence. It is clear that a "splendid first strike" that knocks out Iraqi nuclear weapons with conventional ordnance would be best, but it seems improbable that the military can provide a high-confidence option to do this. Thus, the coalition's strategy must be to wage a conventional war with Iraq while deterring its resort to nuclear weapons. Deterrence depends on capability and will. There is no doubt that the United States and its allies command massive conventional and nuclear capability. It ought not to be impossible to explain to Iraq what the United States can do in response to an Iraqi use of nuclear weapons. The problem is convincing Saddam Hussein that the United States *will* retaliate in particularly horrible ways if he employs nuclear weapons. Only diplomacy can convince Saddam Hussein that U.S. will is stronger than his.

The United States must persuade Iraq that the United States cares more about the liberation of Kuwait than Iraq cares about holding it. Deterrent diplomacy is exactly that; explanations matter as much or more than military aspects. The problem is difficult. Iraq is the local power, has developed a "historical" (albeit trumped up) claim to the land, has tried to seize the country once before, can claim (with some legitimacy) a number of injuries inflicted by Kuwait, and has just seized the country. Iraq probably perceives that its bargaining position is very strong. Heretofore, the United States has cared about Kuwait largely for economic reasons. Kuwait has a lot of oil, but there is plenty more in Saudi Arabia and in the Gulf states, and those are now well protected by the forces entering the region for Desert Shield. Why should Saddam believe that nuclear threats will not stop a coalition offensive? And, if need be, why should not a single nuclear explosion cause the United States to reconsider its entire position? The United States, and its many allies, obviously must disabuse Iraqi leaders of this belief.

The United States and its friends must explain to Iraq just how terrible the political world appears to them, if Iraq is allowed to enjoy its conquest. The method for accomplishing this is to explain over and over U.S. expectations about the grave security situation that will result if Iraq is allowed to deter the coalition from liberating Kuwait. This story is somewhat analogous to the "we don't deal with terrorists and hostage-takers" position that many Western countries, including the United States, have taken. (Of course, the United States did, to its great discredit, deviate from this position.) Countries say that they will not deal with terrorists because it leaves them open to future exploitation. Failure to counter Iraqi aggression would effectively do the same, and much more.

The task of explaining this position to Iraqi leaders, especially Saddam Hussein, is not a simple one, and must be pursued with enormous energy and creativity. Both public repetition and private messages must be employed. Numerous senior U.S. figures, representatives of other governments, and even representatives of international organizations such as the United Nations, will have to repeat these arguments publicly. Go-betweens, such as the Russians and the Jordanians, must convey these messages privately. In addition, the appropriate historical analogies must be made publicly. While the "Hitler analogy" is already proving a useful public relations tool within the United States, perhaps the "Cuba analogy" could prove useful in communicating the seriousness of U.S. intent to Saddam Hussein.

Given the Cold War experience with coercive diplomacy, it may be reasonable to move some nuclear weapons toward the theater to enhance the credibility of the threat of ferocious retaliation in the event of an Iraqi nuclear attack. There is no particular *technical* utility to such action. It would be purely a diplomatic signal. Some joint coalition planning for the retaliatory use of such weapons probably ought to occur. Rumors of such consultations would have an additional diplomatic effect. At the same time, such consultations would lend credibility to the message that this is a contingent threat. If Iraq does not use nuclear weapons, neither would the allies.

INFERENCES FROM THE ACTUAL CRISIS

The issue of "intrawar deterrence" did in fact arise in a muted way duringthe planning for Desert Storm. U.S. Secretary of State James Baker, in his January 9, 1991, meeting with Iraqi Deputy Prime Minister Tariq Aziz, carried a letter from President Bush that threatened underspecified but nevertheless horrible consequences if Iraq used chemical or biological weapons.[31] Indeed, Aziz thought that these were nuclear

---

31. The Bush letter reportedly read, "The United States will not tolerate the use of chemical or biological weapons, support of any kind of terrorist actions, or the destruction of Kuwait's oilfields and installations. The American people would demand the strongest possible response. You and your country will pay a terrible price

threats.[32] Though Iraq had plenty of chemical weapons, it did not employ them.[33] It seems plausible that even more explicit retaliatory threats would have been leveled against a nuclear threat.

Oddly, the private message to Saddam Hussein was preceded by what appears to have been a deliberate public message indicating that the use of nuclear weapons was not under consideration. A January 7, 1991, *Washington Post* story quotes several named and unnamed senior civilian and military officials to the effect that nuclear use was not under consideration. Ironically, this message was probably an effort to quell the fears of allies that the United States was considering the use of nuclear weapons, fears that may have been created by deterrent diplomacy. For example, U.S. Secretary of Defense Richard Cheney said during a visit to the Gulf on December 23, "Were Saddam Hussein foolish enough to use weapons of mass destruction, the U.S. response would be absolutely overwhelming and it would be devastating."[34] Both from the *Post* story, and from other accounts, it appears that a variety of oblique yet ferocious messages had been sent.[35] I suspect that the administration suddenly

---

if you order unconscionable action of this sort." Lawrence Freedman and Ephraim Karsh, *The Gulf Conflict: 1990–91* (Princeton, N.J.: Princeton University Press, 1993), p. 255.

32. Aziz informed Rolf Ekeus, the chief UN investigator of Iraq's nuclear, biological, and chemical weapons programs, that Iraq decided not to employ biological weapons because of a strong, if somewhat ambiguous, warning received from the Bush administration on January 9, 1991, which Iraqi leaders interpreted as a nuclear threat. In a four-hour meeting with Aziz in Geneva, Secretary of State James A. Baker III threatened "a U.S. response that would set Iraq back years by reducing its industry to rubble." R. Jeffrey Smith, "U.N. Says Iraqis Prepared Germ Weapons in Gulf War," *Washington Post*, August 26, 1995, p. A1. Other accounts suggest that Baker threatened the overthrow of the Iraqi government in the event of a use of chemical or biological weapons. See Freedman and Karsh, *The Gulf Conflict*, p. 257; Baker says that he *did* hope to create the impression that nuclear weapons might be employed. James Baker with Thomas DeFrank, *The Politics of Diplomacy* (New York: Putnam and Sons, 1995), p. 359.

33. On current views of the extent of Iraq's chemical weapon arsenal, see Christopher Wren, "U.N. Arms Inspector Firm on Iraq Nerve Gas," *New York Times*, June 25, 1998, p. A10.

34. Jeffrey Smith and Rick Atkinson, "U.S. Rules Out Gulf Use of Nuclear, Chemical Arms," *Washington Post*, January 7, 1991, p. A1.

35. According to Lawrence Freedman, "In the 1991 Persian Gulf War, Britain and France were ready to rule out nuclear use prior to the conflict: the Americans were more hesitant, although they worked to deter Iraqi chemical-weapon use largely through threats to the regime in Baghdad." "Great Powers," p. 41. See also Molly Moore, *A Woman at War* (New York: Scribner's Sons, 1993), p. 318. "During his speaking engagements, Boomer [Lieutenant General Walter Boomer, the U.S. Marine

became concerned that deterrent diplomacy had perhaps gotten out of control. Perhaps certain coalition partners were beginning to believe that the United States did intend to employ nuclear weapons.

Nuclear weapons were apparently discussed more frequently than policymakers admitted at the time. In November of 1990, Cheney is quoted as having told a proponent of nuclear weapons use that nobody had suggested such a thing in the entire three months of the crisis.[36] In an interview in 1996, former President George Bush asserted that nuclear use "was not something we really considered at all."[37] Yet General Colin Powell says in his memoirs that Cheney had asked him to take a look at tactical nuclear options in early October 1990.[38] General Scowcroft stated in a 1995 interview that the administration did indeed discuss nuclear weapons, but determined that their use would be unnecessary.[39] Sometime in "late fall," General Schwarzkopf suggested to General Powell a diplomatic demarche that the United States would use nuclear weapons if the Iraqis used chemicals.[40] *Newsweek* magazine reported in its January 14, 1991, issue that Schwarzkopf had earlier asked for permission to use a nuclear weapon to generate an electromagnetic pulse to wreck Iraqi

---

Corps commander in Desert Storm] talked rather candidly about how the media had been used to intimidate Saddam Hussein. As I watched from the back of one conference room, Boomer told the audience that reporters like Molly Moore . . . frequently asked him during the buildup: 'What will be the American response if Saddam uses chemical weapons?' Boomer said his standard reply was: 'It's going to be something worse, something terrible,' implying that perhaps nuclear weapons would be considered. Boomer smiled when he admitted to the audience, 'I just made that up; I didn't know what the hell we were going to do'." According to two RAND analysts, an unnamed British general told reporters that "chemical use by Iraq would mean a nuclear response by the coalition forces," though they provide no citation. Paul K. Davis and John Arquilla, "Deterring or Coercing Opponents in Crisis, Lessons from the War with Saddam Hussein" (Santa Monica, Calif.: RAND, National Defense Research Institute), p. 59, n. 35.

36. Jeffrey Smith, "U.N. Says Iraqis Prepared Germ Weapons in Gulf War," *Washington Post*, August 26, 1995, p. A1.

37. See "A Gulf War Exclusive: Talking With David Frost," Transcript No. 51, January 16, 1996, p. 5, cited in Payne, *Deterrence in the Second Nuclear Age*, pp. 138–139.

38. Powell, *My American Journey*, p. 486.

39. General Brent Scowcroft, "Meet the Press," NBC News, August 27, 1995, transcript. Mr. McManus: "Did you deliberately give the Iraqis the impression that you were ready to use nuclear weapons?" Answer: "We kept it very ambiguous, which I think was the right thing to do . . . . I'm just telling you what we had privately decided. But, no, we never ruled out the use of nuclear weapons and should not, I don't think."

40. Rick Atkinson, *Crusade: The Untold Story of the Persian Gulf War* (Boston: Houghton Mifflin, 1993), p. 86.

radars and communications devices at the outset of a campaign, and that the Pentagon actually hired some consultants to examine the option. And it was suggested during the planning discussions for attacks on Iraqi biological weapons storage sites that nuclear weapons would be the surest way to generate enough heat to actually kill the viruses with high confidence. Thus, the use of nuclear weapons seems indeed to have been considered by the U.S. Secretary of Defense, and by the Commander-in-Chief of U.S. forces in the Persian Gulf, and not solely for reasons of deterrence. Nuclear weapons use was not ruled out a priori. Because nobody in authority at that time believed that Iraq had a nuclear weapons capability, or even a chemical capability that could seriously disrupt U.S. conventional operations, these nuclear ruminations did not go anywhere. Indeed, one of the reasons most often suggested for the absence of serious consideration of a nuclear response was confidence that coalition forces could effectively operate even if Iraq employed chemical weapons.[41]

Powell himself was drafting an explicitly non-nuclear deterrent message to Iraq on January 15 listing a host of conventional attacks the United States and its coalition partners would mount in the event of a chemical or biological attack. Specific targets such as the merchant fleet, the road and rail network, ports, oil facilities, commercial airports, and even dams on the Tigris and Euphrates were on the list.[42] These targets, particularly the dams, are similar to those suggested the preceding autumn by the principal air commanders in the theater as possible responses to an Iraqi chemical attack.[43]

I argue above that dealing with a nuclear-armed Iraq would have demanded a political and military strategy of intrawar deterrence. Such a strategy depends both on will and capability. Will, in particular, must be systematically communicated to the adversary. Evidence from the actual crisis on the matter of Iraqi chemical weapons use shows that such a strategy was followed. Numerous messages, public and private, formal and informal, were deliberately sent to Iraq. These messages threatened a massive response if Iraq employed chemical weapons. The threats were sufficiently frequent and ferocious that Iraqi officials claim after the fact that they were taken as nuclear threats. Circumstantial evidence suggests

---

41. See especially Powell, *My American Journey*, p. 468. "The Iraqi chemical threat was manageable. Our troops had protective suits and detection and alarm systems. In battle, we would be fast-moving and in the open desert, not trapped as civilians might be. A chemical attack would be a public relations crisis, but not a battlefield disaster."

42. Powell, *My American Journey*, p. 504.

43. Atkinson, *Crusade*, p. 86.

that even some U.S. allies may have been concerned that the United States was contemplating a nuclear response. The only thing missing from the record is evidence that a comprehensive explanation accompanied the threat to Saddam. The closest thing to such an explanation is the passage in President Bush's January 9 letter to Saddam, "The American people would demand the strongest possible response," a proposition reiterated verbally to Tariq Aziz by Secretary of State Baker. For purposes of my argument, one would have liked to see a statement such as "We will have no choice but to retaliate massively, because we must expect to encounter adversaries in the future who are armed with these weapons, and who must be made to understand the grave consequences of using them against the United States." Nevertheless, it seems fair to conclude that the concept of intrawar deterrence and the mechanics of deterrent diplomacy were well understood by the Bush administration, and were employed. This lends credibility to the proposition that the kind of strategy I outlined above would have fallen on a receptive audience of high-level policymakers.

The Conduct of the War

THE VIEW FROM 1990–1991

Intrawar deterrence is one key means to control the risks of a large-scale conventional war with nuclear-armed Iraq. A second potential means is by limiting the coalition's war aims, and by constraining the coalition's military operations consistent with these aims. In stark terms, intrawar deterrence aims to discourage Iraq from employing nuclear weapons out of calculation. We must also consider how to avoid provoking the Iraqis into using their nuclear weapons out of desperation.

What should coalition war aims be? It is the coalition's irreducible requirement to eject Saddam Hussein from Kuwait. Moreover, if this must be done by force, it is reasonable to strive for the maximum destruction of Iraqi military capabilities in and near Kuwait. The Iraqis must be taught that aggression does not pay. And it is a good idea to take this opportunity to whittle down Iraqi military power to help stabilize the region.

There are other plausible and tempting objectives beyond these, but it may be imprudent to pursue them. It would be wonderful to be rid of Saddam entirely. If Saddam's regime cannot be eliminated, it would be beneficial not only for the region, but for the future of U.S. nonproliferation policies, to eliminate Iraq's weapons of mass destruction. Unhappily, for the sake of avoiding Iraqi resort to nuclear weapons out of desperation, the coalition probably must forego these objectives. The goals of the war needed to be "depersonalized." The coalition should not even hint that Saddam Hussein is the target of the war; indeed, it should explicitly say that neither he, nor the sovereignty of Iraq within its prior boundaries, is at issue. Only Kuwait is at issue. A cornered rat with a nuclear weapon is a pretty dangerous animal. Leave it an escape route.

Can Iraq's nuclear weapons program safely be included as a military objective of the coalition's conventional military attack? Since Iraq already has a bomb, this project seems too dangerous; it might put Iraq in a "use or lose" situation. During the Cold War, it was generally believed that U.S. and Soviet nuclear forces were the "family jewels." Though each side targeted the other's nuclear forces, neither side's leadership doubted the grave risks associated with the initiation of a counterforce campaign. Though Iraq is new to the nuclear club, and possesses only a few weapons, it is reasonable to assume that it places a high value on these weapons. The coalition ought not to count on Saddam Hussein's accepting his gradual nuclear disarmament. It is plausible that he would fear that once the nuclear weapons were gone, the coalition would feel free to do anything it wanted—perhaps even use a nuclear weapon on him. Though the last possibility might seem slight, an attempt to topple the regime would seem quite plausible. Saddam might calculate that the use of a single nuclear weapon, perhaps only as a demonstration, would be enough to cause the coalition to cease its military campaign to disarm him.

The coalition's military strategy should be amended to help limit escalatory risks. In particular, we must ensure that conventional operations do not create unintended damage that would be construed either as an attack on the regime, or as an attack on the weapons of mass destruction that Saddam Hussein may view as his secure second-strike capability. If U.S. military planners could be very sure that these capabilities could be quickly destroyed, then it might be reasonable to risk such an attack. But it seems unlikely that U.S. military planners could offer such assurances.

To ensure that Saddam does not overestimate the hostile intent of the coalition, considerable restraint in military operations seems in order. Most notably, the coalition will have to forego the bulk of the "strategic" bombing campaign now under consideration. The coalition's political leadership will need to draw some notional lines to constrain its military operations, perhaps the boundaries of what is now called the "Kuwait Theater of Operations" (KTO), Kuwait and Southern Iraq. Intense military operations will be permitted inside the KTO; only limited operations would be permitted in the rest of Iraq.

Because Saddam will not be forced to pay a cost for his aggression in terms of damage to his country, the full cost would be extracted from Iraqi field forces. Annihilation of Iraqi combat forces in the theater should be an explicit military goal. The troops can go home, but the equipment stays. The message to Saddam, and to those who might follow, is that the forces one invests in aggressive war simply are lost.

INFERENCES FROM THE ACTUAL CRISIS

What is the likelihood that these notions of "limited war" would have found their way into actual coalition behavior? In the aftermath of Vietnam, the generally negative appraisal of constraints placed both on military means and political ends in that war has made "limited war" a very problematical concept in security policy circles, both civilian and military. U.S. planning for conventional war with the Soviet Union did not demonstrate any great sensitivity to the question of whether conventional threats against nuclear forces might be escalatory.[44] Yet, both the objectives and conduct of Desert Storm were wrapped in political constraints. Positive and negative civilian intervention into military operational planning on both diplomatic and domestic political grounds occurred. Indeed, Secretary of Defense Richard Cheney went to considerable lengths to educate himself to facilitate such intervention.

A limited war aim was the principal form of restraint practiced during the war, though the precise political reasons for this limitation are murky. Specifically, it was understood from the beginning that the coalition would not invade and occupy Iraq with an eye toward changing its government. Former Secretary of State James Baker states that a "march to Baghdad" was never on the table, because everyone understood it was costly and unnecessary.[45] General Powell suggested that the "Hitler rhetoric" be toned down because it could imply that Saddam would in fact be ousted, which was not a primary war aim.[46] At the February 27, 1991, Oval Office meeting where the decision to end the war was made, Secretary of State Baker argued, "We have achieved our aims. We have gotten them out of Kuwait."[47] Powell believed that the war had a specific and limited objective, and in the controversial end-game supported a rapid termination of the ground war since this limited objective—the liberation

---

44. Barry R. Posen, *Inadvertent Escalation, Conventional War and Nuclear Risks* (Ithaca, N.Y.: Cornell University Press, 1991). See especially chaps. 2 and 4.

45. James Baker with Thomas DeFrank, *The Politics of Diplomacy* (New York: Putnam and Sons, 1995), p. 437.

46. Powell, *My American Journey*, p. 491.

47. Gordon and Trainor, *The General's War*, p. 416.

of Kuwait—had been achieved.[48] The decision to end the war was made even though Saddam Hussein was still in power. Moreover, though the key decision-makers were more optimistic about the damage to Iraqi weapons of mass destruction than ultimately proved warranted, they knew then that Iraq still possessed chemical weapons stocks.[49]

Initial accounts of the war suggested that the United States was somehow restrained from intervention inside Iraq by its regional coalition partners. Two of the more thorough scholars of the war argue the opposite; both Saudi Arabia and Turkey supported intervention during the Iraqi civil war that followed Desert Storm.[50] Therefore, in spite of the ambitious rhetoric that provided the public rationale for the war, and local support for direct intervention in Iraq, it seems reasonable to conclude that the United States was not willing to pay very much to overthrow Saddam Hussein. The U.S. war aims were limited by its sensitivity to costs.

If aims were limited, so were means. Secretary of Defense Richard Cheney went to great lengths to develop his own ability to assess professional military advice. He asked for, and received, what was essentially a short course in military operational planning.[51] He apparently told Paul Wolfowitz that though he did not intend to micromanage the preparation of the war plan, he did "intend to own it when it's finished."[52] And he was not shy about making his own suggestions. The initial briefings by the Central Command (Centcom) planners for a possible counteroffensive into Kuwait disturbed him. "Colin, I have been thinking about this all night. I can't let Norm do this high diddle diddle up-the-middle plan, I just can't let him do it."[53] Cheney proceeded to make his own inquiries and became an advocate of a large-scale two-division raid into western Iraq. Though this suggestion was ultimately deflected by Schwarzkopf and Powell, it did contribute to the development of the ultimate left-hook plan.

Limitations were also suggested on the air war. President Bush was particularly concerned that targets of special historical or cultural significance not be struck. He raised his concerns at two briefings, one in

---

48. Powell, *My American Journey*, pp. 519–527.

49. Gordon and Trainor, *The General's War*, p. 414.

50. Ibid., pp. 454, 456.

51. Woodward, *The Commanders*, p. 317.

52. Ibid., p. 316.

53. Gordon and Trainor, *The General's War*, p. 141.

October and one in January.[54] A different kind of discussion arose over the issue of direct attacks on presumed Iraqi biological weapons storage facilities. Out of fear of the collateral damage that could have been produced had the viruses been inadvertently released, Cheney wanted to be quite sure that there was a very high probability of complete destruction.[55]

The actual conduct of the crisis does suggest that "limited war" concepts were not entirely alien to the thinking of the principal decision-makers. Constraints were placed on both ends and means. Nevertheless, there is no evidence of fear that attacks on any of Saddam's weapons of mass destruction could conceivably produce "use or lose" behavior.

Thus, there is only qualified support for the proposition that decision-makers would have followed the advice outlined above, and avoided attacks on Iraqi nuclear weapons in order to avoid escalation. Iraq's nuclear, biological, and chemical weapons programs and its missile delivery capabilities were a military objective of the coalition. Had U.S. decision-makers understood that these had only been partially damaged by the forty-three-day air war, it is possible that the air campaign would have continued after the liberation of Kuwait. Yet it *was* then understood that some chemical weapons and missiles had survived the war. The costs or risks of additional attacks upon them were apparently deemed to have exceeded the potential benefits. That this tradeoff was made suggests that decision-makers would at least have considered seriously the costs and risks of direct attacks on Iraqi nuclear weapons.

War Outcome

How would the war have gone if the coalition had fought a limited war against a nuclear-armed Iraq? Would Iraq have used nuclear weapons?

---

54. After the October 11 briefing to the president on air and ground war plans, "Bush told Powell to make doubly sure that no targets of religious or historic significance were on the target list." He apparently did not want to anger unnecessarily the Iraqi people or Arabs throughout the Middle East. Ibid., p. 136. On the evening of January 13, Cheney reviewed the target list with the president, to ensure that "Bush was aware of potential points of controversy." The president asked that one group of targets be dropped, "statues of Saddam and triumphal arches thought to be of great psychological value to the Iraqi people as national symbols." Woodward, *The Commanders*, p. 353.

55. Gordon and Trainor, *The General's War*, pp. 191–193. The authors say that the sites in question were not in fact biological weapons sites; Atkinson, *Crusade*, pp. 88–90, suggests that there was a six-week debate on whether or not to attack these targets. Both Schwarzkopf and General Horner, his air commander, favored the attack and ultimately prevailed with a technical argument on how it could be done successfully.

Would the conventional victory have been as splendid? What message would have been sent to other aggressors? What message to those contemplating a nuclear weapons program?

I doubt that Iraq would have used nuclear weapons under these conditions. There would have been no political or military incentive, and many disincentives.[56] Saddam did not use chemical or biological weapons under conditions of conventional disaster; why would he have used nuclear weapons?[57] The more important question, however, is whether Western leaders would have believed that they could deter nuclear escalation by Iraq. It is difficult to judge. My argument is that once they had taken reasonable steps to control the risks, the costs and risks of inaction would have seemed greater than the costs of action.

The war would probably have gone about as well as it did, maybe better. But this is a statement that benefits entirely from hindsight. It is now clear that the bombing of Iraq proper did not play the decisive role in the ejection of Iraqi troops from Kuwait. Bombing in the KTO did play a decisive role. Iraqi command and control was degraded but not severed.[58] Since Iraq had many months to prepare the theater, it turned out that the interdiction of supplies into the theater did not matter much. There were plenty of weapons, munitions, food, fuel, and troops in southern Iraq and even Kuwait. The hungry troops were at the front; they were hungry because the supply lines to their in-theater depots were severed, not because the supply lines to Baghdad were severed. The Iraqi air force was so bad that it would have been blown out of the sky had it tried to affect the ground battle. It is not clear that killing the air force on

---

56. Aharon Levran shrewdly observes that during the war with Iran, Saddam did not employ chemical weapons indiscriminately; they were employed mainly in situations where the risk of decisive Iranian ground attacks was high, and largely on Iraqi soil. Thus, even against an adversary with no in-kind retaliatory capability, Saddam saw it in his interest to observe some limits, and employed chemicals as weapons of last resort. See Lerran, *Israeli Strategy After Desert Storm: Lessons of the Second Gulf War* (Frank Cass: London, 1997), pp. 68–71.

57. The Pentagon concedes that there was one inadvertent release of chemical agents immediately following Desert Storm as a result of U.S. demolition of an Iraqi weapons dump. A White House panel appointed to study the question found substantial evidence that some U.S. Marines were exposed to chemical agents during the war. Philip Shenon, "Panel Says Pentagon Ignored Signs of Poison Gas," *New York Times*, October 31, 1997. Jonathan Tucker, a researcher dismissed from the White House panel, believes that "considerable evidence suggest that the Iraqi forces engaged in sporadic, uncoordinated chemical warfare during the gulf war." Philip Shenon, "Weapons Expert Tells of Possible Iraqi Gas Attacks in Gulf War," *New York Times*, April 25, 1997.

58. Thomas A. Keaney and Eliot A. Cohen, *Gulf War Air Power Survey: Summary Report* (Washington, D.C.: U.S. Department of the Air Force, 1993), p. 70.

the ground, in its shelters, was preferable to killing it in the air.[59] Moreover, it was the air force's inability to compete in the air that drove it to hide in shelters. Attacking the shelters simply drove aircraft to Iran, a bizarre and not wholly beneficial outcome. An irony here is that many sophisticated aircraft and munitions would have been diverted from their early campaign against bunkers, bridges, and aircraft shelters, to the more serious business of wrecking Iraq's theater forces; this might have hastened their collapse.

Offensive air attacks on Iraqi Scud missiles destroyed few if any of them.[60] The pressure of sustained air attacks probably did complicate and hasten launch procedures, eroding the already poor accuracy of the system. Moreover, to avoid detection and destruction, Scud crews seem to have launched less frequently than they might have otherwise. And salvo firing was inhibited. In the end, this reduced but did not eliminate the effectiveness of Scud missiles as terror weapons.

An Iraqi nuclear weapons program would have complicated Saddam's calculations about conventional Scud launches. Scuds might have been his preferred nuclear delivery system. Would he have wanted to bring down Western air power on them if he did not have to do so? Perhaps concerns about the survival of nuclear Scuds would have discouraged him from firing conventional Scuds. On the other hand, he might have felt safe behind his nuclear shield, and launched the conventional Scud attacks in the same fashion as actually occurred in the Gulf War. He seems to have hoped to provoke Israeli counterattacks that would in turn detach the Arab members of the coalition. The United States would have had to retaliate in some way against conventional Scud launches to prevent the Israelis from doing so. This would have undermined a "limited war" strategy, but it would have been the lesser evil.

---

59. One could argue, however, that attacks on the airforce bases all over Iraq exerted debilitating effect on Iraqi offensive air capability in excess of what one might infer from the total count of aircraft destroyed or flown to Iran, about half the total inventory. Air war planners would have argued that it was unduly risky to allow the Iraqis a sanctuary where they might plan and organize massed, chemically armed air attacks. A possible compromise would have been to attack and suppress the Iraqi airfields within effective unrefuelled, tactical, range of the KTO, particularly those airfields hosting the most effective Iraqi attack aircraft. Naturally, air defense and command and control assets protecting these bases would also need to be suppressed.

60. Keaney and Cohen, *Gulf War Air Power Survey*, write, "Although Iraq's average weekly launch rate of modified Scuds during Desert Storm (14.7 launches per week) was lower than it had been during the 1988 'war of the cities,' and while launch rates generally declined over the course of the Gulf War, the actual destruction of any Iraqi mobile launchers by fixed-wing Coalition aircraft remains impossible to confirm" (p. 83). See also pp. 89–90.

Important lessons about what nuclear weapons are *not* good for would have emerged if the coalition had fought a limited war against a nuclear-armed Iraq. They are not shields for conventional conquests. Where important great power interests are engaged, they are not the great equalizer. Moreover, the often expensive conventional forces that one must bet in wars of conquest face a great risk of annihilation. This lesson might affect those who are meant to take orders, not merely those who give them. Field commanders might see the odds of a successful coup at home to be better than the odds of surviving a war with the United States.

Nevertheless, a lesson would have been taught about what nuclear weapons *are* good for. They are great instruments to deter threats against one's homeland, even conventional threats, even by a great power, even when you are in the wrong. Moreover, by invoking nuclear deterrence, the coalition would itself have demonstrated in the broadest sense the continued utility of nuclear weapons in international politics.[61] Thus, any hopes of "delegitimizing" these weapons, or sustaining a confidence game regarding their supposed inutility, would be dashed. Some damage would have been done to the cause of nonproliferation, but not nearly as much as would have been done by a failure to act.

The Future

It can be argued that all I have done here is to tell a theoretically informed alternative story, a different story than is usually implied by the question "What if Iraq had had nuclear weapons?" Given the nature of my methodology this is unavoidable. Nevertheless, the analysis does help us think about the future.

The United States is indeed working energetically to limit the number of nuclear powers in the world. Policymakers struggled to ensure that only one nuclear power emerged from the wreckage of the Soviet Union, with apparent success. They have worked with considerable effectiveness to dismantle Iraq's nuclear weapons program, and have at least limited North Korea's nuclear materials production capability.

Some U.S. policy analysts call for the abolition of nuclear weaponry, in part to strengthen the legitimacy of the NPT. The problem of fulfilling the Article VI treaty pledges of the nuclear weapons states to seek com-

---

61. This has apparently occurred. Commenting on the 1995 revelations by Tariq Aziz that Iraq was deterred from chemical and biological weapons use by what its leaders perceived to be a nuclear threat, a French defense official averred, "Did Tariq say they were deterred? Whatever the reason, we like that kind of testimonial." Joseph Fitchett, "Nuclear States See Vindication: Threat of Annihilation Deterred Iraq, They Say," *International Herald Tribune*, September 12, 1995.

plete nuclear disarmament was a key issue of the 1995 NPT review conference.[62] The U.S. government hopes that there will be further accessions to the treaty, especially by India and Pakistan, which so dramatically revealed their possession of such weapons in 1998, and by Israel, which is widely suspected to have nuclear weapons.

The Pentagon has a "counterproliferation initiative," which includes among other projects research and development on various offensive and defensive measures to use conventional means to neutralize the nuclear weapons of other countries in the event of crisis.[63] These means must be conventional rather than nuclear in order to protect the "devaluation" of nuclear weaponry now presumed to be necessary to support nonproliferation diplomacy.[64] Some hint that these capabilities could be employed

62. George Bunn, Roland Timerbaev, and James Leonard, "Nuclear Disarmament: How Much Have the Five Nuclear Powers Promised in the Non-Proliferation Treaty?" (Washington, D.C.: Lawyers Alliance for World Security, Committee for National Security, and the Washington Council on Non-Proliferation, June 1994), pp. 1–15, provides an exhaustive discussion of Article VI, which reads, "Each of the Parties to the Treaty undertakes to pursue negotiations in good faith on effective measures relating to cessation of the nuclear arms race at an early date and to nuclear disarmament, and on a treaty on general and complete disarmament under strict and effective international control." The authors tend toward a literal interpretation of the clause.

63. See Cambone and Garrity, "The Future of United States Nuclear Policy," pp. 87–90, for a summary of the counterproliferation initiative.

64. See, for example, former Secretary of Defense William Perry's statement, "I can't envision the circumstances in which the use of nuclear weapons would be reasonable or prudent military action." Meet the Press transcript, NBC News, April 3, 1994, pp. 7–8, cited in Payne, *Deterrence in the Second Nuclear Age*, p. 139. New Pentagon guidelines for the use of nuclear weapons, however, seem to include options for "retaliating against smaller 'rogue' states that might use chemical or biological weapons." See Steven Lee Myers, "U.S. 'Updates' All-Out Atom War Guidelines," *New York Times*, p. A3; see also Hans Kristensen, "Targets of Opportunity," *Bulletin of the Atomic Scientists* (September/October 1997), pp. 22–28. A subtle change can also be observed between the 1996 and 1997 versions of the Pentagon's *Proliferation: Threat and Response*. In the 1996 version the possibility of nuclear retaliation against an adversary that employed weapons of mass destruction against the United States is only hinted at in the preface, and, as far as I can discern, discussed not at all in the body of the report. Office of the Secretary of Defense, *Proliferation: Threat and Response 1996* (Washington, D.C.: U.S. Government Printing Office [GPO], April 1996). In the 1997 version, however, a more explicit statement is offered in the body of the text. In the introductory passages of the section "DOD Response," the following appears: "DOD is undertaking a variety of programs and activities to deter the use of NBC weapons against U.S. and allied forces, as well as against the territories of the United States and its friends and allies. The effectiveness of these efforts will depend on the perceptions and assessments of potential aggressors who possess NBC weapons regarding *the resolve of the United States to deal with such threats. Indeed the knowledge that the United States has a powerful and ready nuclear capability is a significant deterrent to the use of these weapons.* Effective deterrence will depend on a range of nuclear and conventional

preventively, to ensure that nascent nuclear weapons states never complete their programs. Operational and tactical adaptations that might reduce the vulnerability of U.S. military forces to weapons of mass destruction are also under study. The coalition diplomacy of confrontations with new nuclear powers has also received some attention.[65]

Although a combination of international legal instruments and unilateral military means may do much to reduce the proliferation problem, it is unlikely that the success will be total. North Korea's future cooperation is not a sure thing. New nuclear hopefuls will surely arise, with Iran as perhaps the most likely candidate. Though Ukraine's program to relinquish the nuclear weaponry it inherited from the Soviet Union is complete as of this writing, its substantial civil nuclear energy industry provides a foundation for a future weapons program. The NPT nuclear weapons states will continue to reduce the number of nuclear weapons they hold, but real progress to complete nuclear disarmament is unlikely. Nuclear weapons are the sole source of modern Russia's great power status, and the sole bulwark of the security of eastern Russia against 1,200,000,000 Chinese. It is fatuous to imagine that Russia would support complete nuclear disarmament. Thus, the complaints of bad faith by the NPT non-nuclear states against the nuclear states are likely to increase rather than diminish, degrading somewhat the legitimacy of the treaty.

The counterproliferation initiative will undoubtedly yield some interesting technology and weapons. But it is improbable that a high-confidence capability to neutralize even small nuclear forces with conventional means will emerge. Coalition air power destroyed few if any mobile Scud missiles in Desert Storm, suggesting that a moderately well-organized mobile ballistic missile force can be incredibly elusive.[66]

---

response capabilities, as well as active and passive defenses and supporting command, control, communications, and intelligence" (my italics). See Office of the Secretary of Defense, *Proliferation: Threat and Response 1997* (Washington, D.C.: U.S. GPO, November 1997), http://www.defenselink.mil/pubs/prolif97/secii.html# prevention.

65. See, for example, Thomas J. Hirschfeld, "The Impact of Nuclear Proliferation: Final Report," Center for Naval Analysis, July 1995, which devotes considerable attention to force structure and operational requirements of confrontations with nuclear-armed countries. It pays less attention to alliance diplomacy.

66. Keaney and Cohen, *Gulf War Air Power Survey,* pp. 78–90. The report estimates that Iraq had, "at most, thirty or so mobile [Scud] launchers at the start of the war . . . few mobile Scud launchers were actually destroyed by Coalition aircraft or special forces during the war . . . Coalition air power does not appear to have been very effective against this militarily insignificant target category" (pp. 89–90). "Roughly

The Patriot missile's erratic performance in Desert Storm also provides some cause for doubt about the likely success of future defensive programs, though it was not designed for this role, and purpose-built theater ballistic missile systems would likely do better.[67]

Finally, a series of preventive wars to neutralize nascent nuclear programs is unlikely; the United States would not have waged such a war against Iraq in the absence of its attempted conquest of Kuwait. The United States probably will not do so against Korea. Even if it does, one doubts that the United States can muster the political will to do so repeatedly.

One day, the United States will probably face a crisis caused by aggression of a nuclear-armed regional power against an important U.S. interest. The United States would be unlikely to have a combination of offensive and defensive capabilities that could eliminate an adversary's nuclear forces. Thus, the United States will face a defining moment of the kind I have outlined. If so, it will have to assess the very factors discussed here. While the highly competitive political world that could follow U.S. inaction may be perceived as uninviting, more thinking must be done about how the United States might live in that world. Are any of the three alternative strategies outlined above—preventive war, "little NATOs," or isolationism—acceptable alternatives for U.S. foreign and security policy? Perhaps I have overlooked other diplomatic and military strategies that would be preferable to the risks of a nuclear confrontation. If the United States chooses to act, however, it is difficult to see any other foundation upon which to rest its action but intrawar nuclear deterrence, and limited war. These expedients require as much attention and analysis as the rest of the proliferation agenda, but there seems to be no natural constituency for this work in the policy community. Intrawar deterrence raises the political salience of nuclear forces, and thus weakens the nuclear devaluation tactic that is now an element of U.S. nonproliferation policy. Limited war strategies, especially limits on military means, go against the grain of military wisdom, and exacerbate civil-military conflict in the United States. If both the costs and risks of inaction, and the costs and risks of action seem unpalatable, then diplomatic strategies must be devised that offer some hope of redefining the situation to make inaction appear

---

1500 strikes were carried out against targets associated with Iraqi ballistic missile capabilities." Perhaps another 1000 "Scud Patrol" sorties dropped on other targets (pp. 83–84).

67. Ballistic missile designers may devise simple "penetration aids" as well, complicating the task of Anti-tactical Ballistic Missile systems more advanced than Patriot.

differently than how I have portrayed it.[68] This would seem unusually creative diplomacy.

---

68. For example, when Iraq invaded Kuwait, Saddam Hussein argued that Kuwait was a lost province of Iraq, which he was merely reclaiming. Few would credit this assertion. But if the United States wished neither to risk nuclear escalation in a military campaign of liberation, nor to risk the lessons that putative aggressors might draw from the episode, U.S. diplomacy might simply have found it expedient to accept, indeed to argue, the legitimacy of the Iraqi claim. In this way, the basic principle that nuclear weapons deter attacks on one's own territory, but do not protect conquests, would have been sustained. Such a diplomatic campaign might have limited the lessons other states might have drawn from the U.S. concession of Kuwait to Iraq, and thus reduced the nuclear proliferation damage of the concession. This might not work more than once.

Chapter 7

Containing Rogues and Renegades: Coalition Strategies and Counterproliferation

Stephen M. Walt

Since the end of the Cold War, U.S. defense planners have emphasized the threat from a set of countries whose values and ambitions appear to be sharply at odds with U.S. goals and international norms. In addition to a deep-seated hostility toward the United States, so-called rogue or "backlash" states such as Iraq, Iran, North Korea, and Libya are believed to harbor aggressive designs toward one or more of their neighbors and are regarded as especially willing to use violence to advance their foreign policy goals.[1] Each of these states has taken steps to acquire weapons of mass destruction (or WMD, usually defined as nuclear, biological, or chemical weapons), with varying degrees of success. The threat from these potential "nuclear rogues," which was first acknowledged in the Department of Defense's *Bottom-Up Review* in 1993, led to the Defense Counterproliferation Initiative that same year and continues to guide U.S. efforts to isolate these regimes and retard or reverse their nuclear programs.[2]

---

1. While serving in the Clinton administration, former U.S. National Security Adviser W. Anthony Lake characterized "rogue" (or "backlash") states as undemocratic regimes that promote radical ideologies, repeatedly violate human rights, and whose behavior is "often aggressive and defiant." He also described them as increasingly isolated and guided by a "siege mentality." See Lake, "Confronting Backlash States," *Foreign Affairs,* Vol. 73, No. 2 (1994), pp. 45–46. Current U.S. defense plans call for military forces sufficient "to help defeat aggression in two nearly simultaneous major regional conflicts," specifically mentioning North Korea and Iraq in this context. See *A National Security Strategy for a New Century* (Washington, D.C.: The White House, 1998); and for background, Michael T. Klare, *Rogue States and Nuclear Outlaws* (New York: Hill and Wang, 1995).

2. In 1993, Secretary of Defense Les Aspin identified four main threats to U.S. security and stated that "the one that most urgently and directly threatens Americans at home and American interests abroad is the new nuclear danger. . . . [It] is perhaps a handful

The emergence of rogue states armed with WMD is believed to be especially dangerous for at least four reasons.[3] First, such regimes are believed to be more likely to use these capabilities than other states would be, either because they are ideologically committed to altering the status quo or because they are less sensitive to the human costs that the use of such weapons might entail.[4]

Second, a rogue state armed with WMD might be able to deter other states from intervening against it, thereby facilitating its efforts to coerce or conquer its neighbors. In particular, some U.S. officials suggest that the threat of retaliation with WMD might deter the United States from using its superior conventional capabilities to counter conventional aggression, thereby placing current U.S. allies in jeopardy. According to former U.S. Secretary of Defense Les Aspin, for example, "Today, the United States is the biggest kid on the block when it comes to conventional military forces and it is our potential adversaries who may attain nuclear weapons. So nuclear weapons may still be the great equalizer; the problem is the United States may now be the equalizee."[5]

---

of nuclear weapons in the hands of rogue states or even terrorist organizations." See Secretary of Defense Les Aspin, "The Defense Department's New Counterproliferation Initiative," Address to National Academy of Sciences Committee on International Security and Arms Control, Washington, D.C., December 7, 1993; Aspin, *The Bottom-Up Review: Forces for a New Era* (Washington, D.C.: U.S. Department of Defense, 1993); and Office of the Deputy Secretary of Defense, *Report on Non-Proliferation and Counterproliferation Activities and Programs* (Washington, D.C.: U.S. Department of Defense, May 1994). More recently, Secretary of Defense William Cohen has warned that "the United States faces a heightened prospect that regional aggressors, third-rate armies, terrorist cells and even religious cults will wield disproportionate power by using—or even threatening to use—nuclear, biological, or chemical weapons against our troops in the field and our people at home." See Office of the Secretary of Defense, *Proliferation: Threat and Response* (Washington, D.C.: U.S. Government Printing Office [GPO], 1997).

3. The potential dangers created by the spread of weapons of mass destruction are summarized in Peter R. Lavoy, "The Strategic Consequences of Nuclear Proliferation," *Security Studies*, Vol. 4, No. 2 (1995).

4. For example, an official in the Defense Department's Office of Counterproliferation Policy has argued that "many states anxious to acquire nuclear, biological, chemical weapons and missiles (NBC/M) . . . are led by undemocratic, unpredictable leaders whose motivations are largely opaque to outside observation and understanding." Citing North Korea, the same official noted that "the relatively opaque character of the North Korean leadership increases the difficulty of persuading the regime not to acquire NBC/M, to roll back its capabilities, or not to use them." See Paul R.S. Gebhard, "Not by Diplomacy or Defense Alone: The Role of Regional Security Strategies in U.S. Proliferation Policy," *Washington Quarterly*, Vol. 18, No. 1 (1994), pp. 170–171.

5. Similarly, U.S. Senator John McCain once warned that a nuclear North Korea might be more inclined to attack South Korea, because its nuclear arsenal would deter the

Third, and following from the second point, the emergence of rogue states armed with WMD might make threatened states more reluctant to join alliances against them, for fear of becoming the victim of a highly destructive attack. Such fears could inhibit U.S. efforts to contain these regimes and could defeat multilateral efforts to moderate their international conduct. For example, Roger Molander and Peter Wilson argue that "a regional predator will find a small nuclear arsenal a powerful tool for collapsing regional military coalitions that the United States might craft to oppose such a future opponent," and former Defense Department official Zalmay Khalilzad suggests that rogue states such as Iraq and Iran might use WMD to "deter the United States and its allies from [acting] . . . [or to] intimidate the GCC [Gulf Cooperation Council] states into not inviting U.S. or other Western forces to intervene," thereby facilitating renewed aggression in the Persian Gulf.[6]

Finally, many experts argue that a rogue state's unconventional arsenal would lack adequate safeguards and command-and-control devices, thereby increasing the risk of theft, accidents, or unauthorized use. Even if rogue leaders proved more rational than many believe, the danger of inadvertent attacks would grow as more states acquire WMD capabilities.[7]

United States from counterattacking. See Aspin, "Defense Department's Counterproliferation Initiative"; and McCain, letter to the *New York Times*, March 28, 1994, p. A10. In the same spirit, John Arquilla suggests that the acquisition of nuclear weapons by rogue states would create a situation similar to Europe in the 1970s and early 1980s, "when it began to grow clear that the U.S. threat to use nuclear weapons to defend against Soviet conventional aggression was hollow." See Arquilla, "Bound to Fail? Regional Deterrence after the Cold War," *Comparative Strategy*, Vol. 14, No. 2 (1995), p. 133. Even Kenneth Waltz, who is usually sanguine about the possible spread of nuclear weapons, has conceded that "if weak countries have some it will cramp our style. Militarily punishing small countries for behavior we dislike would become much more perilous." See Kenneth N. Waltz and Scott D. Sagan, *The Spread of Nuclear Weapons: A Debate* (New York: Norton, 1995), p. 111.

6. See Roger C. Molander and Peter A. Wilson, "On Dealing with the Prospect of Nuclear Chaos," *Washington Quarterly*, Vol. 17, No. 3 (1994); and Zalmay Khalilzad, "The United States and the Persian Gulf: Preventing Regional Hegemony," *Survival*, Vol. 37, No. 2 (1995), pp. 98–99, 105–106; and also Marc Dean Millot, Roger Molander, and Peter A. Wilson, *"The Day After . . ." Study: Nuclear Proliferation in the Post–Cold War World* (Santa Monica, Calif.: RAND Corporation, 1993); Aspin, *Bottom Up Review*, p. 5; and Steven David, "Why the Third World Still Matters," *International Security*, Vol. 17, No. 3 (Winter 1992/93), pp. 145–155.

7. See especially Scott Sagan's contributions to Waltz and Sagan, *Spread of Nuclear Weapons;* and also Peter Feaver, "Neooptimists and Proliferation's Enduring Problems," *Security Studies*, Vol. 6, No. 4 (1997). Important challenges to this view include Bradley Thayer, "The Risk of Nuclear Inadvertence: A Review Essay," *Security Studies*,

Is all this concern justified? The fear that "nuclear rogues" would have profoundly destabilizing effects on world politics lies at the heart of present efforts to fashion ambitious military responses, ranging from sophisticated disarming attacks to elaborate types of strategic defense. Even so, many strategic planners seem pessimistic about the prospects for containing these regimes, and warn that the spread of WMD would have corrosive effects on important U.S. security commitments.

This chapter offers a different assessment. The central question is straightforward: will the acquisition of WMD by so-called rogue states encourage or impede the formation of defensive coalitions to contain, coerce, or in extreme cases, disarm them? Contrary to much of the conventional wisdom, I argue in this chapter that it will not be especially difficult to forge effective alliances against so-called rogue states, even if they were to acquire WMD. Indeed, overstating the difficulties could be self-fulfilling, if it encouraged potential partners to expect more protection than is necessary or if it convinced U.S. leaders that the prospects for success were not worth the costs and risks.

The first section of this chapter examines the characteristics of so-called rogue states and the difficulties involved in forecasting their behavior. The second section draws on the theoretical literature on alliance formation to explain why states join military coalitions, and identifies the main obstacles that can impede the formation of effective alliances and undermine their value as a deterrent or their performance in war. The third section examines the specific problems that rogue states armed with WMD pose to the formation and effectiveness of coalitions, and argues that a strategy of containment will minimize these obstacles most effectively. Finally, the conclusion outlines several ways to facilitate efforts to contain these states, assuming that this remains an important U.S. interest.

What Is a Rogue State? Why Are They Dangerous?

While experts disagree about the speed with which WMD will spread, there is a widespread consensus that many states could acquire these weapons if they so desired.[8] In spite of renewed efforts to limit the

---

Vol. 3, No. 3 (1994); David Karl, "Proliferation Pessimism and Emerging Nuclear Powers," *International Security*, Vol. 21, No. 3 (Winter 1996/97); and Jordan Seng, "Less is More: Command and Control Advantages of Minor Nuclear States," *Security Studies*, Vol. 6, No. 4 (1997).

8. See, among many others, Steve Fetter, "Ballistic Missiles and Weapons of Mass

spread of WMD (symbolized by the agreement to extend the Nuclear Non-Proliferation Treaty), the ability of even weak states to acquire some type of WMD has been enhanced by the growing diffusion of technical knowledge; the break-up of the Soviet Union and the resulting uncertainty about custodianship of nuclear materials there; the permeability of existing international safeguards; and the willingness of several states to provide key components of nuclear, biological, and chemical weapons technology. The recent nuclear tests by India and Pakistan and the reconstitution of Iraq's WMD capability despite the intrusive United Nations Special Commission (UNSCOM) inspections regime confirm that these capabilities are within the reach of many states. Thus, part of the concern about these rogue states stems from the recognition that there may be little that outside powers can do to prevent a state from acquiring WMD if it wants them badly enough.[9]

The acquisition of WMD by rogue states seems dangerous because these states are believed to harbor particularly aggressive goals, an assessment that rests on a set of core assumptions about their motivations and likely behavior. First, virtually all analysts regard such regimes as fundamentally hostile to the United States. Second, rogue states are assumed to have inherently revisionist aims, meaning that they seek to alter the status quo for reasons other than a desire to improve their own security. Third, they are often regarded as especially willing to take risks, *hmm.* because they are highly autocratic, inspired by a radical ideology, or

---

Destruction: What is the Threat? What Should be Done?" *International Security*, Vol. 16, No. 1 (1991); Brad Roberts, "From Non-Proliferation to Antiproliferation," *International Security*, Vol. 18, No. 1 (1993); Mitchell Reiss, *Without the Bomb: The Politics of Nuclear Non-Proliferation* (New York: Columbia University Press, 1988), p. 23; William C. Martel and William T. Pendley, *Nuclear Coexistence: Rethinking U.S. Policy to Promote Stability in an Era of Proliferation* (Montgomery, Ala.: U.S. Air War College, 1994), pp. 7–9; Aspin, "New Counterproliferation Initiative"; Peter D. Zimmermann, "Technical Barriers to Nuclear Proliferation," *Security Studies*, Vol. 2, Nos. 3–4 (1993); and "Proliferation: Bronze Medal Technology is Enough," *Orbis*, Vol. 38, No. 1 (1994); and Mark D. Mandeles, "Between a Rock and a Hard Place: Implications for the U.S. of Third World Nuclear Weapon and Ballistic Missile Proliferation," *Security Studies*, Vol. 1, No. 2 (1991).

9. The motivations for nuclear acquisition are examined in Bradley A. Thayer, "The Causes of Nuclear Proliferation and the Utility of the Nuclear Non-Proliferation Regime," *Security Studies*, Vol. 4, No. 3 (1995); Zachary S. Davis and Benjamin Frankel, eds., "The Proliferation Puzzle: Why Nuclear Weapons Spread (and What Results)," special issue of *Security Studies*, Vol. 2, Nos. 3–4 (1993); Stephen M. Meyer, *The Dynamics of Nuclear Proliferation* (Chicago: University of Chicago Press, 1984); and Martel and Pendley, *Nuclear Coexistence*, pp. 18–27.

prone to make decisions in an "irrational" fashion. As a result, such regimes are believed to be especially difficult to deter.[10] In short, the perceived threat from rogue states rests on the belief that they are unusually bellicose, and would become even more aggressive were they to acquire weapons of mass destruction.[11]

This characterization should not be accepted uncritically. Portraying rogue states as a looming international threat obscures their relative weakness, international isolation, and internal vulnerabilities.[12] It also discounts the possibility that their aggressive behavior stems largely from insecurity rather than ideological conviction or a desire for glory or material gain.[13] States that see their security as precarious are often more willing to run risks (including the use of force) to improve their positions; one need not approve of the actions of Wilhelmine and Nazi Germany, Imperial Japan, or Ba'thist Iraq to recognize that intense perceptions of threat lay at the core of their expansionist behavior. German and Japanese elites saw themselves as tightly encircled, and believed their regimes were destined to decline unless they took action to reverse the trend. Similarly, Iraq's attempt to seize Kuwait probably arose less from Hussein's megalomania or from Ba'thist ideology than from the profound vulnerabilities facing the Iraqi regime. In addition to the serious long-term threat from Iran (which possesses far more latent power), Iraq was burdened by the debts it incurred during the Iran-Iraq war, the falling price of oil, the OPEC (Organization of Petroleum Exporting Countries)

---

10. As U.S. Senator John McCain has noted, "the rapids of internal politics—and external rivalries—in broad stretches of the developing world confound any confident projections regarding 'stable and restrained,' as contrasted with 'unstable and potentially unrestrained' actors on the proliferation stage." See his "Proliferation in the 1990s: Implications for U.S. Policy and Planning," *Strategic Review*, Vol. 17, No. 3 (1989), p. 12.

11. The *Bottom-Up Review* emphasized the danger from states "*set on regional domination through military aggression* while simultaneously pursuing nuclear, biological, and chemical weapons capabilities." See Aspin, *Bottom-Up Review*, p. 1 (emphasis added).

12. In 1996, the U.S. population was approximately 268 million, its gross national product (GNP) was $7.6 trillion, and its defense expenditures were roughly $265 billion. By contrast, Iraq, Iran, Libya, and North Korea have a *combined* population of 120 million, a combined GNP of $128 billion, and their combined defense spending was $11.4 billion. Thus, the U.S. population was 2.22 times larger than the combined population of the four main "rogue" states, its GNP was nearly sixty times greater than theirs, and it spent over twenty-three times as much on defense as they did. These calculations are based on data in International Institute for Strategic Studies, *The Military Balance: 1997–98* (Oxford: Oxford University Press, 1997).

13. Bradley Thayer argues that insecurity is the principal motive behind the Iraqi, Iranian, and North Korean nuclear programs. See his "Causes of Nuclear Proliferation," pp. 495–496.

states' refusal to adhere to prior price agreements, and deep internal divisions. Seizing Kuwait promised a quick fix to these problems, and the apparent absence of an outside commitment to protect the sheikhdom made the temptation impossible to resist.[14]

The point is not to defend Iraqi expansionism, of course; the point is *Really?* simply to recognize that insecurity is often a major source of aggression. If insecurity is the main cause of a rogue state's bellicose behavior, then the acquisition of WMD could make a seemingly "rogue" state less dangerous. By contrast, if the desire to expand arises from other sources (such as religious or ideological convictions, long-standing territorial irredenta, etc.), then obtaining WMD might easily make a state more bellicose. This suggests that efforts to deal with rogue states should be informed by a systematic and dispassionate analysis of their underlying motivations.[15]

Finally, the assumption that such regimes are either irrational or insensitive to costs is problematic. Although rogue regimes often treat their own populations with great brutality, that is quite different from taking steps that are likely to lead to the destruction of the regime itself. Even if a rogue state were somewhat more likely to use WMD than a "non-rogue" state, the likelihood of use is still quite low. Thus, it is by no means obvious that nuclear rogues pose a special sort of danger.

Indeed, the alleged threat from so-called "rogue states" is not new. The Soviet Union was the first modern "rogue" insofar as its ruling ideology and foreign policy behavior revealed a deep-seated commitment to altering the global status quo.[16] U.S. policymakers were alarmed by the Soviet acquisition of nuclear capabilities in 1949, and their rhetoric was strikingly similar to the language now used to describe the threat from contemporary rogue states. In 1949, for example, the National Security Council Memorandum "U.S. Objectives for National Security"

---

14. See Efraim Karsh and Inari Rautsi, "Why Saddam Hussein Invaded Kuwait," *Survival*, Vol. 33, No. 1 (1991); and Lawrence Freedman and Efraim Karsh, *The Gulf Conflict, 1990–91: Diplomacy and War in the New World Order* (Princeton, N.J.: Princeton University Press, 1993), chap. 2.

15. Robert Blackwill and Ashton Carter recommend "understanding the strategic personality of a new proliferator," noting that "a wider variety of proliferator personalities is possible than is sometimes acknowledged in writings about nuclear strategy." See Blackwill and Carter, "The Role of Intelligence," in Robert D. Blackwill and Albert Carnesale, eds., *New Nuclear Nations: Consequences for U.S. Policy* (New York: Council on Foreign Relations, 1993), p. 236.

16. The Soviet Union was treated as a pariah state by the Western powers throughout the 1920s, and the United States did not even establish diplomatic relations with Moscow until 1934.

(NSC-68) warned that Soviet possession of nuclear weapons would provide it "with great coercive power for use in time of peace" and would discourage "the victims of its aggression from taking any action . . . which would risk war."[17] Similar fears accompanied the Chinese acquisition of a nuclear capability in 1964, an event that led both U.S. and Soviet leaders to consider preemptive strikes against Chinese nuclear facilities.[18] As with today's "rogue states," U.S. concerns about a nuclear China were exacerbated by the unwarranted belief that the communist government in China was highly aggressive, inherently unpredictable, and therefore difficult to deter. As U.S. Secretary of State Dean Rusk warned in 1966, "a country whose behavior is as violent, irascible, unyielding, and hostile as that of Communist China is led by leaders whose view of the world and of life itself is unreal." Such rhetoric could be applied to contemporary "rogue states" virtually unchanged.[19]

Not only are rogue states not a new phenomenon, but the historical record suggests that the existing nuclear states acquired these weapons largely for defensive purposes (i.e., in response to legitimate security fears), or for reasons of status, and have not behaved more recklessly since acquiring these capabilities.[20] The introduction of nuclear weapons into South Asia ended the costly cycle of conventional war between India and Pakistan, for example, and both regimes recognize that these weapons are useful only in a deterrent role.[21] Iraq's decision to refrain from

17. See "U.S. Objectives for National Security (NSC-68)," in Thomas H. Etzold and John Lewis Gaddis, eds., *Containment: Documents on American Policy and Strategy, 1945–1950* (New York: Columbia University Press, 1978), p. 398.

18. See Gordon H. Chang, *Friends and Enemies: The United States, China, and the Soviet Union, 1948–1972* (Stanford, Calif: Stanford University Press, 1990), chap. 8; and Henry A. Kissinger, *White House Years* (Boston: Little, Brown and Co., 1979), pp. 183–187.

19. See Franz Schurmann and Orville Schell, eds., *The China Reader, vol. 3: Communist China* (New York: Vintage Books, 1967), p. 508. On China's nuclear program, see Avery Goldstein, "Understanding Nuclear Proliferation: Theoretical Explanation and China's National Experience," *Security Studies*, Vol. 2, No. 3–4 (1993); on its rather cautious approach to the use of force, see Allen S. Whiting, "The Use of Force in Foreign Policy by the People's Republic of China," *Annals of the American Academy of Political and Social Science*, Vol. 402 (1972), pp. 55–66.

20. The Soviet acquisition of nuclear weapons seems to have had little effect on Stalin's decision to approve the North Korean invasion of South Korea in 1950. See Sergei N. Goncharov, John W. Lewis, and Xue Litai, *Uncertain Partners: Stalin, Mao, and the Korean War* (Stanford, Calif.: Stanford University Press, 1993), chap. 5.

21. For example, Indian Defense Minister George Fernandes remarked in 1998 that "it is an established fact that any country that has nuclear weapons cannot use them. By definition, they can be used only as a deterrent. If we had to go nuclear, it was for the purpose of possessing a nuclear deterrent that would enable us to tackle some of the threats that we faced, only that." Quoted in *New York Times*, June 18, 1998, p. A6.

using chemical weapons during the Gulf War suggests that even a leader as risk-acceptant as Saddam Hussein can be deterred by a powerful retaliatory capability.[22] Thus, there seems to be little reason to regard today's rogue states as an especially novel or grave danger.

The problem, of course, is that one cannot be sure what a regime would do were it to attain WMD. Furthermore, motivations and intentions can change without warning, and weapons that were acquired for one purpose could be used for other ends later. Thus, even if today's "rogue states" are probably not as dangerous as the more alarmist rhetoric suggests, there are still good reasons to worry about their weapons programs and foreign policy objectives, and good grounds for thinking about how the United States and its allies should respond. In particular, how will the acquisition of WMD by a perceived "rogue" state affect the willingness of other states to form alliances against it? To answer this question, let us begin by considering why states form alliances in the first place.

When Do States Form Alliances?

Alliances are formed so that their members can combine their capabilities in ways that further their respective interests. Alliances can take many forms (offensive or defensive, symmetrical or asymmetrical, formal or informal, etc.), but the common thread in virtually all such arrangements is each member's desire to increase its security by collaborating with at least one other state.

BALANCING BEHAVIOR

The principal reason to form an alliance is to oppose an external threat.[23] In particular, states form military coalitions in order to balance against the most threatening state or coalition that they face. The rationale for this tendency is straightforward: because no supreme authority exists to

---

22. During the Gulf War, several U.S. officials made statements that were clearly designed to deter Iraq from using its chemical arsenal against the coalition's forces. See Freedman and Karsh, *Gulf Conflict*, pp. 257, 288–289.

23. This discussion is based on Stephen M. Walt, *The Origins of Alliances* (Ithaca: Cornell University Press, 1987), esp. chap. 2. See also Stephen M. Walt, "Testing Theories of Alliance Formation: The Case of Southwest Asia," *International Organization*, Vol. 43, No. 2 (1988); "Alliances, Threats and U.S. Grand Strategy: A Reply to Kaufman and Labs," *Security Studies*, Vol. 1, No. 3 (1992); "Multilateral Collective Security Arrangements," in Richard Shultz, Roy Godson, and Ted Greenwood, eds., *Security Studies for the 1990s* (New York: Pergamon-Brassey's, 1993); and "Why Alliances Endure or Collapse," *Survival*, Vol. 39, No. 1 (1997).

protect states from each other, states facing a possible threat will join forces with others in order to amass sufficient power to deter or defeat an attack.

This explanation is often framed in terms of *power*; states with lesser capabilities are presumed to combine against a stronger state in order to prevent the stronger power from dominating.[24] But it is more accurate to say that states form coalitions to balance against *threats*, and power is only one element in their calculations (albeit an important one). In general, the level of threat that states face will be a function of four distinct factors: aggregate power, geographic proximity, offensive capabilities, and aggressive intentions.

AGGREGATE POWER. Other things being equal, the greater a state's total resources (population, industrial capacity, military strength, etc.), the greater harm it can inflict on others. Thus, the greater a state's *aggregate power*, the larger the potential threat. Recognizing this, the traditional aim of U.S. grand strategy has been to prevent any single power from controlling the combined resources of Eurasia, because such an agglomeration of power could pose a serious threat to U.S. security. England's traditional reliance on a balance of power policy reflected similar motivations, and contemporary concerns over China's economic and military growth betray much the same sensitivity. A state's aggregate power is not the sole determinant of threat, of course, but it is always an important one.

GEOGRAPHIC PROXIMITY. Because the ability to project power declines with distance, nearby states are usually more dangerous than those that are far away.[25] As a result, states are more likely to form coalitions in response to threats from their neighbors, and will prefer allies that are geographically separate. This factor helps explain why, during the Cold War, regional powers usually preferred to rely on support from a distant superpower rather than on cooperation with other regional actors. First, the superpowers could do more to help. Second, allying with a neighbor can be dangerous if the neighbor becomes too strong. This motive also

---

24. For classic expressions of this view, see Hans J. Morgenthau, *Politics Among Nations* (New York: Alfred A. Knopf, 1967); and Kenneth N. Waltz, *Theory of International Politics* (Reading, Mass.: Addison-Wesley, 1978).

25. On the so-called power-distance gradient, see Kenneth Boulding, *Conflict and Defense: A General Theory* (New York: Harper Torchbooks, 1962); and Albert Wohlstetter, "Illusions of Distance," *Foreign Affairs*, Vol. 46, No. 2 (1968). See also Kautilya, "Arthasastra," in Paul Seabury, ed., *Balance of Power* (San Francisco: Chandler Publishing, 1965).

explains why the superpowers' efforts to recruit regional clients during the Cold War were only partly successful. Although Moscow and Washington sought allies primarily to counter each other, their clients in the developing world were usually more worried about regional or internal threats and relatively unconcerned about the global balance.[26]

OFFENSIVE CAPABILITIES. States with large offensive capabilities—defined as the ability to threaten the sovereignty or territorial integrity of other states—pose a greater threat than states whose capabilities are designed primarily to defend their own territory. As a state's offensive capabilities increase (either because its military expenditures are growing or because its armed forces are being tailored for offensive warfare), other states will be more inclined to ally against it. The ability to undermine other regimes via propaganda or subversion can be a potent source of threat as well, which explains why revolutionary states that see themselves as a model for others are usually regarded as dangerous, even when they are militarily weak.[27]

OFFENSIVE INTENTIONS. States with aggressive aims are obviously more dangerous than states that are satisfied with the status quo. Even a relatively weak state is likely to face a defensive coalition if other states view its intentions as especially bellicose. Other things being equal, therefore, the belief that a state harbors offensive intentions will increase the likelihood that a countervailing coalition will form against it.

Together, these four factors explain why potential hegemons like Napoleonic France, Wilhelmine Germany, and Nazi Germany eventually faced overwhelming opposing coalitions: each of these states was a great power lying in close proximity to others, and each combined large offensive capabilities with extremely aggressive aims. The same factors explain why the Soviet Union was badly overmatched during the Cold War: it was the second most powerful state in the system, it was geographically close to the medium powers of Europe and Asia, it possessed a large military establishment whose forces and doctrine were tailored for offensive wars of conquest, and it never abandoned its public commitment to world revolution. Although the United States had greater overall capabilities (including a large network of overseas bases and a significantly

---

26. See Walt, *Origins of Alliances*, pp. 158–165, 266; and Steven David, *Choosing Sides: Alignment and Realignment in the Third World* (Baltimore, Md.: Johns Hopkins University Press, 1991).

27. See Stephen M. Walt, *Revolution and War* (Ithaca: Cornell University Press, 1996); and Stephen M. Walt, "Revolution and War," *World Politics*, Vol. 44, No. 2 (1992).

greater capacity for global military intervention), its distance from the other centers of world power made it the perfect ally for the medium powers that were most directly threatened by the Soviet Union.

"Balance-of-threat" theory also explains why the United States was able to assemble such a large and cohesive coalition during the Gulf War. Although Iraq was not a great power, its military forces were growing, its conquest of Kuwait threatened Western oil supplies, and Saddam Hussein's behavior suggested that he had extremely aggressive ambitions and was willing to run impressive risks to achieve them. Adroit diplomacy helped bring the coalition into existence, but the essential preconditions for an effective coalition were certainly present.[28] The theory also tells us why it has been more difficult to maintain the Gulf War coalition after its smashing victory over Iraq. Because Iraq's power is greatly diminished, other states feel less need to join forces to oppose it. Moreover, states like Russia and France are now worried about excessive U.S. influence, and so are increasingly reluctant to support U.S. policy wholeheartedly.

BANDWAGONING BEHAVIOR
Although the desire to balance threats is the main reason why states form coalitions, some states may prefer to "jump on the bandwagon" and join forces *with* the dominant (or most threatening) power rather than attempt to form a coalition against it. This may be done either for defensive or offensive reasons: some states may seek the stronger side to appease it; others join the strong side to profit from its victory.

Several factors will affect a state's propensity to bandwagon. First, weak states are more inclined to bandwagon than are strong states; they can do little to affect the outcome and must choose the winning side at all costs.[29] The likelihood of bandwagoning also increases when a state faces an imminent threat and potential coalition partners are unavailable. Even if such a state would prefer to balance, it may be forced to bandwagon (or adopt a neutral position) if it cannot find adequate support. Third, bandwagoning is more likely to occur in the latter stages of a war, when the outcome is clear and states rush to curry favor with the victors

---

28. See David Garnham, "Explaining Middle Eastern Alignments During the Gulf War," *Jerusalem Journal of International Relations*, Vol. 13, No. 3 (1991); and Freedman and Karsh, *Gulf Conflict*, chaps. 6, 7, 25.

29. Eric Labs suggests that even weak states are reluctant to bandwagon, though they will do so as a last resort. See his "Do Weak States Bandwagon?" *Security Studies*, Vol. 1, No. 3 (1992). See also Michael Handel, *Weak States in the International System* (London: Frank Cass, 1981); and Robert L. Rothstein, *Alliances and Small Powers* (New York: Columbia University Press, 1968).

or to extract spoils from the vanquished. Lastly, revisionist states will be more inclined to bandwagon, because they are especially interested in reaping the benefits from victory and must take pains to pick the likely winners.[30] Mussolini's behavior following the German attack on France in 1940 offers a nice example of all three factors; Italy was a weak state with revisionist aims, and Mussolini waited until France's defeat was apparent before declaring war.

Although great powers often fear that potential allies will align with the strongest or most threatening state, this fear receives little support from international history. Every modern attempt to achieve hegemony in Europe has been thwarted by a powerful balancing coalition. More recent examples of balancing behavior include the coalition that fought the Gulf War against Iraq, the resurgence of the Association of Southeast Asian Nations (ASEAN) following the Vietnam war, the Sino-U.S. rapprochement in the 1970s, the formation of the Gulf Cooperation Council and the U.S. tilt against Iran during the Iran-Iraq war, and the coalition of Front-Line States against South Africa during the 1970s and 1980s.[31] This tendency should not surprise us; balancing is usually preferred for the simple reason that no national leader can be completely sure what others will do. Bandwagoning is dangerous because it increases the capabilities available to a threatening power and requires that the weaker state trust in the continued benevolence of the stronger state.[32] In effect, the state must subordinate itself to the will of another power (and at worst completely lose its sovereignty), and most leaders, irrespective of their relative power, regime type, or personal inclinations, will be loath to embrace this possibility if other options are available. Because perceptions of intent are often unreliable and intentions can change overnight, it is usually safer to balance against potential threats rather than take their continued forbearance on faith.

IMPLICATIONS FOR THE UNITED STATES

The ability of the United States to forge and manage future coalitions will be primarily a function of the external security environment. States do not join alliances or fight wars for altruistic or sentimental reasons—such

30. See Randall K. Schweller, "Bandwagoning for Profit: Bringing the Revisionist State Back In," *International Security*, Vol. 19, No. 1 (1994).

31. For other examples, see Walt, "Testing Theories of Alliance Formation."

32. For this reason, partners in an offensive coalition are prone to quarrel once victory is achieved, often over the division of spoils. Examples include the victorious coalition of Serbia, Bulgaria, Macedonia, and Greece after the First Balkan War, or the Nazi-Soviet agreement to divide Poland in 1939.

vital decisions ordinarily reflect careful considerations of self-interest. Although shared values and similar political systems can facilitate alignment and smooth potential rifts, U.S. leaders should not expect assistance from other states unless these governments believe it is in their interest to help. If other states do not see themselves as threatened, it will be extremely difficult for the United States to entice them into a military coalition.[33]

The U.S. ability to form and lead future coalitions will depend on its own capacity to contribute to them. Although states generally prefer to balance against threats, they will find this option less appealing if they believe that the effort is unlikely to succeed. Thus, even if the United States comes to rely less on formal alliances such as NATO and more on ad hoc coalitions, it must still possess sufficient military assets to make itself an attractive partner. In short, coalition support can be an adjunct to U.S. military power but not a substitute for it.

OBSTACLES TO EFFECTIVE ALLIANCE FORMATION

When the balancing process works smoothly, and when states' tendency to form balancing coalitions is widely understood, aggression will be discouraged because even highly ambitious leaders will expect opposition to any aggressive moves. For example, the rapid establishment of the Western alliance in 1949–50, which made it clear to the Soviet Union that any overt act of aggression would be met by a combined Western response, enhanced the security of the West and helped preserve peace between East and West for more than four decades. It is also likely that Saddam Hussein would not have acted as he did in August 1991 had he known that the seizure of Kuwait would generate a vast coalition of powers seeking to expel his forces and disarm his regime.

By contrast, aggressors are harder to deter when balancing takes place slowly or inefficiently, and the probability of war will perforce increase. For instance, although Britain and France did balance against Nazi Germany in the 1930s, their efforts to do so were neither efficient nor swift. Not only were Britain and France distracted and diverted by threats in several separate theaters (which made it hard to focus their attention on Germany), but Adolf Hitler was a master of dissimulation who carefully cloaked the scope of his ambitions. As a result, his future opponents did not recognize the true magnitude of the threat until very

---

33. Because clever aggressors try to conceal their aims until their preparations are well advanced, status quo powers may want to coordinate possible responses in advance, to facilitate the formation of a defensive coalition in the shortest possible time.

late. Efforts to form a countervailing coalition were also impaired by ideological antipathies between Britain and France and the Soviet Union, by rivalries among Germany's East European neighbors, and by the East European states' reluctance to ally with Stalin (whose embrace they correctly judged to be as dangerous as Hitler's). For these and other reasons, the effort to balance against Germany fell short of the level necessary for successful deterrence, especially given Hitler's extraordinary desire for war.[34]

The efficiency of the balancing process also plays an important role in determining success in wartime. The ability to coordinate strategy and maintain an adequate effort will maximize the likelihood of victory, while disputes over wartime priorities and attempts to pass the costs of the war onto one's allies will undermine any coalition's efforts and may lead to an unnecessary defeat. In the wars of the French Revolution, for example, the first and second coalitions against revolutionary France failed in part because the member states found it difficult to coordinate their military activities and because they focused on maximizing their individual gains instead of concentrating on the common aim of defeating France. By contrast, although the victorious coalitions in World Wars I and II repeatedly quarreled over strategy and relative levels of effort, these disagreements did not undermine their battlefield performance significantly.[35]

These examples reveal that one cannot assume that balancing coalitions will always form or will always operate well enough to succeed. What are the main obstacles to the formation of effective and efficient alliances?

CONFLICTING INTERESTS. States are unlikely to form a coalition if their interests are directly opposed, or if they interpret a particular strategic situation in very different ways. Obviously, forming an effective coalition will be impossible if one state regards another as especially dangerous but other states do not share its view. For example, European allies of the United States did not support its effort to overthrow the Sandinista

---

34. On these points, see Walt, "Alliances, Threats and U.S. Grand Strategy."

35. On the tensions within the first and second Coalitions, see John M. Sherwig, *Guineas and Gunpowder: British Foreign Aid in the Wars with France* (Cambridge, Mass.: Harvard University Press, 1969); Paul Schroeder, *The Transformation of European Politics* (Oxford: Clarendon Press, 1994); and Steven Ross, *European Diplomatic History, 1789–1815: France against Europe* (Garden City, N.J.: Anchor Books, 1969), pp. 87–92. On the Grand Alliance in World War II, see William H. McNeill, *America, Britain and Russia: Their Cooperation and Conflict, 1941–1946* (New York: Johnson Reprint Co. reprint of 1953 ed., 1970); and Robert Beitzell, *The Uneasy Alliance: America, Britain and Russia, 1941–1943* (New York: Alfred A. Knopf, 1972).

regime in Nicaragua, largely because they did not regard the Sandinistas as a serious threat. Alternatively, other states may recognize that a particular power is dangerous, but conclude that their own interests are not affected. For example, U.S. efforts to isolate Iran during the 1980 hostage crisis were weakened by the unwillingness of key U.S. allies to go beyond token sanctions, largely because they did not regard the seizure of the U.S. embassy as a threat to their own interests. More recently, U.S. efforts to isolate Tehran and Baghdad have also been undermined by the fact that medium powers such as Germany and France did not share the U.S. appraisal of Iran and Iraq and were interested in pursuing commercial opportunities with both states.[36] Lastly, states may refrain from joining a coalition simply because they prefer to support the other side. During the Cold War, for example, the United States and Soviet Union interpreted most international events in a strictly zero-sum fashion. Because their interests were in conflict on so many issues, they rarely saw each other as prospective coalition partners.[37] In general, the greater the divergence in state interests among a set of potential allies, the more difficult it will be to form a coalition among them.

COMPETING PRESCRIPTIONS. Even when states agree that a particular power is dangerous and that they share an interest in countering it, they may disagree over how to respond. If a particular state appears to have aggressive aims, for instance, some states may favor a strategy of accommodation while others advocate efforts to contain or overthrow it. For example, French officials at the Paris Peace Conference in 1919 repeatedly advocated joint Western intervention against the Bolshevik regime in Russia, but were unable to persuade British and U.S. leaders that such a step was necessary or advisable.[38] Similarly, although the United States and the members of the so-called Contadora Group (Mexico, Venezuela, Colombia, and Panama) generally agreed that the Sandinista regime in Nicaragua posed a threat to regional stability, they disagreed completely on both the magnitude of the threat and the best way to deal with it. Similar dynamics can be found in the efforts in the 1990s to formulate

---

36. See Charles Lane, "Germany's New Ostpolitik," *Foreign Affairs*, Vol. 74, No. 6 (1995).

37. One of the clearest signs of the Cold War's passing was Russia's willingness to acquiesce in the United Nations' action during the Gulf War, despite its long-standing alignment with Iraq and its reservations about U.S. policy toward Baghdad.

38. See John Thompson, *Russia, Bolshevism and the Versailles Peace* (Princeton, N.J.: Princeton University Press, 1966), chap. 6.

a unified policy toward North Korea's nuclear weapons program or in U.S. and European differences over the best approach to take toward the civil war in Bosnia, and the problem has been especially stark in the protracted debate over how to respond to Iraq's repeated efforts to evade the UNSCOM inspections regime. Because many U.S. allies believed that a punitive response (such as airstrikes) would be ineffective or counter-productive, they were unwilling to support U.S. efforts to compel Iraq to comply fully with the sanctions regime. In each case, the question is less one of conflicting interests than competing prescriptions; the potential partners generally want the same thing, but do not agree on how to get it.[39]

COLLECTIVE ACTION. The problems just noted are compounded by the familiar problem of collective action.[40] Even if all members of a coalition agree on what the main threat is *and* on how they should respond, their efforts to do so may be undermined by disagreements over who should bear the burden. The danger, of course, is that each member's efforts to "pass the buck" to its partners may lead the coalition to do too little to achieve its common objective, even when the coalition possesses greater overall resources than its adversaries. This problem clearly afflicted the coalitions against revolutionary France and Napoleon, whose members were constantly wrangling over subsidies, spoils, and strategy. Similar problems weakened the effort to balance against Hitler during the 1930s.[41] The familiar debates over burdensharing in NATO are another

39. In the February 1998 confrontation between Washington and Baghdad, only Great Britain and Kuwait supported the use of military force to compel Iraqi compliance with the sanctions regime. Turkey refused to let the United States use its bases to conduct air strikes, France and Russia openly opposed U.S. policy, and the Arab League (including Egypt and Saudi Arabia) opposed any use of military force. The general problem of competing prescriptions is discussed in Charles T. Allan, "Extended Conventional Deterrence: In from the Cold and Out of the Nuclear Fire," *Washington Quarterly*, Vol. 17, No. 3 (1994), pp. 222–223.

40. See Mancur Olson, *The Logic of Collective Action* (Cambridge, Mass.: Harvard University Press, 1965); and Mancur Olson and Richard Zeckhauser, "An Economic Theory of Alliances," *Review of Economics and Statistics*, Vol. 48, No. 3 (1966).

41. On the French case, see Schroeder, *Transformation of European Politics*, and Paul Schroeder, "The Collapse of the Second Coalition," *Journal of Modern History*, Vol. 59, No. 2 (1987). On the 1930s, see Barry R. Posen, *The Sources of Military Doctrine: France, Britain and Germany between the World Wars* (Ithaca, N.Y.: Cornell University Press, 1984); and Thomas J. Christensen and Jack Snyder, "Chained Gangs and Passed Bucks: Predicting Alliance Patterns in Multipolarity," *International Organization*, Vol. 44, No. 2 (1990).

example of this problem, and U.S. complaints about the distribution of costs during the Gulf War eventually convinced Japan to contribute roughly $12 billion to the allied war effort.

As the latter two examples indicate, the problem of collective action is not insurmountable. But resolving issues of collective action are likely to be more difficult when the members of a coalition face different strategic problems—some risk annihilation while others do not—and so cannot adjust the burdens by the relatively simple expedient of transferring funds.

THE PROBLEM OF UNCERTAINTY. A final obstacle to effective alliance formation is uncertainty. Inefficient balancing is more likely when the threat is ambiguous, since it is more difficult for states to agree on what the main threat is and on how they should respond. After the Bolshevik revolution, for example, British Prime Minister David Lloyd George argued against allied intervention in Russia by warning that the facts required to make such a decision "had never been ascertained and were probably unascertainable."[42] As noted earlier, it was difficult to form an effective coalition against Nazi Germany in part because the medium powers of eastern Europe could not tell whether Nazi Germany or the Soviet Union posed the greater danger, and Great Britain could not decide whether the main danger to its interests was Japan in the Far East, Italy in the Mediterranean, or Germany on the continent. More recently, a lack of information has plagued efforts to respond to the North Korean nuclear program and to Iraq's efforts to evade the UNSCOM inspections regime. Because potential coalition members disagreed about the circumstances in each case and could not obtain definitive information either about current conditions or the likely consequences of alternative courses of action, it was more difficult to assemble multilateral support. By contrast, the German occupation of the rump of Czechoslovakia in March 1939 and Iraq's invasion of Kuwait in August 1991 provided unambiguous evidence of their aggressive aims and facilitated the formation of strong countervailing coalitions.

Uncertainty is a constant feature of world politics; fortunately, the lack of perfect information is never an absolute barrier to an effective

---

42. Similarly, U.S. President Woodrow Wilson told a confidant that his impressions of Russia were based on "indefinite information," and his doubts helped convince him that the Allies should "leave it to the Russians to fight it out among themselves." Quoted in Richard H. Ullman, *Anglo-Soviet Relations, vol. 2: Britain and the Russian Civil War* (Princeton, N.J.: Princeton University Press, 1968), pp. 96–97; and Frederick S. Calhoun, *Power and Principle: Armed Intervention in Wilsonian Foreign Policy* (Kent, Ohio: Kent State University Press, 1986), pp. 232, 238.

alliance. Still, ambiguity about the nature or extent of the threat will make it more difficult to bring a coalition into being and will fuel disputes among the members, lowering its deterrent power and impairing its wartime performance.

SUMMARY

While each of the obstacles identified above impedes the formation of an effective alliance, each barrier is likely to shrink as the level of threat increases. As rival states become more powerful or aggressive, the danger they pose overrides other interests and encourages states that are otherwise deeply suspicious to join forces, as the United States and the Soviet Union did in World War II or as the United States and Syria did during the Gulf War. Similarly, disagreements over how to respond decline as the threat becomes manifest, if only because a strategy of appeasement will be exposed as infeasible and firmer measures (including the use of force) will come to be seen as necessary. The dilemmas of collective action operate with less force when states are facing a truly grave threat, because guaranteeing victory will become more important than trying to minimize one's own contribution.[43] Finally, although ambitious aggressors may be able to disguise their ambitions for some time, they cannot hide them forever. Thus, uncertainty about the level of threat is likely to diminish as the level of threat increases, thereby facilitating the formation of a countervailing coalition.

Containing Rogue States: Implications and Obstacles

Do these same dynamics operate when states face a threat from a rogue state armed with WMD? I answer this question below.

WILL OTHER STATES BE WILLING TO BALANCE?

Concerns about the possible emergence of rogue states armed with WMD rest on the belief that other states will be reluctant to stand up to them for fear of inviting a devastating response. Such fears are compounded by the worry that great powers such as the United States will be more reluctant to help out, thereby allowing rogue states to commit acts of aggression or intimidation with impunity, and reinforcing the temptation

---

43. An important exception arises when the threat is clearly confined to only one member of a coalition. Such circumstances tempt that state's allies to remain aloof so as to avoid the costs of fighting, possibly in the hope that a war will leave both antagonists worse off. Obviously, such opportunistic behavior makes sense only when one is confident that the aggressor will either be weakened by defeat or sated by its victory.

to bandwagon. Such fears are almost certainly overstated. Rogue states that acquire WMD are likely to face formidable balancing coalitions. Although these countervailing alliances will have to overcome the obstacles identified in the previous section, doing so should not be especially difficult, for four reasons.

First, rogue states are invariably regarded as threatening even before they acquire WMD, usually because they are believed to harbor highly revisionist intentions; neighboring powers will already be on their guard. As a result, it will require little or no effort for the United States to convince them that the danger is real. The acquisition of WMD will merely increase the rogue state's capabilities (thereby increasing the level of threat) and provide an additional incentive to balance. The rogue's acquisition of WMD is also likely to strengthen other states' belief that the rogue state holds aggressive ambitions, further reinforcing their desire to join forces against it.[44]

Second, neighboring powers have little incentive to cave in to a rogue state's threats, even when the enemy acquires WMD. Its threats will be effective only if the victims are convinced that the rogue state is willing to launch an unprovoked attack, and the historical record suggests that this sort of blackmail rarely succeeds.[45] Acceding to such a threat would invite additional demands, and in effect means relinquishing one's own sovereignty and independence without a fight. Few ruling elites would be willing to accept this sort of subordination, particularly when it would leave them at the mercy of a regime that had shown a tendency to brutalize its own citizens. Would *anyone* voluntarily accept Saddam Hussein's authority while other options were available, particularly if he were trying to compel compliance by threatening annihilation?[46]

---

44. These effects might not be as large as is sometimes supposed, however. Nuclear weapons are not as usable as conventional capabilities, and would not be effective instruments of conquest unless they were married to a large conventional military capability. Accordingly, they will not increase a rogue state's *offensive power* as much as a comparable increase in conventional military capabilities would. Similarly, because rogue states are by definition regarded as having highly revisionist aims, the discovery that they had obtained WMD would at most reinforce the existing belief that they were aggressively inclined. The historical record supports this view: U.S. relations with Great Britain, France, and Israel remained positive after these states obtained nuclear weapons, just as U.S. relations with the Soviet Union and China were hostile both before and after they joined the nuclear club.

45. See Richard K. Betts, *Nuclear Blackmail and Nuclear Balance* (Washington, D.C.: Brookings Institution, 1987).

46. The precarious existence of the Iraqi Kurds provides a sobering reminder of what subordination to Iraqi hegemony entails.

Third, a rogue state's threats to use WMD against its neighbors will not be very credible. Those who fear that rogue states will use WMD aggressively adopt a very pessimistic double standard. They assume that outside powers like the United States will be deterred by a rogue state with a handful of unreliable weapons of mass destruction. They also assume that the rogue state's leaders will be utterly unfazed by the vast nuclear and conventional capabilities of the United States.[47] The credibility of a rogue state's nuclear threats is further reduced by the possibility that an attack on a neighbor would inundate its own territory with damaging radioactive fallout, a problem heightened by the lack of a long-range missile capability. Even if these states do obtain WMD, in short, these weapons will be no more usable than the superpower arsenals were.

Finally, although neighboring states might be forced to bandwagon if they were utterly isolated, the great powers (and especially the United States) have an enduring interest in preventing other states (and especially relatively weak states) from using WMD for blackmail, coercion, or as a shield for conventional aggression.[48] Nuclear weapons have been a profoundly *defensive* force throughout the nuclear age, and this tendency has been an important cause of peace between the major powers.[49] To permit a rogue state to alter the status quo through the use of nuclear blackmail would reverse this equation, and would send a powerful signal that these weapons had become an effective instrument of expansion or aggression. The consequences for world politics would be tremendous: incentives to proliferate would grow apace, revisionist powers would be quick to repeat their efforts to intimidate others, and war as a result of miscalculation would become much more likely. Neither the United States nor the other great powers has any interest in allowing such a world to emerge, which means that they have ample reason to bolster states facing threats from a well-armed rogue.[50]

---

47. On this point see Waltz, *Spread of Nuclear Weapons*, p. 40.

48. This line of argument follows Barry Posen's chapter in this volume.

49. On the defensive nature of nuclear weapons, see Robert Jervis, *The Meaning of the Nuclear Revolution: Statecraft and the Prospect of Armageddon* (Ithaca, N.Y.: Cornell University Press, 1989) and *The Illogic of U.S. Nuclear Strategy* (Ithaca, N.Y.: Cornell University Press, 1984); and Kenneth N. Waltz, "Nuclear Myths and Political Realities," *American Political Science Review*, Vol. 85, No. 4 (1991).

50. One could even argue that the United States has a general interest in punishing the use of WMD for aggressive purposes, even if its own specific interests were not affected.

Thus, states facing a rogue state's blackmail should be able to obtain U.S. backing.[51] And because helping to thwart nuclear blackmail is in the larger U.S. interest, it should not be too difficult for U.S. leaders to convince both potential allies and possible challengers that its pledges are genuine.

Since nearby states have little reason to kowtow to a rogue state's blackmail, and little reason to fear that they will be utterly abandoned, there is also little reason to fear that a rogue's acquisition of WMD would immediately undermine existing security commitments or trigger a cascade of capitulations. Once again, the history of the nuclear age provides reassuring evidence on this point. U.S. leaders once worried that Soviet acquisition of nuclear weapons would allow the Kremlin to conduct "piecemeal aggression against others, counting on our unwillingness to engage in atomic war," and cause our allies to "lose their determination."[52] Similar fears accompanied the growth of Soviet nuclear and conventional forces in the 1970s, which many experts believed cast doubt on U.S. credibility and placed NATO in danger of collapse.

In fact, most non-nuclear states have been more than willing to balance against hostile nuclear adversaries. For example, non-nuclear powers such as West Germany, Belgium, Holland, Italy, and Japan joined defensive alliances against the Soviet Union throughout the Cold War, just as Cuba and Nicaragua aligned themselves with Moscow to gain protection from the United States. Pakistan strengthened its alignment with China following India's "peaceful" nuclear explosion in 1974, and Israel's not-so-secret development of nuclear weapons did not deter a number of Arab states from allying against it. Indeed, Syria and Egypt launched a conventional attack on Israel in October 1973 despite their awareness that Israel already possessed a number of nuclear bombs. Syrian troops opposed the Israeli invasion of Lebanon in 1982 despite their conventional inferiority and Israel's nuclear arsenal, and Iraq's chemical weapons capability did not deter the United States or its coalition partners from ousting it from Kuwait in 1991.[53] Thus, the historical record is admirably clear: an opponent's possession of WMD does not make other states unwilling to defend their own vital interests, or deter them from using force when necessary.

---

51. It is worth noting that the United States has been willing to undertake a variety of new military commitments in the late 1990s—including Bosnia, Macedonia, and the expansion of NATO—even though its own security was not directly at stake.

52. See "U.S. Objectives for National Security (NSC-68)," p. 414.

53. On the latter example, see Quester and Utgoff, "No-First-Use and Proliferation," pp. 109–110.

Two caveats should be noted at this point. First, the willingness of non-nuclear states to balance against nuclear powers during the Cold War may have been partly an artifact of bipolarity, insofar as the U.S.-Soviet rivalry made each superpower's commitments to its allies more credible. Because each superpower had an obvious interest in containing expansion by the other, other states could be fairly confident that their patron's "nuclear umbrella" would remain open. With the passing of bipolarity, the credibility of existing commitments may wane, and it may be somewhat harder for the United States (or other great powers) to convince threatened states that it will be willing to help. But as noted above, the United States has both a general interest in deterring nuclear-backed aggression and specific interests in a number of regions. Although it may take a bit more effort to reassure potential victims, both their own desire to remain independent and the U.S. interest in preventing nuclear coercion should combine to provide a sufficient level of credibility.

Second, the acquisition of WMD could enhance a rogue state's ability to deter other states from trying to conquer and overthrow it. Yet because such a regime would not be as fearful of losing power should a war go badly, its WMD might make it more adventurous than it would otherwise be. However, the rogue state armed with WMD might actually be *less* willing to use force to advance its foreign policy objectives, out of a fear that inadvertent or deliberate escalation might trigger an overwhelming nuclear response. Thus, the net effect of WMD on a rogue state's behavior is not obvious.

States are unlikely to view a rogue state's acquisition of WMD with equanimity, if only because they cannot be sure how these new capabilities might be used in the future. States facing a threat from a well-armed rogue will prefer to balance against them, and they should be able to count on adequate allied support. What goals might such an alliance seek, and how will the main obstacles to effective alliance formation affect its cohesion and effectiveness?

OBJECTIVES

Coalitions formed in response to a rogue state's acquisition of WMD could pursue several different objectives. First, other states might join forces to persuade the rogue state to abandon its WMD ambitions by threatening it with diplomatic isolation, economic sanctions, or military encirclement. By increasing the political, economic, and strategic costs of acquiring WMD, this strategy seeks to convince the rogue state that it will be better off if it forgoes them. Such a policy could also provide a lesson to other states that might consider obtaining their own WMD arsenals.[54] The problem, of course, is that efforts to pressure a rogue state are likely to reinforce its sense of insecurity and increase its desire for a more powerful deterrent.[55]

Second, states may ally against a rogue state to deter it from taking aggressive actions in the future. Here the goal is one of containment; instead of trying to halt or reverse the acquisition of a WMD arsenal, threatened states would seek to persuade the rogue state not to use its new capabilities to intimidate others or to gain tangible benefits (such as additional territory).

Third, if containment failed to work (or was not attempted), a defensive coalition might try to reverse the outcome of the rogue state's aggression, as the United States and its allies did following the Iraqi seizure of Kuwait. This objective is more demanding than containment (particularly if achieving it requires direct attacks against the homeland of a country that possesses WMD), but might be necessary either to protect specific interests or to reinforce the norm that WMD cannot be used as a shield for conventional aggression.

Finally, in extreme cases, two or more states could join forces to conduct a preventive strike against the rogue state's WMD arsenal. Such a policy is likely to be adopted only when there is strong evidence that the rogue state might actually use its new capabilities. As we shall see, this goal would place the greatest strains on an alliance's solidarity.

POSSIBLE OBSTACLES

Which obstacles to an effective alliance will operate most powerfully when states seek to join forces against a rising rogue state, and which objectives are likely to minimize these difficulties?

CONFLICTING INTERESTS. In general, the potential threat from a rogue state's WMD should reduce conflicts of interest among its potential opponents. However, the interests of allies are unlikely to be fully congru-

---

54. Most states will weigh the costs and risks of obtaining WMD before they begin such a program, however, and only a regime that was highly motivated is likely to make such a decision in the face of high costs and risks. Unfortunately, persuading a highly motivated adversary to abandon this goal will not be easy. Interestingly, although several states (e.g., Sweden, Brazil, Argentina, and South Africa) appear to have abandoned their nuclear programs voluntarily, outside pressure seems to have played little role in their decisions. See Mitchell Reiss, *Bridled Ambition: Why Countries Constrain Their Nuclear Capabilities* (Washington, D.C.: Woodrow Wilson Center, 1995).

55. Pakistan's experience supports this point, insofar as U.S. efforts to punish Pakistan by withholding conventional weaponry probably encouraged it to continue its nuclear program.

ent, if only because geography places potential partners in different circumstances. States that are near the rogue state will have a greater interest in containing its expansionist tendencies, but they will also be more vulnerable to conventional or nuclear retaliation by the rogue power and less willing to take direct action against it. States outside the region will be less concerned by the spread of WMD (and thus less willing to act to stop it), especially because none of the current so-called rogue states is a global power or about to become one. Thus, anti-rogue alliances will inevitably exhibit certain conflicts of interest, even when all parties are aware of the threat.

Alliances against a rogue state may also be impaired by other differences among the potential partners. Although the end of the Cold War has made it easier to isolate contemporary rogue states, which can no longer turn to the Soviet Union, other political divisions could undermine efforts to contain or coerce these regimes. For example, support from China helped North Korea resist U.S. efforts to pressure it into abandoning its nuclear program, and U.S. efforts in the 1990s to contain Iran and Iraq simultaneously have been undermined by its own reluctance to use either one as a counterweight to the other.[56]

Another potential conflict of interest could arise over the goal of nonproliferation itself. In the Far East, for example, efforts to reverse the North Korean nuclear program were undermined by the ambiguous motivations of South Korea and Japan. Not only do these states worry about a possible military confrontation in the region, but both may be ambivalent about the North Korean nuclear program itself. South Korean strategists may hope to inherit the North Korean program after reunification, while some Japanese officials may regard the North Korean program as a pretext that could eventually allow Tokyo to acquire a nuclear arsenal of its own. Thus, although neither Japan nor South Korea is comfortable with the idea of a nuclear North Korea, they have been reluctant to do much to prevent it.[57]

These conflicting interests highlight the problem of credibility. Local powers will prefer to balance against a rogue state, especially if they believe that other states are willing to help. If they fear that they will be left in the lurch, however, they may be more inclined to adopt a position of neutrality. The question of credibility was a recurring (albeit exagger-

---

56. The United States did tilt toward Iraq during the 1980s, but has clearly been unwilling to do so since the Gulf War.

57. Evidence of conflicting Korean, Japanese, and U.S. interests can be found in Leon V. Sigal, *Disarming Strangers: Nuclear Diplomacy with North Korea* (Princeton, N.J.: Princeton University Press, 1998), especially pp. 9, 58–59, 74.

ated) issue throughout the Cold War, but was mitigated by the United States' obvious interest in helping contain Soviet power. Because none of the current rogue states poses a similar threat to the United States, it will have more difficulty convincing potential allies that Washington will still fulfill its commitments.[58]

In short, although many states are likely to regard a rogue state's acquisition of WMD as a dangerous development, estimates of the seriousness of the threat will vary greatly. A clever rogue government could do much to exacerbate this problem, either by offering concessions to mollify potential opponents, by masking its true ambitions, or by concealing the full extent of its weapons programs. As the Iraqi case illustrates all too clearly, it is difficult to organize extensive counterproliferation activities unless or until a rogue state commits clear and unmistakable acts of aggression. In 1991, Iraq's naked incorporation of Kuwait made it relatively easy to assemble a large coalition for the purpose of expelling an aggressor. In 1998–99, by contrast, Iraq's repeated efforts to thwart the UNSCOM inspectors did not pose as grave a threat and thus did not provoke a united allied response. Current and future rogue states are likely to learn from Hussein's unhappy experience; they will avoid overly provocative actions until their WMD programs are further advanced.

COMPETING PRESCRIPTIONS. Even when several states agree that a particular rogue state is a serious threat, they may disagree on how the threat should be met. If the rogue state has not yet obtained WMD, uncertainty about its actual capabilities will discourage some states from taking forceful action, based on the belief that the use of force may not be necessary.[59] Once such a state does acquire WMD, the fear that preventive action might not be completely successful could undermine allies' cohesion and render a forceful response impossible.[60] Here again, the confrontation between the United States and Iraq over the UNSCOM mission is revealing. Because even the proponents of a military response questioned

---

58. See Marc Dean Millot, "Facing the Emerging Reality of Regional Nuclear Adversaries," *Washington Quarterly,* Vol. 17, No. 3 (1994), p. 57.

59. See Michèle A. Flournoy, "Implications for U.S. Military Strategy," in Blackwill and Carnesale, *New Nuclear Nations,* pp. 155–156.

60. Israel's preventive attack on the Iraqi nuclear facility is an exception that proves the rule, as the operation was conducted unilaterally. By contrast, the international community failed to respond to repeated reports about Iraq's renewed nuclear program until Baghdad foolishly provided an opportunity by seizing Kuwait.

whether airstrikes would eliminate Iraq's WMD capacity or persuade Hussein to permit the UNSCOM inspectors to return, it proved impossible to assemble widespread international support for the use of force.

A policy of containment minimizes these problems. Although some members of the opposing coalition may want to do more, any state that regards the rogue state as dangerous is likely to agree on the need to prevent it from expanding. States within the region have an obvious interest in containing potential aggressors as they are the most likely objects of attack, while the United States and some other great powers have a general interest in deterring aggression, especially when it involves the threat or use of WMD. By placing the onus for starting a war on the rogue state itself, a strategy of containment also minimizes the danger of escalation and the potential risk to each member of the coalition, thereby reducing incentives to defect.[61]

What if containment were to fail? Would an anti-rogue coalition be able to summon the political will to reverse the results of aggression, especially if the enemy already possessed WMD? The most likely answer is "yes." A deliberate act of aggression by a rogue state would remove any lingering doubts about its bellicose aims and its willingness to take risks. Such an act would strengthen both its neighbors' desire to weaken or overthrow it and the great powers' interest in punishing a forceful challenge to the status quo. Thus, it should not be too difficult to convince a defensive coalition to force an aggressor to relinquish its gains, although the risk of escalation would probably make some members reluctant to overthrow the rogue government itself.

By the same logic, it will be very difficult for an anti-rogue coalition to take preventive or preemptive action. As the recent confrontations with North Korea and Iraq reveal, such a policy requires a consensus that the rogue state is sufficiently dangerous to justify starting a war, with all its attendant risks. Not only are democratic states generally disinclined to fight preventive wars, but allies are likely to face different levels of risk and will probably disagree about the level of threat and the probability of success.[62] For all these reasons, therefore, a strategy of containment will generally be the preferred response.

---

61. There is another reason why states may favor different responses. If taking action against a rogue state would require a large foreign military presence, and if such a presence would have worrisome domestic repercussions, then local powers are more likely to favor strategies of deterrence or accommodation rather than prevention.

62. See Randall K. Schweller, "Domestic Structure and Preventive War: Are Democracies More Pacific?" World Politics, Vol. 44, No. 2 (1992).

COLLECTIVE ACTION. Even if the United States and its allies agreed on the need to confront a rogue state *and* were united on the strategy to follow, their efforts to contain, compel, or disarm it might be undermined by each state's effort to pass the costs and burdens of action onto its allies. These conflicts cannot be met simply by negotiating more equitable financial arrangements (as the United States and its NATO allies were able to do), because the costs of membership in an anti-rogue coalition will be difficult to allocate evenly. Nearby states will face a greater risk of retaliation than allies who are far away, while outside powers (and especially the United States) will have to provide the bulk of the military forces needed to deter or reverse aggression. When each member's contributions are difficult to compare, the danger grows that each member will feel it is bearing an unfair burden. It will be difficult even to obtain agreement for modest steps like economic sanctions, given most states' unwillingness to forego profits and market share simply because an economic partner has chosen to acquire WMD.[63]

Because a policy of containment will usually be the least expensive, this approach will minimize quarrels over burden sharing. Neighboring states will have to bear the social and economic costs of joint exercises or the presence of foreign troops, and the United States will have to provide the bulk of the military muscle, but these burdens should not be too onerous for either side. Moreover, because a rogue state's neighbors will face a more direct threat than the United States, they will need its help more than it needs them. As a result, the U.S. ability to insist on adequate allied contributions should be substantial.

If the alliance adopts more ambitious objectives, however, the temptation to free-ride will inevitably increase. All members might have an interest in disarming a dangerous rogue, or in reversing a deliberate act of aggression, but most members are likely to prefer that someone else bear the costs and run the risks. This problem can be reduced by measures designed to protect the alliance members from retaliation (such as theater defense systems), and by strategies designed to share the political and military burdens. For example, the United States might be reluctant to take preventive action unilaterally, for fear of acquiring a reputation

---

63. Responses to the Indian and Pakistani nuclear tests are instructive. Although the United States and others have vowed to impose sanctions on both sides, the actual penalties are likely to be modest. For example, although the Arms Control Export Act of 1994 requires the U.S. government to cut off trade credits and loan guarantees to states that conduct nuclear tests, the Clinton administration quickly requested an exception for wheat sales by U.S. farmers. As Senator Slade Gordon warned, every day that sanctions are in place "we'll lose sales, and they are sales we can't recover." See *New York Times,* June 15, 1998, p. A17.

for bullying weaker states. But this political cost would be reduced if other states were actively involved, even if their participation were largely symbolic. Similarly, neighboring states might be reluctant to take direct action if U.S. troops were not in the front lines, because the fear of a massive U.S. response would probably discourage the opponent from using its WMD in the course of a conventional war.

The bottom line is that the problem of free-riding will be most serious when a coalition contemplates offensive action against a rogue state, and will be lowest when implementing a strategy of containment.

THE IMPACT OF UNCERTAINTY. Each of the obstacles just discussed would be exacerbated by the inherent uncertainty involved in appraising the actual threat that a rogue state represents. Because such regimes will go to great lengths to conceal the extent of their achievements, it will be difficult to estimate their capabilities. In particular, rogue states may adopt a policy of deliberate ambiguity regarding their capabilities (sometimes referred to as a strategy of "opaque" proliferation) to reap the benefits of a WMD deterrent while minimizing the political or military costs.[64] Even when other states are convinced that the regime has acquired WMD, the rogue is likely to conceal the size, location, and capabilities of its arsenal in order to complicate efforts to destroy it. And the less that outsiders know, the harder it will be for them to agree on what to do.

Uncertainty about intentions will be equally troublesome, even when the rogue state appears extremely bellicose. As noted earlier, the problem is the inherent difficulty of determining how the acquisition of WMD will affect the new regime's propensity to use force. Not only would the possession of a reliable deterrent reduce the insecurity that encourages expansionist behavior, but achieving such a capability might invite preemption or retaliation if war broke out. Thus, even seemingly irrational or bellicose rogues will probably behave with great caution as their capabilities increase, which means that efforts to contain them stand a good chance of succeeding.[65]

Overall, uncertainties about both capabilities and intentions will make it difficult for a coalition to reach agreement on offensive action against a proliferating rogue state. The intelligence requirements for a military solution would be extremely demanding, given the need to

---

64. See Benjamin Frankel, ed., *Opaque Nuclear Proliferation: Methodological and Policy Implications* (London: Frank Cass, 1991); and Thayer, "Causes of Nuclear Proliferation," pp. 508–517.

65. This argument has been made by a number of scholars, most notably Kenneth Waltz. See his contribution to Waltz and Sagan, *Spread of Nuclear Weapons*.

locate virtually all potential weapons sites and production and research facilities and the relative ease with which these targets can be concealed.[66] As a result, coalition support for a strategy of preventive war is likely to be forthcoming only when there is unambiguous evidence that a nascent rogue was planning to attack. Unfortunately, history warns that this sort of evidence will rarely, if ever, be available. Indeed, the West's poor track record in gauging the scope of North Korea's and Iraq's WMD programs and its failure to anticipate India's resumption of nuclear testing merely underscore the inherent difficulty of acquiring information that foreign governments are eager to conceal. Given these uncertainties, it would be folly to expect other states to agree to preempt a rogue state merely on the basis of U.S. intelligence estimates.[67]

These difficulties will be compounded by the need for secrecy. A preventive strike is more likely to succeed when it is conducted without warning (as Israel did in 1981), but the advantage of surprise is lost if an extended diplomatic campaign is necessary in order to round up allied support.[68] And a clever rogue can exacerbate virtually all of these problems by masking its intentions, downplaying its radical beliefs, and refraining from overt acts of aggression until its weapons are completed and dispersed.[69]

Taken together, these obstacles suggest that containment should be the primary objective for any anti-rogue alliance, with reversing aggression as a secondary goal should events require it. Containment will reflect the members' shared interests most accurately; is most likely to generate consensus within the alliance and to minimize collective action problems; and will require the least amount of reliable information. Reversing conventional aggression will be more difficult but still feasible—assuming that the act of aggression is unambiguous—but preventive military action is unlikely to garner significant allied backing.

---

66. It is worth noting that the U.S.-led coalition apparently destroyed no more than a handful of Iraq's Scud missile launchers, despite an intensive air campaign against them. See Thomas A. Keaney and Eliot A. Cohen, *Gulf War Air Power Survey: Summary Report* (Washington, D.C.: U.S. GPO, 1993), pp. 78–90.

67. On the failure to gauge Iraq's WMD programs, see Freedman and Karsh, *Gulf Conflict*, p. 320; and David Albright, "Masters of Deception," *Bulletin of the Atomic Scientists*, Vol. 54, No. 3 (1998).

68. See Millot, "Facing the Emerging Reality," p. 48.

69. Once again, the Gulf war provides an apt contrast. The formation of the coalition was facilitated by Iraq's overt aggression against Kuwait, which gave clear evidence of Iraq's revisionist aims. Moreover, the coalition's war plans did not rely heavily on secrecy regarding the timing of the attack or whether it would take place at all.

Recommendations for U.S. Policy

The aquisition of WMD by any rogue state will highlight its threatening nature and make its containment seem more necessary. At the same time, it also makes steps beyond containment more dangerous. These insights lead to two broad conclusions and a number of specific recommendations.

First, despite the various obstacles to coalition formation, it should not be difficult to organize *defensive* coalitions against contemporary rogue states, even if they were to obtain WMD. Indeed, containment is by far the most feasible response to a rogue state that is in the process of obtaining these capabilities. Containment places the onus for conflict on the rogue state, lowers the risk that the United States or its partners will be dragged into a war that might otherwise be avoided, and minimizes the various obstacles to an effective alliance. It also responds to the fear that rogue states will act more aggressively once they obtain WMD, while leaving the door open to improved relations should the rogue state feel more secure once it acquires its own deterrent. And such a policy might help deter the acquisition of WMD in the first place, by forcing potential proliferators to weigh the benefits of possession against the political and military costs of isolation.

Second, efforts to organize *offensive* coalitions to compel or disarm a rogue state are unlikely to succeed. The U.S. efforts to pressure North Korea and Iraq into abandoning their WMD ambitions are quite revealing on this point, because widespread allied support for military action was lacking in both cases. The obstacles identified above will be especially severe should the United States seek allied support for preventive military action; even when potential allies secretly favor such a step, the temptation to free-ride will be especially high. Among other things, it should not surprise us that there was strong allied support to reverse Iraqi aggression in 1991, or that support for the UNSCOM regime has deteriorated since then.

RECOMMENDATIONS
In light of the various incentives and obstacles identified in this chapter, U.S. efforts to organize military coalitions against a rogue state should emphasize several elements.

REEMPHASIZE INTELLIGENCE. Timely and accurate intelligence is a prerequisite for forming an effective coalition; the United States cannot expect others to help if they do not believe that a rogue state is dangerous.

Similarly, mounting a credible deterrent to aggression by a rogue state armed with WMD will require accurate information about its present capabilities and the likely objects of its revisionist aims.[70] In addition to monitoring rogue states' capabilities, therefore, U.S. intelligence efforts should also focus on identifying the underlying sources of each rogue's ambitions, and their susceptibility to change. In particular, efforts to ascertain how the acquisition of WMD is likely to affect the rogue state's behavior will be especially valuable as a safeguard against either complacency or paranoia.

SHARE THE COSTS AND RISKS. Assuming that other states share similar perceptions of the threat, the United States will have to find ways to demonstrate its own commitment and to share the costs and risks. Potential allies need to be confident that they will not be left in the lurch should deterrence fail. As discussed above, this step will not be as difficult as many experts fear, in part because the United States has specific interests in most regions of the world and also because the United States has a general interest in reinforcing the defensive character of WMD. Obviously, reiterating these interests will be an important element of U.S. diplomacy toward potential coalition partners.

In addition, efforts to contain rogue states are likely to require some tangible expressions of U.S. commitment, even if only of a symbolic nature. Joint military exercises nicely reinforce both commitment and capability, and can be bolstered by public declarations, naval port calls, diplomatic visits, and in extreme cases, the permanent deployment of a modest number of U.S. troops. By ensuring that the United States will be involved should force be used, such measures diminish the odds that a rogue state would take action in the first place.

This sort of forward policy would also give the United States a bargaining chip in its subsequent dealings with the rogue state; for example, the United States could offer to reduce its military presence if the rogue state abandoned its WMD programs. If a large U.S. presence is infeasible, pre-positioned stocks and other preparations for rapid reinforcement can convey commitment and enhance military effectiveness.[71] Another possibility is the deployment of theater ballistic missile defense systems, which can both protect U.S. military forces in the area and

---

70. See Henry Sokolski, "Fighting Proliferation with Intelligence," *Orbis*, Vol. 38, No. 2 (1994); and Blackwill and Carter, "The Role of Intelligence."

71. There is a potential tradeoff here; a large stockpile of pre-positioned equipment might be an inviting target in the event of war. On the other hand, destroying that stockpile would probably ensure that the United States would become involved.

reassure potential victims who may fear a WMD attack, thereby reinforcing containment in two mutually reinforcing ways.[72]

Over time, public support for an active counterproliferation policy will depend on the willingness of other states to contribute to the common goal. Most vital U.S. interests are only indirectly affected by contemporary rogue states (because none of them is in a position to attack the United States directly), and U.S. taxpayers will not support a major effort to contain rogue states if the states it seeks to protect are unwilling to aid its efforts. As in any alliance, in short, success will depend on each member's ability to convince its partner or partners that they are not being exploited.

IDENTIFY ALLIES CAREFULLY. The importance of maintaining credibility and distributing costs suggests that the United States needs to think clearly about which states actually merit a U.S. commitment and why. Should the United States try to protect *any* state that is threatened by an aggressive rogue state, or should it do so only when U.S. interests are more directly engaged? If the latter, then the United States would commit itself to aid Saudi Arabia against Iraq or Iran but might not commit itself to oppose Libyan interference in Chad. The need to set priorities will become more acute as U.S. military capabilities decline. The general sense that rogue states are potentially dangerous does not mean that they are equally threatening or that the United States must respond to each one.[73]

DON'T OVERSTATE THE THREAT. U.S. leaders should be careful not to exaggerate the danger posed by rogue regimes. Ironically, efforts to limit proliferation will be less effective if rogue states conclude that developing WMD is the best way to get a lot of international attention or to get the United States to focus on regional security issues. Potential allies might conclude that these states are so powerful or dangerous that it would be foolish to stand up to them. As I have already argued, today's rogue states are rather weak and insecure regimes whose nuclear ambitions are based primarily on a deep sense of insecurity.[74] There seems to be little reason

---

72. See Utgoff and Quester, "No-First-Use and Proliferation," pp. 107–108.

73. This criterion is complicated by the general U.S. interest in discouraging the use of WMD for aggressive purposes, which might lead the United States to punish *any* state that uses WMD to alter the international status quo even when the state in question is not hostile to the United States itself. But this problem does not alter the value of carefully identifying places where the United States should make a special effort to contain rogue states.

74. The worst case would be if a rogue state's sense of insecurity were so great that

to fear that these states would be suicidally bellicose even if they did acquire such capabilities, which suggests that prompt efforts to contain them are both necessary *and* likely to succeed. In seeking allies, therefore, U.S. strategic planners should strive to portray the threat as accurately as possible, without indulging in the hyperbole that all too often accompanies efforts to rally public or international support.[75]

COMBINE DETERRENCE AND REASSURANCE. The West won the Cold War by containing the Soviet Union while avoiding all-out war, until a combination of external pressures and internal developments brought the fall of communism. Similar objectives should guide U.S. behavior toward most rogue states: it should strive to contain them while emphasizing that its aims are limited. No matter how objectionable such regimes may be, it will rarely be in the U.S. interest to back them into a corner: States that see others as irrevocably hostile are prone to extremely risky behavior. Unless the United States is willing to take more forceful action, containment should be tempered with contingent forbearance. In other words, the United States should communicate that it will not try to overthrow these regimes so long as they abstain from aggression; but will act to oust them if they attack their neighbors, and especially if they use WMD.

CREATE "FIREBREAKS" AGAINST THE USE OF WMD. The United States should consider how to erect "firebreaks" against the use of WMD in the event that deterrence does break down.[76] One way is to reemphasize the taboo against nuclear weaponry, making it clear that the United States and its allies would regard the use of such weapons as a qualitatively different act and would adjust their evaluations of the regime and their conduct toward it accordingly.[77] Continued reductions in the U.S. arsenal will reinforce this norm, and the ratification of the chemical and biological

---

it felt compelled to expand even after it acquired WMD, and attempted to use its new capabilities to accomplish that objective.

75. However tempting it is to magnify external threats in a era of budgetary pressure, the strategy can be counterproductive if it suggests that the problem is so grave as to allow for no solution.

76. On the nature and construction of such thresholds, see Thomas C. Schelling, "Bargaining, Communication, and Limited War," and "Nuclear Weapons and Limited War," in *The Strategy of Conflict* (Cambridge, Mass.: Harvard University Press, 1960).

77. For this reason, proposals that the United States respond to this threat by developing "micro"-nuclear capabilities would probably be counterproductive, as they would merely serve to blur the distinction between conventional and nuclear weapons and open the way for large-scale nuclear use by others. For an example of this sort of

weapons conventions could begin to expand the norm to cover these technologies as well. Finally, the United States should also remind the international community that using a weapon of mass destruction against U.S. territory, military forces, or allies will inevitably invite overwhelming U.S. retaliation. Because the United States now possesses overwhelming conventional superiority (and is likely to retain it for some time), it is in its interest to keep WMD as "unusable" as possible.

MAINTAIN ADEQUATE U.S. DEFENSE CAPABILITIES. Any serious effort to combat the spread of WMD and to contain rogue states will require the United States to maintain a robust set of military capabilities. No state can expect to wield international influence without possessing substantial power, and efforts to forge anti-rogue coalitions will be more difficult if the United States lacks the strength that has made it a valuable ally in the past. Although the programmed decline in U.S. military forces could make the United States more dependent on its allies' support, its ability to obtain this support would evaporate if its capabilities erode too far. To repeat: allied support can be an invaluable adjunct to U.S. power, but it is not a substitute for it.[78]

FINAL THOUGHTS

If the main arguments advanced here are correct, then the alleged danger from "nuclear rogues" is overstated. Instead of enabling otherwise weak and unpopular regimes to expand with impunity—secure in the belief that neither their neighbors nor the great powers will dare resist—the acquisition of WMD is more likely to spark the formation of countervailing coalitions committed to containing or reversing any attempt by the rogue to expand. Although these coalitions will face the usual obstacles to effective balancing, these barriers can be overcome with fairly modest efforts. In particular, costly programs to develop theater defenses or to

---

argument, see Thomas W. Dowler and Joseph S. Howard II, "Stability in a Proliferated World," *Strategic Review,* Vol. 23, No. 2 (1995).

78. Of course, one could argue that the absence of a major great power challenger means that the United States has few if any overseas interests at present, and thus little need either for large military forces or an extensive alliance network. By this logic, states that are directly threatened should bear the costs of containing today's rogue states, and the United States can abandon its efforts to promote international order and concentrate on domestic issues. This position is defensible (although I do not share it), but it requires a willingness to abandon any effort to shape the future evolution of the international system. International influence cannot be had "on the cheap"; if the United States wants to wield substantial leverage, it will have to pay the price.

acquire the capacity for a disarming conventional attack are probably unnecessary and might even be counterproductive.

This conclusion does not imply that the United States can retreat to isolationism, secure in the belief that other states will do all that is needed. U.S. military power will be an important component in these defensive coalitions, and will be especially necessary to deter the actual use of WMD against states that lack their own deterrent.

Indeed, how the United States responds to the spread of WMD could have far-reaching effects on the future of world politics. Although the United States cannot prevent some states from acquiring WMD, it can do much to affect the implications of this trend, and thus its pace and scope. If the United States does not act to contain rogue states armed with WMD, or if it allows them to use these weapons to coerce or to conquer other states, then it will be helping reinforce the belief that these weapons are a potent instrument for altering the international status quo. The demand for WMD will grow apace, and the danger of war is likely to rise, at least in the short run. By contrast, if the United States arrays itself against both the spread of WMD and their use for coercion or conquest, it will help discredit them as instruments of aggression and diminish incentives to acquire them.

As the Cold War fades into history, therefore, the United States stands at a crossroads. Down one path lies a world where WMD remain scarce and unusable; down the other lies a world where such weapons are plentiful, where they are seen as potent sources of political influence, and where they are somewhat more likely to be used. Because the latter world is clearly less desirable, the United States should try to prevent it. Such a strategy is likely to garner considerable allied backing. The more pertinent question is whether U.S. citizens are willing to support active involvement in international affairs, even when it costs large sums of money, and occasionally the lives of its citizens as well.

Chapter 8

The Response to Renegade Use of Weapons of Mass Destruction

George H. Quester

How the United States responds to an adversary that used nuclear, biological, or chemical weapons to attack U.S. forces at bases abroad, a U.S. city, or the city of a major ally would play a critical role in determining how such weapons are seen afterward. The U.S. reaction to an attack would influence whether other rogue states would make attacks with weapons of mass destruction (WMD). Indeed, if the United States lets other states know what its response would be, the prospect of U.S. retaliation could deter the use of WMD, if the response is of the proper form and magnitude.

This chapter explores two models for how the United States could respond to a renegade's use of WMD—a law enforcement model, as illustrated by the Allies' actions after World War II, and the Cold War deterrence model. I argue that unconditional surrender is the appropriate objective if a country such as Iraq or North Korea uses nuclear weapons. In contrast, the Cold War theory of deterrence does not apply to a rogue state that has used WMD. If a renegade state uses one or several such weapons, it should not assume that the rest of its WMD arsenal can ward off extreme punishment. The treatment that was applied to Hitler's regime after it had killed millions of people is the appropriate treatment for a rogue state that uses WMD today.

The next section examines the law enforcement model: for what reasons does society imprison criminals? The second section examines how well these motivations explain U.S. policies toward Germany and Japan after World War II. The third section surveys the usefulness of Cold War–era ideas about deterrence and the argument that no surrender is truly unconditional. The conclusion offers policy recommendations.

From the Korean War to the response to the Iraqi invasion of Kuwait, and as far back as U.S. persistence in the Pacific during World War II, the United States has failed to advertise the vehemence with which it will respond to foreign aggression. As doubts are expressed today about U.S. commitments after the end of the Cold War, it would be wise for the United States, and its adversaries, to take account of this pattern.

The Law Enforcement Model

Analogies with domestic law enforcement can help to illuminate U.S. military choices at the international level, since, as Thomas Schelling noted in the 1950s, a criminal justice system applies parallel strategic approaches.[1] The law enforcement model is also valuable in that it highlights the role of justice and legitimacy; it will always be easier to deter a potential renegade, and to enlist cooperation in punishing such a renegade, when other countries share the U.S. sense of justice and outrage.

Why do we "retaliate" against criminals—murderers or embezzlers— by placing them in prisons that are quite costly to the taxpayer? There are at least four reasons, each of them relevant to how the United States should respond to the use of WMD by a renegade state.

One reason for putting criminals into prisons is to disarm them, to render them incapable of committing similar crimes again. Murderers are a threat to citizens besides those they have already killed. At the minimum murderers must be denied weapons; since kitchen knives and other deadly weapons are so easy to come by, this may require locking such people in prisons. (This reasoning applies much more to a murderer than to an embezzler; once the identity of an embezzler is known, banks will not hire him or her again, and the monetary assets of the rest of society would be fairly safe, even if the individual were never to be imprisoned.)

Since so many tools, such as axes and kitchen knives, can also be instruments of crime, we cannot ban all weapons; we must wait until someone commits a crime. Once someone is identified as an axe-murderer, he or she will have to be denied all use of axes or other potential weapons. A similar approach may be useful with the "dual-use" technologies of chemistry, biology, and nuclear physics, where what serves civilian purposes also produces weapons. For example, Iraq cannot be allowed the same facilities as states with less of a history of aggression.

A second reason to imprison criminals is to "make an example" of the violator, thus deterring other potential law-breakers. Imprisonment is a way to punish the criminal. (This, to use the jargon of military strategy, is a "countervalue" rather than a "counterforce" argument for imprison-

1. Some interesting discussion of analogies between domestic justice systems and international strategy can be found in Thomas Schelling, *The Strategy of Conflict* (Cambridge, Mass.: Harvard University Press, 1960).

ment.) For this purpose, being in prison must be made a decidedly less pleasant experience than being free.

There are limits to how painful U.S. citizens and other civilized people want their prisons to be. The horror stories of southern state prisons of the early twentieth century are not something we would want to bring back, in part because this is too close to "torture" or "cruel and unusual punishment," and also because it may conflict substantially with the goal of reforming the criminal. Yet, the U.S. public also objects whenever penitentiaries come to be labeled "country club prisons," because the prospect of a penitentiary with color television sets and gymnastic equipment and good food may not deter a potential law-breaker who is living in an environment of urban poverty.

The major reason we imprison people who embezzle large amounts of money may thus be to make an example of them, in the hope that others will be deterred from stealing money if they see that this can lead to the discomfort of imprisonment.

Just as we cannot let a law-breaker "get away with" a crime in domestic society, lest everyone else feel free to commit the same offense, countries must respond to violations of their rights in the international arena; North Korea will be watching to see what Iraq can get away with, and vice versa.

A third argument for imprisoning criminals is justice. Most people grow up with an elementary sense of justice, by which wrongs should be retaliated, simply to rectify the score. We are angry when our rights are violated; we become calmer, we feel right about the world again, if we are able to inflict punishment on the violators.

"Justice" is an important word in most cultures, and enormous arguments (and wars) can erupt about exactly what is "just."[2] Indeed, some of the worst conflicts are those in which each side feels itself the victim of an injustice inflicted by the other, and where each thirsts for retribution. It is very human, in such debates about justice, for each side's sense of rightness to intensify. Thus, conflicts where each side brings along its own version of history, such as that between Croats and Serbs in the former Yugoslavia, are exacerbated.

The goal of "justice" may seem closely related to the goal of "setting an example," but it is not identical. Justice is an end in itself, something sought not because of any conscious calculation of consequences, such as deterring a potential adversary from misbehavior. The relatives of a murder victim typically feel relief when the murderer is convicted and

---

2. On the manifold disagreements of definitions of justice, see Carl J. Friedrich, *Justice* (New York: Atherton Press, 1963).

given a stiff sentence, not so much because they are concerned about their own immediate safety, or because they are concerned to deter other murderers in the abstract, but because they wish for revenge.

The instinct for revenge is very human. As an assurance that the guilty will be punished, the existence of such instincts are indeed valuable for deterring crimes and aggressions. The common instinct for revenge helps to make the *prospect* of retaliatory punishment credible.

But taking revenge may not always make sense after a crime has been committed or after an aggression has occurred. Such an abstract psychological sense of retribution, however emotionally important, cannot be the only consideration in domestic law enforcement decisions, or in the international response to a WMD use. Feelings of justice are grounded in deep emotions, which tend to confuse rational calculations. Justice may come at too high a price in other, more practical, considerations.

A fourth purpose for prisons is to reform criminals, to change their behavior, to teach them that they can support themselves without violating the laws. U.S. citizens worry that their prison systems are not very successful at rehabilitating inmates, but many convicts do indeed leave prison and never violate the law again. (Of course, the process of aging reduces some of the energies required for criminal activity. If law-breakers are held out of reach of society until they "mature," they are less of a risk thereafter. This rationale combines arguments one and four for imprisonment.)

Reform is certainly among the goals pursued internationally. The United States is gratified by the change in Germany and Japan since 1945, and hopes for a similar change as the former Soviet Union becomes a democratically governed Russia.

In setting sentences and in designing prisons, it is often difficult to determine which approach—disarming, reforming, or punishing the criminal, or establishing a clear deterrent precedent for others—is the most feasible.

In addition, each of the goals of imprisonment can conflict with each of the others. What exacts the greater punishment may not always achieve the maximum of disarmament. (The 1943 bombing of Hamburg provides a very clear example in the field of military strategy, as *more* German workers showed up at factories after the bombing, because the delicatessens, etc., that had exempted them from war work had been leveled.) For example, excessive punishment, or excessive security to ensure that inmates cannot arm themselves, may get in the way of reforming the criminals. And the abstract pursuit of justice or revenge could come at too high a price for our more practical goals.

The End of World War II

The United States has responded to international instances of "mass destruction" in the past. In World War II, the Nazi regime in Germany killed millions of people in concentration camps.[3] The Japanese army, after the capture of the Chinese capital at Nanking in 1937, reportedly killed several hundred thousands of people by bayonet.[4] The lethal gas used in gas chambers and bayonets are not what we normally classify as WMD, but such weapons facilitated destruction of the magnitude that we might experience if rogue proliferators used WMD today.

How did the outside world respond to these acts of mass destruction? What lessons can we learn from what happened to the Nazis or the Japanese military leaders, and to their countries?

DISARMING THE RENEGADE

The Allies' original intention after World War II was that the Axis powers should be totally disarmed, that is, rendered incapable of ever again militarily threatening and inflicting mass destruction on their neighbors. Japan was denied any armed forces in the constitution drafted for it by the staff of U.S. General Douglas MacArthur, and the Allies agreed that Germany was not to have any military forces.

For a time, such a total disarmament was indeed applied. The Japanese and German military forces were disarmed and disbanded, and the Axis states lost their sovereignty to the Allied occupying powers.

After some years, however, the exigencies of the Cold War led in Japan to the emergence of three "Self-Defense Forces," Ground, Air, and Naval, in place of an Army, Air Force, and Navy.[5] In the German case, an escalatory race emerged; the Western allies established a Federal Border Police while the Soviets established a "People's Police," each of which looked remarkably military. They were followed by openly acknowledged armed forces on each side.[6]

---

3. On the Holocaust, see Raul Hilberg, *The Destruction of the European Jews* (New York: Holmes and Meier, 1985).

4. See Edward Russell, *The Knights of Bushido* (London: Corgi, 1960) for an account of Japanese behavior in Nanking and elsewhere in Asia.

5. The emergence of the Japanese "Self-Defense Forces" is discussed in Michael Schaller, *The American Occupation of Japan: The Origins of the Cold War in Asia* (New York: Oxford University Press, 1985).

6. The escalatory process of East and West German rearmament is discussed in Hans Speier, *German Rearmament and Atomic War* (Evanston, Ill.: Row, Peterson, 1957).

If the victims of Axis aggression had expected that Germany and Japan were to be totally and permanently disarmed, they would have been disappointed. But from their very creation, the West German military forces were tied closely to NATO, and the Germans were not to have their own general staff. And the constraints on the Japanese "Self-Defense Force" also amount to a compromise of sovereignty greater than most nations would accept.

In one most important aspect the Axis powers have *not* been allowed to rearm at all: with weapons of mass destruction.

If one asks the basic question today about the Nuclear Non-Proliferation Treaty (NPT)—why some countries are allowed to possess nuclear weapons and others are not—the answer is often evasive, amid assurances that a total nuclear disarmament of all the powers of the world is the goal. The more honest answer might be that the non-use of nuclear weapons may require that there be at least two nuclear forces deterring each other, given the inability of the world to forget how to produce such weapons, and given the inherent difficulties of the safeguards required to assure each country that no other country is cheating.

The five states that had weapons when the NPT was drafted may be more than are necessary for mutual deterrence, for "mutual assured destruction," but this is where, by accident of history, the nonproliferation barrier was drawn.

To justify this barrier further, one sometimes hears references to the political instability of third world countries, of countries such as Libya or Iraq; but note is then taken that such instability has appeared also in Beijing and Moscow, while it has not in Sweden or Australia. It is dangerous to base the nonproliferation argument on the premise that Moscow and Washington are "more qualified" to handle nuclear weapons, while New Delhi and Islamabad are not, for this will simply unite the South Asians in indignation at the slur on their character.[7]

Even fifty years after World War II, many still feel that Germany, Italy, and Japan, the parties guilty for the barbarism that was at the heart of World War II, are particularly disqualified to handle nuclear, biological, or chemical weapons.

The Federal Republic of Germany was required to renounce nuclear, biological, and chemical weapons by the Western European Union Treaty that legally sanctioned its rearmament in 1955. It grudgingly signed the

---

7. For this author's more extensive view of South Asian attitudes on nuclear proliferation, see George H. Quester, *Nuclear Pakistan and Nuclear India: Stable Deterrent or Proliferation Challenge?* (Carlisle, Penn.: Army War College Strategic Studies Institute, 1992).

NPT at the end of the 1960s, renewing this commitment to forego nuclear weapons. Japan has maintained a nuclear taboo, even while its "Self-Defense Forces" had become a substantial conventional army, navy, and air force by the end of the 1980s.

PUNISHING THE RENEGADE

In addition to disarming the Axis powers, in 1945 the Allies intended to impose some punishment. In practice, there was indeed an important setting of an example here, so that anyone contemplating emulating Hitler or Tojo would have to think twice.

One cannot forget, of course, that nuclear destruction was imposed on Japan in 1945. It is more often forgotten that the physicists and engineers who worked so feverishly to produce nuclear weapons in the Manhattan Project visualized them as being readied for use against Germany, more than against Japan.[8]

The focus on Nazi Germany stemmed, of course, from fears that the Germans might be racing to acquire nuclear weapons themselves. The bomb might thus be used to defeat Germany before Hitler could use such weapons against the West. But another reason for the focus was that so many of the nuclear scientists were direct or indirect victims of Nazi policies, who had been driven to flee Germany or Europe, guessing or knowing that their relatives were being murdered in concentration camps.[9] If the atomic bomb had been ready in time, it almost surely would have been dropped on German cities before Japanese cities. Instead, German cities were exposed to deadly conventional bombardment, with the casualties in Hamburg and Dresden reaching into the hundreds of thousands.

Of course, these attacks were primarily intended to force the Germans and the Japanese to surrender, rather than to set an example for other aggressors in the future. That is, it was an attempt at intrawar compellence of the Axis, rather than future deterrence of others. Germany did not surrender because of the punishment of World War II aerial bombardment, while Japan most probably did.

Still, some of the motivation for punishing the Axis populations went beyond speeding the Allied victory, and this motivation persisted after the surrenders. The Soviets surely punished the Germans (and even the Japanese, with whom they were at war for a mere three days) through

---

8. For a contrary interpretation, see Arjun Makijani, "Japan Always the Target," *Bulletin of the Atomic Scientists*, Vol. 51, No. 3 (May/June, 1995), pp. 23–27.

9. On the motives driving the Manhattan Project, see Thomas Powers, *Heisenberg's War* (New York: Knopf, 1992).

their treatment of prisoners of war—a great number did not survive to return to their homes, and the rest were held in forced labor for many years.[10] Some of the U.S. plans for the occupation of Germany after the war, for example the Morgenthau Plan, likewise amounted to punishment or revenge, albeit these were also labeled as "disarming" the Germans by eliminating their heavy industry.[11] Policies of "nonfraternization" and the maintenance of austere food rations could similarly be justified less by the logic of disarmament and more by that of punishment. One did not have to execute leaders like Goering or Tojo to render them incapable of launching wars again; rather, their execution achieved a feeling that "justice was done" and "a good precedent had been set." The war crimes trials in Nuremburg and in Tokyo could be advertised, like any other judicial process, as the setting of an enduring legal example; critics of their legality claimed that these processes simply allowed the victors to take revenge.[12]

Thus there was *some* punishment of the Germans and Japanese, whether for the sake of justice or for setting an example, but much less than these defeated peoples might have had reason to expect, especially given how their own armies and governments behaved. Both of the defeated Axis powers were on short rations immediately after the war, but so were the rest of Europe and East Asia. It took time to overcome the economic disruption that the war inflicted.

The punishments were surely greater while these countries were still holding out against surrender, and less very soon thereafter. This suggests that the punishments were being inflicted on the Axis states because they were still engaging in behavior threatening to the outside world. Similarly, the punishments a judge might impose on someone who is refusing to testify when he has been given immunity are often much more severe and open-ended than the punishments imposed on a robber who has been arrested and disarmed, and so can no longer engage in robbery.

But the reduction in the level of punishment after 1945 also shows how the desire for revenge can cool, at least in a democracy, with its humane instincts. For example, in a public opinion poll taken among U.S. citizens in November 1945, immediately after the atomic bombings and Japan's surrender, some 22 percent of the respondents regretted that the

---

10. Soviet treatment of prisoners of war is detailed in Helmut Fehling, *One Great Prison* (Boston: Beacon Press, 1951).

11. On the Morgenthau Plan, see Warren Kimball, *Swords or Ploughshares?* (Philadelphia: Lippincott, 1976).

12. The Nuremberg trials are discussed in Telford Taylor, *The Anatomy of the Nuremberg Trials* (New York: Knopf, 1992).

United States had not inflicted more nuclear destruction on the Japanese.[13] Most U.S. citizens would today be shocked by this 1945 expression of opinion.

Passions also cooled after 1946 because of the new drives of Cold War competition between Moscow and the West. But, in the U.S. and British zones of Germany, and in Japan, occupied entirely by the United States, the occupying forces by and large did not have their heart in being punitive, even if it still seemed quite plausible that the German and Japanese crimes should be blamed not only on the leaderships, but on the peoples as a whole.

Indeed, within two years the United States was donating substantial economic assistance to its former enemies through the Marshall Plan and related aid efforts.[14] Perhaps this was because the economic situation in the former war zones had deteriorated to the extent that no additional punishment was necessary, but the aid also demonstrated an unusual generosity, where the winner in a war delivers gifts to the loser.

The Allied bombings of Germany substantially exceeded the volume of the German bombings of Britain, and hence did more than enough to achieve justice for Coventry and London.[15] But they were not enough, by any "eye for an eye, tooth for a tooth" reasoning, to make up for the concentration camps and the Holocaust and the totality of the Nazi occupation of Europe, and all the deaths and destruction suffered in ordinary combat.

It is hardly surprising that most Jewish people in the United States are less inclined to turn the page on German history than are other citizens, or that the Chinese and Koreans, and indeed almost all East Asians, are much more bitter than are the average U.S. citizens toward Japan.

The more direct victims of wartime atrocities might thus feel bitter that less than justice was achieved. Perhaps too many war criminals received light sentences or were not punished at all.

Perhaps, where *mass* destruction is involved, it may be difficult to inflict commensurate revenge, unless it is inflicted very rapidly and all at once. A civilized people will not stomach the task of being

---

13. Poll of November 30, 1945, cited in *Public Opinion Quarterly*, Vol. 9, No. 4 (Winter, 1945–1946), p. 530.

14. On the Marshall Plan, see Armand Clesse and Archie Epps, eds., *Present at the Creation* (New York: Harper and Row, 1990).

15. The magnitudes of the German and British-U.S. bombing campaigns are compared in Charles Webster and Noble Frankland, *The Strategic Air Offensive Against Germany* (London: Her Majesty's Stationery Office, 1961).

cruel in retaliation, once the transgressor has ceased to defy the norms of the world—once the immediate compellence task has been accomplished.

Where it is necessary, for longer-term considerations, to "teach a lesson" to deter others, it will not be easy to persist in this retaliation. When the enemy is no longer defiantly engaged in aggression or atrocities, when it is no longer challenging the democratic powers in a contest of wills, the revenge motive is reduced, and the punishments soon cease.

Many analysts argue that there should be no nuclear retaliation for a chemical or biological attack on a U.S. military base or U.S. city, because the robust array of U.S. conventional weapons will allow for more than sufficient retaliation. (And a few analysts, with logical consistency, also suggest that there should be no U.S. nuclear response even if a country such as Iran uses a *nuclear* weapon against the United States.) However, if the United States sees its task as making an example of the transgressor, so as to deter others, conventional retaliation could take too long, and might not be pursued for long enough to impose a commensurate punishment. To respond tellingly to a use of weapons of mass destruction, to really "set an example," it may be necessary to respond with WMD.

The United States may thus have to retain an array of nuclear weapons for a variety of reasons. First, such weapons will be required to respond to another power's use of nuclear weapons. Second, some tactical situations might arise where conventional defenses could not protect an ally; rather than have the ally obtain its own nuclear weapons, there may be a need to maintain U.S. extended deterrence. Third, the threat of a U.S. nuclear response may be needed to deter some renegade state from using chemical or biological WMD.[16]

REFORMING THE RENEGADE

After 1945, the Allies' emphasis shifted to reforming Germany and Japan, instead of very severely punishing the Axis countries, or totally and permanently disarming them.

In 1919, the French government under Georges Clemenceau shared none of U.S. President Woodrow Wilson's enthusiasm and optimism for achieving democracy in Germany, fearing that a democratically governed German neighbor would be just as hostile as one governed by the Kaiser. With the Treaty of Versailles, France was intent entirely on disarming

---

16. For an analysis urging U.S. nuclear retaliation threats against another state's chemical or biological attacks, but not against a purely conventional attack, see David Gompert, Kenneth Watman, and Dean Wilkening, "Nuclear First Use Revisited," *Survival*, Vol. 37, No. 3 (Autumn 1995), pp. 7–26.

Germany, and on punishing the Germans to set a deterrent example for the future.

But in the total defeat and unconditional surrender imposed on the Axis powers in 1945, the Germans and Japanese did not just have their governments defeated and removed; their political cultures were successfully reformed.[17] The decades since 1945 might suggest that it is foolish to pursue revenge or justice (as in Germany after World War I), but that it is possible to inculcate democracy, and that democracies are more peaceful and internationally cooperative than are other forms of government.

Of course, some analysts still voice doubts about whether Germany and Italy and Japan have been reformed. They note that Italy in 1994 elected a governing coalition that included neo-Fascists.[18] Germany experiences neo-Nazi "skinhead" attacks on resident foreigners. The same kinds of marginal anti-social behavior can be found in places like France and Sweden, perhaps with no less frequency on a per capita basis, but it understandably attracts more notice in Germany.[19]

Scholars debate whether Japan's openness and liberalization and democracy are superficial.[20] For example, it is disturbing that the Japanese school system does far less than the German to address the basic guilt of the Axis powers for the outbreak of World War II.[21] Yet, the democratization of these defeated Axis powers now looks like a general success, far exceeding what most U.S. citizens and others might have dared hope for during World War II.

Prisons often fail to reform convicts, which is an argument for increasing the punishments to deter crime, or for disarming such criminals by locking them up, rather than pursuing a "reform-school mentality." By contrast, the handling of the defeated Axis powers after World War II suggests that the reform of governments and their foreign policies is attainable. Thus, reform efforts could be as important as "setting an example" or disarming a defeated enemy. The possibility of reform should not be forgotten in the passion for justice or revenge.

The United States has felt generally successful in its response to the mass destruction perpetrated by the Axis powers in World War II. But

---

17. See John D. Montgomery, *Forced to be Free* (Chicago: University of Chicago Press, 1957).

18. See *New York Times*, April 16, 1994, p. 1.

19. See, for example, *New York Times*, October 14, 1994, p. A-12.

20. See Karel Van Wolferen, *The Enigma of Japanese Power* (New York: Knopf, 1989).

21. See Saburo Ienaga, "The Glorification of War In Japanese Education," *International Security*, Vol. 18, No. 3 (Winter 1993/94), pp. 113–133.

the modes of achieving the success were not what would have been predicted in 1943.

As with law enforcement, it is not easy to predict which of the four basic kinds of response to a transgression will offer success. And difficult choices must be confronted, as one approach will conflict with another.

The Cold War Deterrence Model

The situation that the United States faced in its Cold War standoff with the Soviet Union is very different from the situation it would face if a renegade were to use WMD, in two important ways.

First, had the Soviet Union made an initial use of weapons of mass destruction, the damage could have been infinitely more severe than what any renegade state can launch in the near future. A Soviet missile and bomber surprise attack might have killed most of the U.S. population.

It thus was not really clear *why* the United States should have retaliated to set an example here (for it would be difficult, after this disaster, to imagine any comparable situation emerging). The holocaust that Moscow launched would have brought the United States to the point where many analysts predicted that "the survivors would envy the dead," and retaliation against the cities and civilian population of the Soviet Union would mainly have increased the aggregate calamity suffered by all when "deterrence had failed."[22]

But "everyone knew" that the United States would "of course" retaliate for such an attack, and the leaders in Moscow surely knew this. The infliction of 100 million casualties on the peoples of the Soviet Union would have achieved the satisfaction of revenge, matching the 100 million Americans killed in this "nuclear Pearl Harbor."

Here is one place where the distinction between seeking justice and revenge for its own sake and seeking to "set an example" is very important, since there might not have been enough relevant political leaders around to remember "the example." The United States would indeed have retaliated to achieve some sense of justice, even if there were less other reason to do so. "Everyone knew" that there would be retaliation, and, for this reason if for no other, no such sneak attack ever came.

By contrast, if the rogue proliferators we are addressing here were to utilize WMD, the numbers of Americans killed would be far smaller. In addition to containing our worst-case fears, this also resurrects the more

---

22. For a discussion of why retaliation would allegedly make no sense here, see Philip Green, *Deadly Logic* (Columbus: Ohio State University Press, 1966).

practical motivation for a retaliatory response. In addition to the simple satisfaction of revenge and justice, we would again be "setting an example" for the deterrence of similar acts in the future.

But there is a second, even more important distinction with the years of the Cold War. What if the Soviet Union had decided to wage a *limited* nuclear war, detonating perhaps only three or four nuclear warheads, while holding thousands in reserve? It would have been very difficult for the United States and the rest of the world to respond by pursuing the total elimination of Soviet weapons of mass destruction, or the complete defeat and reform of the Soviet Union.

There surely would have had to be some punitive U.S. response against such a Soviet limited nuclear attack. Moscow could not have been allowed to get away with such a nuclear escalation, lest it feel free to do the same whenever it felt so inclined in the future. Yet the response would have had to be measured, in accord with the dictates of graduated deterrence and intrawar deterrence. It was generally assumed that the United States could never put its adversary into a position of "using or losing" its strategic nuclear forces, the forces that could wipe out the United States.

In contrast, a confrontation with a renegade user of WMD in the next decades will not threaten the national existence of the United States. As in World War II and Desert Storm, there will be more immediate risks for U.S. allies; but, in confronting a seeming lunatic, a state that is already inclined to inflict mass destruction on the world, the main priority will be to disarm that renegade, and disarm it totally.

THE END OF COLD WAR INTRAWAR DETERRENCE

It is easiest to retaliate for a renegade's use of WMD if it has nothing left in its WMD arsenal. The choices are more difficult when the renegade retains even a few additional warheads, and can inflict additional destruction.

While the renegades the United States may face will possess some nuclear warheads or other weapons of mass destruction, they will not have thousands nor even hundreds for a long time to come, and a meaningful counterproliferation program will reduce the destructive potential any such proliferator can deploy. Because they can brandish so much less destructive potential than the former Soviet Union, it would be a poor prescription for policy to project the constraints of the Cold War.

World War II may be much more relevant. What if Hitler or the Japanese had acquired some nuclear warheads at about the same time as the United States, and had used one or two against London or

Chungking? Would the United States have settled for a "containment" of the adversary's forces in such a case, merely pushing the Nazis inside Germany, and the Japanese Army to its home islands? Or would it have demanded "unconditional surrender," as it did? By the Cold War conventional wisdom of limited war, there would have been definite constraints on how the United States could respond to any use of nuclear weapons of mass destruction when the opponent retained any reserve of atomic bombs.

However, I argue that there would be much less restraint on how far the United States felt able to go. A grim analogy is the U.S. and British responses to the early reports of the Nazi death camps.[23] The idea of negotiating with the Nazis to end these operations was rejected out of hand, on the argument that the only acceptable objective was the total defeat of Hitler's regime. Yet millions of people perished as the Nazis continued their program of mass destruction. And a great number of ordinary West and East Europeans perished in the Allied military offensives directed to their liberation. The decision to press for a total defeat of the Axis had taken these losses into account. Thus, whenever an adversary has demonstrated a willingness to inflict horrendous aggressive acts, the threats of *what else* the rogue adversary can inflict until it is defeated should not be allowed to lead to "intrawar deterrence."

The motivation might be simple anger and a thirst for revenge. Or it might be more practical; a Hitler or a Saddam Hussein who had dropped a nuclear bomb could never be trusted not to return to mass destruction, so that disarming the rogue-state would be more relevant than revenge. In any event, the world simply could not tolerate the continuation in power of leaders who had killed so many people using weapons of mass destruction.

If the Axis powers had mounted a nuclear attack, the United States and its allies would have had to push on to Berlin and Tokyo in 1945 so as to decisively eliminate their regimes, even as the few Axis nuclear warheads were still being fired back in retaliation.

UNCONDITIONAL SURRENDER

The issue of constraints during wartime relates closely to the concept of unconditional surrender, an idea that requires our scrutiny.

The Allied demand for unconditional surrender was used by the Nazi and Japanese leaderships to reinforce the fighting will of their forces, on

---

23. The controversial Allied policies toward the Holocaust are given critical attention in Arthur D. Morse, *While Six Million Died* (New York: Random House, 1968).

the basic argument that surrender to the Allies now offered no limits at all to the punishments to be inflicted.

In 1958, Paul Kecskemeti argued that "unconditional surrender" is always a contradiction in terms.[24] If an adversary is to choose to cease fire, it must be promised something in exchange for this, at the least a guarantee that one will also cease shooting, and that the defeated population will not be massacred. Kecskemeti's study detailed how the Italian, German, and Japanese surrenders in World War II, along with the French 1940 surrender to the Germans, all involved some important promises by the winner to the loser.

Kecskemeti and many others drew the logical deduction that the additional retaliation a defeated adversary could still inflict precludes all thought of unconditional surrender. Kecskemeti's argument was received as plausible, in large part because it seemed to fit the mutual assured destruction situation of the Cold War. Since the Soviet nuclear arsenal could destroy the national existence of the United States, and vice versa, neither power could dare to push the other to total defeat, or to ask for unconditional surrender.

But the resources of a renegade like Saddam Hussein would be far smaller than what the Soviets possessed in 1958. The problem with Saddam Hussein will less resemble the Cold War and more the problem with Hitler in World War II, and the problem in law enforcement.

Renegade Use of WMD: The Unconditional Surrender Formula

What should the United States do if weapons of mass destruction are used by Iraq, North Korea, or a similar renegade state in the future?

Two important inferences can be drawn from this chapter. First, rather than being an automatic and unavoidable consequence, the constraining fears of "intrawar deterrence" and "limited war" may *not* take effect if a renegade actually uses WMD and still retains more; it may instead seem too urgent to wrest all these WMD away from the renegade.

Second, rather than being an unreasonable posture and a logical oxymoron, "unconditional surrender" should be the demand of the civilized world after a renegade's acts of mass destruction. A use of WMD could be so outrageous, so much beyond the normal standards of governments around the world, that the offending government would have to be replaced from the top down.

24. Paul Kecskemeti, *Strategic Surrender* (Stanford, Calif.: Stanford University Press, 1958).

If Saddam Hussein had clearly used chemical or biological weapons against UN forces in Desert Storm, the case for advancing all the way to Baghdad might have become overwhelming. And the same logic may apply in the future, if the regimes in Tehran or Pyongyang should use such weapons, or if Iraq should somehow reacquire, and then use, such WMD.

Such a pursuit of unconditional surrender would be driven by one or all three of the major motives for response noted above.

First, perhaps most urgently, it would seem necessary to render the regime incapable of repeating its acts of mass destruction. Military conquest as far as the capital may not be the *only* way to accomplish this; the hope of President George Bush and the rest of the UN coalition in 1991 was that the peace terms imposed on Saddam Hussein at the end of Desert Storm would suffice to deny Iraq any WMD. But a total occupation of the guilty country, and the supplanting of its government, might seem the best way of achieving disarmament.

The most appropriate response to any renegade use of WMD might thus simply be a maximum of disarmament. The United Nations Security Council resolutions on Iraq (when Saddam Hussein had of course not yet used his WMD, but had engaged in blatant conventional aggression) established a useful legal precedent.

There would be a need to strive for an even more rigorous version of the nuclear disarmament imposed on Iraq, going beyond normal International Atomic Energy Agency (IAEA) procedures under the NPT, and for a removal of most or all of the dual-use nuclear and biological and chemical infrastructure that lends itself to weapons of mass destruction. Given the inherent problems posed by such dual-use technologies, an unconditional surrender might be required to put them all under effective control.

The second major focus of a policy of unconditional surrender, after a renegade state's use of WMD, would be as important: to show any similar state around the world that using such weapons leads to punishments that outweigh the gains.

In addition to striving for the same disarmament imposed on Germany and Japan, it will be important to set a precedent. For example, punishing the leaders in a war crimes trial would make a great deal of sense, not so much to reform the regime, which may remain difficult, but to set an example that would be watched by others.

The United States might also consider "counterpower" nuclear targeting—an idea that emerged during the Reagan administration about how best to deter the Soviet leadership. It was argued that the proper targeting objective—one that would *truly* punish the Kremlin leaders, and more importantly deter them from aggression in the first place—was not to kill millions of Soviet citizens, but rather to undermine the Communist Party's control over the Soviet Union.[25]

It is reasonable to assume that Kim Jong-Il or Saddam Hussein or the ayatollahs in Iran also care about power; the best way to punish any use of WMD by such leaders might be to eliminate their regimes. For North Korea, this would mean merging it with South Korea (something long overdue on the merits of the wishes of the Korean people, in any event). For Iraq, this would mean breaking it into pieces, as Kurds in the north and Shi'ites in the south would be allowed to secede, and in Iran it would perhaps mean imposing a secular and democratic government. All of this might be undertaken not so much to produce peace-loving governments, but to confront other potential WMD-wielding renegades with the punishment they fear the most, losing power.

The third possible goal of demanding unconditional surrender would be to reform the regime, perhaps to install a Western-style democracy, or at least to eliminate the militarism or fanaticism that caused such a threat to the world. The lesson of Germany and Japan is that such reform is possible, and that it is accomplished by a total takeover of power, rather than by stopping short of an occupation of the country.

Given how unwilling the North Korean and Iraqi regimes have been to risk facing opposition in any free elections, there is no reason to suppose that someone like Saddam Hussein or Kim Jong-Il really speaks for his people. But the question remains: If one frees the Kurds and Shi'ites, and arrests Saddam Hussein to be tried for war crimes, would the successor regime be easier to deal with? And if the punishment for Pyongyang's use of WMD were the reunification of Korea, would a united Korea renounce nuclear weapons?

Yet one must remember how unsure the United States was in 1945 about the prospects for reforming and democratizing Germany and Japan. When asked in November 1945 whether they expected Germany or Japan to give up ideas of ruling the world (or Asia), 63 percent of U.S. respondents said that Japan would "try again," and 46 percent felt this way about Germany; but, as a sign of a trend, 60 percent had been pessimistic about Germany on the same poll question in July of 1945.[26]

---

25. On "counterpower" theories of nuclear targeting, see Bernard Albert, "Constructive Counterpower," *Orbis*, Vol. 19, No. 3 (Summer 1976), pp. 343–366.

26. Polls for July 27, 1945, cited in *Public Opinion Quarterly*, Vol. 9, No. 3 (Fall 1945), p. 386; polls for November 30, 1945, cited in *Public Opinion Quarterly*, Vol. 9, No. 4 (Winter 1945–46), p. 534.

The question of Germany or Japan trying to conquer the world or Asia is so settled today that no public opinion poll would pose it.

In sum, though the Allies were unsure in 1943 what their policy of unconditional surrender, in response to the mass destruction inflicted by the Axis, was to achieve, in retrospect the policy made sense and was a success.

In face of an act of mass destruction in 2003, demanding an unconditional surrender might again make sense—and again, it might be easier to identify the policy than to specify exactly *in what way* it would succeed. But by one logic or another, an axe-murderer must be denied access to tools; by one logic or another, the rogue user of WMD may have to be forced to surrender.

The primary effect of an aggressor's use of WMD will not be to convince us that we *must* live with this, but instead that we *cannot* live with it. The constraints on political goals that seemed so self-evident during the Cold War would no longer apply, as an actual renegade's use of WMD might require that the perpetrator's entire political system be overturned and reformed.

Chapter 9

Rethinking How Wars Must End: NBC War Termination Issues in the Post–Cold War Era

Brad Roberts

How might a major theater war involving the use of nuclear, biological, or chemical (NBC) weapons end? How *should* it end? How should the United States think about, plan for, and attempt to shape the termination phase of such wars?

These questions, and speculative answers to them, are generally missing from the debate about proliferation, counterproliferation, and the emerging interstate security order. Yet most everyone in that debate accepts that a major theater war involving the use of NBC weapons would be a watershed event, not just for the nations involved and their immediate neighbors, but also for the nonproliferation project and indeed for the world order such as it now exists. This chapter examines the war termination issues associated with NBC regional wars, both to shed some light on the nature of that watershed, and to identify steps that can be taken to reduce the likelihood of such wars or to halt them before they have escalated to their full and horrible potential.

The chapter begins with a critique of the intellectual inheritance on NBC war termination: the Cold War provided a number of points of reference that continue to surface in analysis of future war termination issues, and the Persian Gulf War of 1991–92 provided some additional points of reference.[1] Unfortunately, this inheritance includes a body of analysis largely suited to a strategic era now past and a collection of sentiment and assumption that ought not be mistaken for systematic analysis.

The second section elaborates a more systematic approach to the war termination subject. To predict what perceptions and interests are likely

---

1. For a comprehensive look at Cold War–vintage thinking on war termination, see William T.R. Fox, ed., "How Wars are Ended," *Annals of the U.S. Academy of Political and Social Science*, Vol. 392 (November 1970).

to shape the thinking of decision-makers in the termination phases of a regional NBC war, it is necessary to project how such a war might unfold, examining the set of decisions associated with the use and non-use of NBC weapons, not the structure and character of the larger conflict. The purpose here is to illuminate the factors of significance to the termination phase of a war in which NBC weapons have been used.

The third section postulates alternative outcomes and weighs them for their likely impact on U.S. interests. It argues that asymmetries of interest, power, vulnerability, and leverage will shape outcomes, but that those asymmetries will be fluid in war and that perceptions of them will change as the war unfolds. The fourth section emphasizes the fundamental differences between achieving battlefield victory and winning the postwar peace.

The chapter closes with a review of the policy implications of the analysis. This includes a discussion of the essential military and political ingredients of successful war termination by the United States.

The Intellectual Inheritance

Start a debate on war termination issues, and most U.S. scholars turn immediately to Herman Kahn and Fred Iklé. Writing in the early 1960s, Herman Kahn sought to come to terms with the particular problems associated with the unique dynamics of wars of mass destruction—wars made possible by the advent of the atomic arsenals then being assembled by the major powers. Such wars involved the risk, indeed the seeming certainty, of Armageddon. Such wars became essentially unthinkable from a political perspective and, as Kahn observed, the task of ending such wars was "truly unthinkable."[2] In a world of mutual assured destruction (MAD), wars terminated themselves with the "spasm" that produced global cataclysm.

Although there was more subtlety to Kahn's argument than presented here, the subtleties have fallen away from the way people utilize Kahn's ideas today. Many people subscribe to the view that wars gone nuclear are wars whose ending cannot be conceived. The problem with this way of thinking about contemporary major theater NBC wars is that such wars, even those waged by an aggressor with nuclear weapons, are

---

2. To quote: "Many people consider nuclear war—how it might start, how it might be fought—as unthinkable. But even more, the problem of how such a war might terminate is 'unthinkable'." See Herman Kahn, "Issues of Thermonuclear War Termination," *Annals of the U.S. Academy of Political and Social Science*, Vol. 392 (November 1970), p. 134. See also Kahn, *Thinking About the Unthinkable* (New York: Horizon Press, 1962), and *On Escalation* (New York: Praeger, 1965).

not MAD wars. The assured destruction is *not* mutual. Such wars do not threaten the United States with assured destruction: only the United States (and potentially its coalition partners) would have the capacity to inflict such damage. The regional aggressor may be able to "tear off an arm" by imposing significant costs, military or human, on its opponents, but not to inflict mass destruction. Of course, these costs make avoiding such wars important. This disparity of destructive power implies that the United States would win any such war, unless it chooses not to for some reason. This obliges the United States to think about what outcomes to such wars it can and cannot tolerate, and what price it is willing to pay to secure its preferences.

Writing a decade after Kahn, Fred Iklé sought to draw lessons from the traumatic effort to extricate the nation from the war in Vietnam. In his book *Every War Must End*, Iklé explores the factors other than the balance of military power that shape the final outcome of wars, including, for example, the war's impact on domestic politics and the degree to which outside powers will intervene. His particular focus was limited wars, in contrast to total wars involving the full application of all power resources available to the belligerents.

While Iklé's arguments have continuing relevance to regional wars in which the United States plays a role, weapons of mass destruction feature hardly at all in his analysis. In a new preface to the book written in 1991, Iklé observes that the strategic concepts of the Cold War era cannot be relied upon in an era characterized by challenges by NBC-armed tyrants. He concludes that "a new strategy [is needed] to supersede the concepts we developed for the bipolar confrontation."[3]

Iklé's legacy is a particular way of thinking about the NBC war termination problem that is widespread today among civilian and military analysts in the United States. It holds that any war that "goes nuclear" moves from the realm of the military to the political, and that its termination is strictly "the president's job," as he or she searches with the adversary for points of compromise and accommodation. By this view, how the next war will end is unknowable. It might go well or badly from the point of view of the interests of the United States. But its outcome and termination will depend primarily on the temperament and skill of the president and the president's immediate political advisers.

After Iklé, the next seminal statement in this debate was by General K. Sundarji, the former chief of staff of the Indian military credited with asserting that "the Persian Gulf War showed that if you are going to take

---

3. Fred Charles Iklé, *Every War Must End*, 2nd ed. (New York: Columbia University Press, 1991), p. xv.

on the United States you had better have a nuclear weapon."[4] This assertion implies that in any regional war involving NBC weapons, the regional power will be able to utilize its mass destruction assets to "win." What does this actually entail? Adherents of this view make three basic arguments. First, a basic asymmetry of interests would characterize any such war—interests would be vital for the regional actor and peripheral for the United States—and the regional power should be able to exploit this asymmetry to its advantage. Second, the United States has shown itself to be very sensitive to casualties; it is deterrable by a state capable of driving human costs beyond what the U.S. public will bear. Third, the Persian Gulf War showed the United States to be unbeatable by conventional military means, but vulnerable to nuclear or biological means. These arguments, correct or not, are used to support the expectation that the first major theater war with NBC weapons will end very badly from the point of view of U.S. interests—that the United States will be coerced into acquiescing to aggression and even perhaps backing down from stated security guarantees.

This view of war termination has generated something of a backlash, especially within the U.S. military. Adherents also offer three basic arguments. First, the United States can dominate at every level of escalation initiated by an aggressor. Second, no regional aggressor possesses a nuclear arsenal capable of inflicting assured destruction on the United States—MAD is gone. Third, U.S. superiority implies that an NBC-armed aggressor could not hope to exploit comparatively paltry assets to deter the United States from intervening and staying the course. Some adherents of this way of thinking, expressing their sense of U.S. supremacy, speak loosely of turning the aggressor's capital into a "glass parking lot." These arguments are used to support the expectation that the first major theater war with NBC weapons will end with unconditional surrender by the regional aggressor.

These three different ways of thinking about regional NBC war termination might be deemed "schools." The term suggests a coherence the views lack, but it accurately conveys the way in which different expectations have begun to coalesce around particular views of the lessons of the past, assumptions about the present, and expectations of the future. The three schools certainly imply very different views of the consequences of the first major theater NBC war. Those in the third school ("the United States has all the trump cards") predict a more orderly world after

---

4. This view was expressed to a conference of the Defense Nuclear Agency in June 1993. See *Proceedings,* Defense Nuclear Agency Second Annual Conference on Controlling Arms, Richmond, Virginia, June 1993.

any NBC war, as the "world's only superpower" extends its security to others. Those in the second school ("the aggressor's nuke is the trump card") expect the first major theater war involving NBC weapons to result in a breakdown of a U.S.-dominated global security structure, as allies lose confidence in the United States and adversaries are emboldened to test new limits.[5] Those in the first school ("it's the president's job") see the future as a mystery, as dictated by whim and personality and the domestic politics of the moment.

These schools share one striking feature: they posit war outcomes as driven by certain inevitable forces of power, history, or personality. Outcomes are foreordained. But of course history does not work that way. Certainly, the history of war is full of surprises—of results that were not predicted ahead of time by one or both sides. Wars unfold in a series of choices by the belligerents, choices made on the basis of available military capabilities; what is seen as militarily and politically necessary, viable, or suicidal; and perceptions of what is at stake in each stage. The combatants' sense of honor can play a decisive role in their decisions.[6] By thinking through these factors and their impact on the dynamic of an NBC war, it is possible to get beyond mechanistic predictions of the outcomes of such wars to a better understanding of how the central decision-makers will conceive their interests and stakes and the means to protect them—and how these will play out in the war escalation and termination phases. This is the purpose of the following section.

NBC Wartime Choices: Why, When, and How?

For purposes of this analysis, the focus here is on three primary choices: the aggressor's choice to initiate NBC warfare, the U.S. choice about how to respond, and the aggressor's choice about how to respond to U.S. actions or inaction.

WHY INITIATE NBC WARFARE?
The first choice is the aggressor's: whether or not to initiate the use of NBC weapons. For many in the U.S. strategic community, the choice to use NBC seems baffling or irrational. But such use might appear to an aggressor as logical and necessary. An aggressor might be motivated to initiate the use of NBC weapons for a variety of reasons. It might believe that such use would cripple U.S. political will to prosecute or escalate a

---

5. This argument is elaborated in Stephen Rosen's chapter in this volume.

6. Donald Kagan, *On the Origins of War and the Prescription of Peace* (New York: Anchor Books, 1995).

conflict by inducing great fear among the U.S. public (and among potential coalition partners). It might hope to achieve prompt battlefield victory over a militarily superior United States by using massively destructive weapons at an early stage of a conflict. Or it might conclude that it can employctrl such weapons to create a fait accompli in regions where the United States is not engaged militarily and lacks historical involvement. These three purposes fit under the general rubric of deterring, defeating, or preempting U.S. action.

If we credit the leaders of NBC-armed regional states with both rationality and at least a modicum of strategic calculus, how might an aggressor think about the risks associated with these uses of NBC weapons?[7] Might the leaders convince themselves that the risk-benefit balance favors their interests? For uses aimed at deterring U.S. action, the aggressor must recognize a possibility that the use of NBC weapons to intimidate the United States might backfire. Rather than inducing accommodation and appeasement, such use could generate anger or acts of reprisal, or strengthen U.S. will to permanently expunge the threat. Coalition partners threatened by NBC weapons might well urge the United States to decisive action, if they come to believe that an aggressor has grown intolerably menacing.[8] The taboo against all these weapons remains strong in the countries that an aggressor might target for intimidation. But an aggressor might well conclude that U.S. interests at stake in the region are peripheral, while its own stake is vital; this is the basis of the expectation that the United States will disengage in response to NBC threats and attacks. Moreover, noting U.S. withdrawal from conflicts in places like Somalia and Lebanon once the human costs began to escalate, an aggressor might expect to manipulate U.S. decision-making processes and institutions by threatening large human costs. The head of an insular and authoritarian regime in particular might well believe that a U.S. public made fearful through shows of indiscriminate violence, and the hint of more to come, will back down. Such a leader might conclude that the risks of rousing potential U.S. coalition partners are outweighed by the likelihood that their vulnerabilities will result in efforts by them to exert leverage on Washington to make compromises that seem to prevent

---

7. For a discussion of the connection between psychopathology and deterrence and a historical review of failures of deterrence, see Keith Payne, *Deterrence in the Second Nuclear Age* (Lexington: University Press of Kentucky, 1996), especially chap. 4.

8. For more on this line of argument, see Brad Roberts and Victor Utgoff, "Coalitions Against NBC-Armed Regional Aggressors: How Are They Formed, Maintained, and Led?" *Comparative Strategy*, Vol. 16, No. 3 (July–September 1997), pp. 233–252.

further loss of life. An aggressor might also calculate that the risks of violating the taboo on biological and chemical weapons are outweighed by the advantages such weapons offer over nuclear weapons. These advantages are their ability to kill invisibly, the fear and revulsion they generate, and the possibility that their use for purposes of coercion might be less likely than the use of nuclear weapons to generate undesirable responses by the United States.

An aggressor that opts to use NBC weapons in pursuit of battlefield victory seems likely to go through a similar risk calculus. One risk is that the use of NBC weapons might not derail U.S. military action. Perhaps the aggressor has overstated U.S. military vulnerabilities or has failed to use its own weapons effectively. Another risk is that even militarily crippling attacks may lead not to the desired disengagement of U.S. forces but to heightened political resolve in Washington to see the war through to unconditional surrender by the aggressor. But an aggressor might convince itself that there is at least a reasonable chance that the United States will not try to reverse the aggressor's gains—or that the United States will not threaten the regime in doing so. Or it might believe that the risks of using NBC weapons are less than the risks of not doing so—the failure to try to take the initiative in a looming conflict with the United States may threaten the survival of the regime or compel it to live with a regional status quo that is itself threatening to the regime or is otherwise intolerable.

An aggressor that opts to try to preempt U.S. action would also confront some risks. The primary risk is that the United States sometimes finds its interests engaged by events in regions where any interests had previously been explicitly discounted. U.S. interests are not always clearly perceived or well defined, but catalytic events can illuminate them and induce the United States to act decisively. Washington's decision in 1950 to come to the defense of South Korea shortly after its explicit exclusion from a catalogue of U.S. interests is a case in point. The use of nuclear weapons anywhere could well lead U.S. leaders and the public to believe that major issues were at stake, which the United States could not allow to play out from the sidelines.[9] An aggressor opting to use NBC weapons must judge the United States as lacking both the interests and the incentive to risk massively destructive war to reverse the aggressor's gains.

---

9. As Lawrence Freedman has argued, even in regions remote from U.S. interests and where the United States lacks deployed military forces, an aggressor's use of WMD is likely to create a compelling national interest for the United States, and for the nuclear weapons states generally. See Freedman, "Great Powers, Vital Interests, and Nuclear Weapons," *Survival*, Vol. 34, No. 4 (Winter 1994–1995), pp. 35–52.

Thus an aggressor might well deem the use of NBC weapons to be strategically sound, if risky. Given those risks, an aggressor opting to use those weapons will likely limit their use to the lowest possible level seen to be likely to achieve the desired result without stimulating an overreaction. Or, if the regime anticipates the impossibility of surviving any confrontation with the United States unless the latter is paralyzed militarily and politically, it might develop a go-for-broke attitude.[10]

HOW MIGHT THE UNITED STATES RESPOND?

If the first choice is the aggressor's, the second is that of the United States: to respond or not to the aggressor's use of NBC weapons and if so, by what means. The United States might choose to withdraw its forces and disengage from the conflict. Or it might strike back in reprisal or retaliation. Its war aims might expand. It might opt to limit its military response to conventional weapons, or to employ nuclear ones. Choices among these options are likely to be shaped by many factors. Seven of these factors are discussed here: the level and character of destruction; the potential for continued destruction by the aggressor; the mood of the U.S. public; congressional debate and action; the international political context; military advice; and the U.S. stake.

THE LEVEL AND CHARACTER OF DESTRUCTION INFLICTED BY THE AGGRESSOR WITH ITS USE OF NBC WEAPONS. If the destruction inflicts few casualties, or is inconsequential for U.S. military operations, the United States may opt to prosecute the war without specific reply or reprisal, while offering perhaps a commitment to remove the offending regime. However, if the aggressor's use of NBC weapons generates high casualties or cripples U.S. and coalition military operations, then the United States may feel compelled to use its own nuclear weapons in response. If the destruction were limited to battlefield targets, the United States might opt to limit its own nuclear response to similar targets. If the destruction reaches into the territories or cities of neighboring states or U.S. allies, the United States may feel compelled to inflict greater punishment on the aggressor state. If the destruction reaches into the United States itself, U.S. leaders

---

10. Such a reading of U.S. will is not unknown in history. For example, Japan risked a politically crippling blow at Pearl Harbor, though the United States was far more capable of war mobilization than Japan. States unlikely to prevail in a long-term war sometimes conclude that war can be made and won in the short term with a decisive attack on the national resolve of the larger power. See Barry Wolf, "When the Weak Attack the Strong: Failures of Deterrence," RAND Note N-3261-A (Santa Monica, Calif.: RAND, 1991).

would likely feel compelled to respond immediately, sharply, and decisively.

THE POTENTIAL FOR CONTINUED DESTRUCTION BY THE AGGRESSOR. If the potential for continued destruction is known to be virtually nonexistent because the aggressor has used most or all of a small NBC force in its initial salvo, the United States might press on with existing war plans and aims. But if some NBC potential remains, a counterforce strike is likely to be seen as an urgent necessity. If a limited potential remains, the United States might opt to reply with a limited strike of its own—limited either in the types and numbers of targets struck or in the weapons chosen. But if such a limited strike is seen as unlikely to deprive the aggressor of the ability to inflict significant damage, perhaps for extended periods of time and beyond the theater of combat, then such restraint may be seen as unwise. Instead, the United States might opt to deliver as hard a blow as possible, perhaps even with nuclear weapons, in the hope that this would eliminate the aggressor's potential for continued destruction. Even if such a strike does not promise complete success, it may be seen as useful for various purposes: breaking the aggressor's will to continue to use NBC weaponry, punishing the first use, demonstrating U.S. willingness to escalate, and reassuring coalition partners that the war will end quickly and with defeat for the aggressor. Such a strike would also be useful for impairing the aggressor's NBC capabilities to the point where active and passive U.S. defenses deprive them of any meaningful operational impact.[11] However, if the end of the confrontation seems sufficiently close, the United States might refrain from reprisal, acting instead to seek a political settlement that induces the aggressor not to use its reserve capability.

THE MOOD OF THE U.S. PUBLIC. The U.S. public's sensitivity to casualties is oft remarked, as is the role of the media in magnifying the impact of casualties on the national mood. The U.S. public may well cringe and shrink in fear after an aggressor's use of NBC weapons, but its sensitivity is often misunderstood. Commentators who expect the U.S. public to cringe in fear as casualties mount seem to take little cognizance of the difference between peripheral and vital interests in the reactions of the U.S.

---

11. For a discussion of the utility of such defenses in influencing the outcome of wars against WMD-armed regional adversaries, see Victor Utgoff and Johnathan Wallis, "Major Regional Contingencies Against States Armed with Nuclear, Biological, and Chemical Weapons: Rising Above Deterrence" (Alexandria, Va.: Institute for Defense Analyses, June 1996).

public and the historically well-demonstrated willingness of the public to suffer casualties in the name of a significant national interest.[12]

Similarly, the antinuclear sentiment prevalent today in the United States suggests that the president could face strong pressures not to use U.S. nuclear weapons except in response to attacks on targets in the United States itself. But once an aggressor breaks the nuclear taboo, and especially if large numbers of U.S. casualties result, an explosion of outrage and anger would likely ensue—especially when television brings home pictures of U.S. victims. The result could be entirely unanticipated by those who make predictions based on the U.S. record in Somalia and Lebanon. Herman Kahn has written of the historic tendencies of a U.S. public inclined to a messianic worldview to apply "extravagant force" when roused to action.[13] Rage could combine with moral fervor to produce a U.S. military campaign to expunge an aggressor seen in the United States as evil, a campaign that would draw on all available military capabilities to seek the most immediate and decisive end to the war at hand and a permanent removal of any threat of future war by this regime.

Moreover, because appeasement has a particularly pejorative meaning in the United States (where it is associated with cowardice, not compromise), the president would likely face strong pressure not to back down from the defense of U.S. interests in the face of threats from what the public might view as a "nuclear pygmy." Nuclear attack on U.S. forces could well produce strong popular pressure for a nuclear reply by the United States, one aimed not just at reprisal or escalation containment, but at immediately ending the conflict and vanquishing the foe. It is useful to recall the private intervention with Saddam Hussein by Japanese Prime Minister Yasuhiro Nakasone, as reported in the press at the time. After U.S. Secretary of State James Baker delivered the famous letter to Iraqi Deputy Prime Minister Tariq Aziz threatening terrible consequences for certain Iraqi actions if war came, Nakasone reportedly sought out Hussein to attest to the fact that the United States is the kind of country that would make good on its promise of retaliation and retribution—as Japan knew better than most.

---

12. See Eric V. Larson, *Casualties and Consensus: The Historical Role of Casualties in Domestic Support for U.S. Military Operations*, RAND Report MR-726-RC (Santa Monica, Calif.: RAND, 1996); Benjamin C. Schwarz, *Casualties, Public Opinion, and U.S. Military Intervention: Implications for U.S. Regional Deterrence Strategies*, RAND Report MR-431-A/AF (Santa Monica, Calif.: RAND, 1994); and Mark Lorell and Charles Kelley, *Casualties, Public Opinion, and Presidential Policy During the Vietnam War*, RAND Report R-3060-AF (Santa Monica, Calif.: RAND, March 1985).

13. Kahn, *On Escalation*, p. 17.

CONGRESSIONAL DEBATE AND ACTION. Members of Congress would certainly have strong views about what to do in a nuclear crisis, views that would have weight not least because of the congressional power of the purse and its ability to shape public opinion. Sharply different views could be present—for every voice of compromise or retreat, there could be others shouting "don't tread on me" and calling for decisive U.S. action. Or Congress might rise up in near unanimity to urge a particular course of action upon the president. Whether Congress would clarify or obfuscate the choices before the United States and its national command authority is an open question. The answer would have much to do with the character of leadership to be found among key committees, their chairpersons, and the political parties. The skill of the executive branch and especially the president in shaping the legislative debate would be critical. The result might be the building of consensus from dissension, as seen in the lead-up to Desert Storm. Or consensus may not be achieved. The president would value consensus, which would strengthen the administration's political hand, but dissension could preserve the freedom to maneuver.

THE INTERNATIONAL POLITICAL CONTEXT. If the United States has assembled an international coalition to face an NBC-armed aggressor, its basic strategic choices would have to be made with the coalition in mind. This might strengthen the tendency to back down in a confrontation, if the coalition is held in check by its most cautious members; or a sense of acute vulnerability among coalition members could stimulate calls for immediate, decisive action. The United States might find that its flexibility to choose among options is limited by its leadership of a coalition in which terms of engagement have been well negotiated as a price of coalition formation, or it might feel less constrained than when acting alone if the international legitimacy offered by the coalition works to expand its options. Deciding how to respond to an aggressor's use of NBC weapons would be especially complex when other coalition members have the capacity to respond independently with NBC attacks of their own. If the coalition were to include states willing and able to use nuclear, biological, or chemical weapons of their own, the United States would find itself in the difficult position of needing to act to forestall or perhaps condone such use at the same time that the weapons of mass destruction (WMD) capabilities of the aggressor are a focus of international fear and condemnation. As in the Persian Gulf War, when the United States persuaded Israel not to retaliate for Iraqi missile strikes on its cities, it seems likely that the United States would try to maintain the maximum degree of control over both the local dynamics of the conflict and the international politics of the confrontation. Furthermore, in the capitals of allied and coalition nations, concern about the durability of U.S. security guarantees and about U.S. credibility more generally would add to the brew. Even in wars where the United States finds itself fighting without a coalition, many of these concerns would nevertheless still be likely to affect the United States as it calculates the long-term consequences of its choices on U.S. alliance relations and its more general reputation and political standing.[14]

MILITARY ADVICE. The influence of a president's military advisers peaks in time of military crisis. Particularly on the issue of a possible U.S. nuclear reply, the precise nature of military advice is difficult to predict. The disposition against nuclear weapons is strong in the U.S. military today, reflecting not least its confidence in its ability to carry out its missions without resort to weapons of mass destruction.[15] But the military's antinuclearism might perish once U.S. forces or interests are attacked with weapons generating mass casualties. In reply, U.S. nuclear use might be deemed necessary and appropriate to punish the enemy, to prevent its further use of NBC weapons, to collapse its military apparatus to prevent further aggression, or to save lives, whether U.S. or allied. Some military advisers might argue that the restoration of the WMD taboo *requires* U.S. nuclear retaliation to punish the aggressor and to substantiate the notion that nuclear weapons are not useful as instruments of aggression or even as ultimate guarantors of the sovereignty of aggressive regimes.

THE U.S. STAKE. The point is not trite. It seems reasonable to assume that the United States would not find itself in confrontations that might involve NBC weapons except in instances of compelling national interest. But, as many have noted, U.S. history is replete with episodes in which, as casualties mount, its level of military engagement has been deemed inappropriate for the stake involved, with a consequent U.S. withdrawal and loss of face (e.g., from Lebanon and Somalia). Thus, the United States

---

14. For more on this theme, see Roberts and Utgoff, "Coalitions Against NBC-Armed Regional Aggressors."

15. This antinuclear disposition has been identified by RAND analysts in follow-up to the "Day After . . . " study of nuclear futures. See Marc Dean Millot, "Facing the Emerging Reality of Regional Nuclear Adversaries," *Washington Quarterly*, Vol. 17, No. 3 (Summer 1994), pp. 41–71; and Roger C. Molander and Peter A. Wilson, "On Dealing with the Prospect of Nuclear Chaos," *Washington Quarterly*, Vol. 17, No. 3 (Summer 1994), pp. 19–39.

might somehow be drawn into an NBC confrontation over issues not of central importance to it. Such a confrontation would likely present U.S. leaders with an unpalatable choice between risking an unpopular war fought with unpopular weapons and trying to cut its losses.

Of course, the NBC component may reveal U.S. interests related to perceptions of its power and credibility that neither it nor the aggressor had conceived. Those interests could lead the United States to take actions not expected by an aggressor that believes the U.S. stake in the conflict to be low. Even if the aggressor only threatens to use NBC weapons, compelling interests could well be generated for the United States. Washington would likely feel that a perception that the United States had backed down in the face of NBC threats from a regional aggressor would have terrible consequences for the United States, both politically and militarily.

LIKELY U.S. PREFERENCES. These seven factors, and others, as magnified by the media and played out through the personalities of the national command authority, would shape the choices made by the United States. Given the variety and contingent character of these factors, simple predictions of U.S. war-time choices, whether by U.S. analysts or by leaders in an aggressor state, seem foolhardy. But likely preferences can be illustrated.

A decision by the United States to back down or withdraw from a confrontation when an aggressor has used nuclear weapons seems unlikely. The president, the U.S. public, and the international community would have to see little benefit in punishing the aggressor. Key constituencies would have to be deeply divided, and the stake widely seen as disproportionately low compared to the costs of continued action. This conclusion seems valid even for the mere threat of use. Many constituencies would likely feel that it was very important not to allow the precedent to stand. Moreover, the president would have to choose to allow blackmail by an aggressor, and it is far from clear that the domestic or international political context would permit this to happen.

A decision to reply with military strikes aimed at annihilating the aggressor state and society is only slightly less unlikely. Key domestic and foreign constituencies might be gripped by rage or fear, and thus might seek such a reply to punish the act of nuclear aggression and remove the threat of continued aggression. Decision-makers might be deluged with pleas to seek an immediate cessation of the conflict in order to protect the broader interests of the United States and the international community. But the president would likely also be cautioned to act with restraint, and could opt to use U.S. force cautiously in the hope that a settlement might become possible without U.S. actions that cause enormous additional casualties.

A decision to ignore an aggressor's nuclear use and to continue the military confrontation on previously existing terms appears likely in certain circumstances. The victims of the first attack must be few, the aggressor's capacity for continued nuclear use must be low, the demands for retribution must be few, and a means of sufficiently expeditious military victory at minimal human cost must be available.

A variant of this option would be the continued utilization of conventional weaponry in pursuit of expanded war aims. All wars unfold within the context of an "agreed battle" by which adversaries accept what is and is not at stake; changes to the "agreed battle" by expanding it to additional fronts, whether military, political, or economic ones, might be alternatives to a U.S. nuclear reply.[16] Such strategies may be successful in imposing new costs upon the aggressor sufficient to secure an outcome acceptable to the United States. But such expansion may not be feasible. If an interventionary force had already thrown its full resources into the conventional battle it would have no further units to deploy, at least in a timely fashion, no new fronts to open, and no new targets to attack. Expansion of war aims to include removal of the regime in power might offer some leverage. But some regimes may already believe that their survival is at stake in a war with the United States, no matter what the U.S. declaratory policy. Moreover, by spelling out plans to remove the regime, the United States may well remove the last barriers of restraint—a regime with nothing to lose may go for broke.

A decision to meet an aggressor's use of nuclear weapons with a limited nuclear reply by the United States seems at least as likely as these last two replies. Especially if victims are numerous, the aggressor's capacity for continued nuclear use remains, publics are angry and vulnerable, and the political groundwork has been done for a concerted coalition response, such limited use appears likely.

The use of nuclear weapons would likely be opposed by those who believe that conventional weapons can do the necessary reprisal, retaliation, and defeat. However, while conventional weapons have taken on many of the strategic targets traditionally assigned to nuclear weapons, and may yet take on more, the speed with which they can achieve U.S. war aims is considerably less than that of nuclear weapons. Carrying on the battle by conventional means alone could come at a high cost, if the aggressor can continue to inflict mass casualties on the forces and popu-

---

16. "Agreed battle" is Max Singer's term, as discussed in Kahn, *On Escalation*, pp. 4–8.

lations of coalition members before the coalition beats its forces on the battlefield.

If the conventional option seems likely to be expensive, the president would face a difficult choice: employ nuclear weapons, hoping that this will reduce the casualties suffered by the coalition, even if their employment might cause many casualties within the aggressor state; or refrain from their use, hoping that overall casualties will be kept to a minimum, even if a disproportionate share is borne by the coalition members. The president would no doubt be advised to do everything possible to protect U.S. forces and to minimize the casualties borne by the regional victims of aggression, even while tallying the possible political costs associated with U.S. nuclear use. Of course, the United States made such a choice in 1945. Faced with a clear choice between using nuclear weapons to try to end a war quickly and relying on slower but still certain non-nuclear means, a U.S. president would probably not want to risk the sacrifice of further U.S. lives, or lives of those within the region already victimized by aggression, to resurrect a non-nuclear taboo that the enemy had already broken.

These predictions are all based on the aggressor's use of nuclear weapons. What about biological and chemical weapons? Can they be expected to shape U.S. perceptions and choices similarly?

Chemical weapons seem unlikely to produce the same effects. Generating large casualties against well-protected forces takes a great deal of sustained effort by a chemically armed aggressor. Moreover, chemical weapons do not have anything like the political symbolism of nuclear weapons. But it is also true that chemical weapons can be used to massively destructive effect. An aggressor equipped with the means to efficiently deliver a chemical agent on unprotected population centers or military targets of campaign significance such as ports, airfields, and other logistic centers could significantly disrupt an intervention force, especially in the early stages of deployment (when such weapons might generate relatively few casualties). Furthermore, a very strong taboo is attached to such weapons in the mind of the U.S. public, which might well feel that strong punishment of such use would be necessary. So a U.S. nuclear reply to an aggressor's use of chemical weapons cannot be ruled out entirely.

Biological weapons seem likely to be closer to nuclear than chemical weapons in their effects on U.S. choices. Their potential for mass destruction of unprotected populations is indisputable, as is their utility in attacks on unprotected military targets of campaign significance. Washington would expect an aggressor that uses them once to use them again, and even the most potent preemptive counterforce attacks are unlikely to remove the fear that the aggressor might be able to inflict significant continued destruction with such weapons. The U.S. public could be particularly enraged by the use of biological weapons, given the deep abhorrence for them. Internationally, there would likely be many calls to reply to and punish such use in order to demonstrate to all states with biolgical weapons programs that such weapons cannot be used success-fully and will bring down the full wrath of the international community. Indeed, the United States will have a very strong interest in ensuring that the "right lesson" is drawn from the first major biological confrontation, a lesson that leads away from and not toward such weapons. On the other hand, biological attacks that generate few casualties and marginal mili-tary or political consequences would be less certain than even limited nuclear attacks of stimulating strong U.S. responses.

In considering possible U.S. responses to an aggressor's use of chemi-cal or biological weapons, due consideration must be given the fact that the United States has forsworn both chemical and biological weapons. Accordingly, it no longer has an option to reply "in kind" to attack on U.S. forces with such weapons.[17] The absence of an in-kind capability may be inconsistent with the desire for maximum U.S. flexibility, but the U.S. military sees no advantage in making in-kind chemical or biological retaliatory strikes to first-use by an enemy, preferring instead a mix of other capabilities, including self-protection, defenses against delivery sys-tems, conventional preponderance, and nuclear threats. A dilemma arises when these latter capabilities might prove inadequate to meet the require-ments created by the seven factors enumerated here. For example, if an aggressor's use of chemical or biological weapons generates huge casu-alties, cripples U.S. or coalition military operations, enrages public and political sentiment, and if the potential for continued destruction is sig-nificant, the U.S. command authority seems likely to consider a nuclear reply. It would then face a clear choice between continuing the war by conventional means, thus suffering large numbers of casualties, and in-itiating nuclear use aimed at bringing that risk to an end.

Thus the United States (and the coalition) could face decisions about the propriety of making the first use of nuclear weapons, albeit not the

---

17. Biological weapons were renounced unilaterally in 1969 as a prelude to a bilateral U.S.-Soviet biological weapons disarmament agreement and ultimately the multilat-eral Biological and Toxin Weapons Convention of 1975. Production ceased and the U.S. stockpile was destroyed by order of the Nixon White House. Chemical weapons were renounced in 1992 in the endgame to the negotiation of the Chemical Weapons Convention. A residual stockpile of chemical munitions remains in the U.S. arsenal but is slowly being destroyed.

first use of weapons of mass destruction. If the aggressor state is a party to the Nuclear Non-Proliferation Treaty (NPT) and is considered not to be in violation of that treaty, U.S. choices would be constrained by the assurances it has made to NPT parties. But questions of legality and of justice may prove separable. Legally, the United States is obliged not to use nuclear weapons against states that are party to the NPT (except for those aligned with a nuclear weapons state).[18] But the failure to use nuclear weapons may be deemed unjust in time of war, if the legal stricture comes to be seen as the only barrier to a quick end to a war of aggression—especially if the aggressor is using other banned weapons to inflict mass casualties on its neighbors and on the members of the coalition formed to reverse the aggression. Both biological and chemical weapons are the subject of a global arms control regime. Of course, if the state is not a party to the NPT or is in violation, the legal constraint disappears, but the presumption against U.S. nuclear use in reply to non-nuclear challenges would remain.

Just as the aggressor's choices are not risk-free, the United States would also face risks as it calculates its choices. It may miscalculate the ability of the aggressor to survive a limited nuclear attack and to continue to inflict damage—or to escalate that damage to the U.S. populace or sustain that damage with acts of vengeance against U.S. leaders or the U.S. public in the months or years after battlefield confrontation ends. The United States may misread the intent or strategic personality of the aggressor, only to find that actions intended to deter escalation instead raise its ambitions. An NBC-armed aggressor with nothing to lose may become even more dangerous, abandoning a strategy of strategic bargaining aimed at escaping war in favor of a strategy of punishment and retribution in the final stage of defeat—or even after losing the war. Biological weapons may be seen as particularly attractive and effective for this purpose. Indeed, the fear that an aggressor might use biological agents surreptitiously after the war could emerge as an important factor in Washington's calculation of an adversary's capacity to inflict continued destruction. It is instructive to recall that the United Nations Special Commission on Iraq (UNSCOM) discovered that Saddam Hussein dispersed biological weapons from their storage facilities as the war began,

---

18. See "Treaty on the Non-Proliferation of Nuclear Weapons," in *Arms Control and Disarmament Agreements: Texts and Histories of the Negotiations* (Washington, D.C.: Arms Control and Disarmament Agency, 1990), pp. 89–102. See also Michael Wheeler, *Positive and Negative Security Assurances*, Paper No. 9, Project on Rethinking Arms Control, Center for International and Security Studies at Maryland (College Park: School of Public Affairs, University of Maryland at College Park, February 1994).

with orders that they be launched against coalition military forces and neighboring states in the event that the coalition destroyed Baghdad.[19]

THE AGGRESSOR'S RESPONSE

Once an aggressor uses NBC weapons and the United States makes its response, what then? Would the aggressor opt to continue the fight with weapons of mass destruction, accept defeat, or something in between?

An aggressor's decision to continue the WMD conflict would likely surprise those in the United States who believe in the dissuasive effect of the overwhelming escalatory potential of the U.S. arsenal. But what might an aggressor that calculates risks and tries to shape the flow of events think? How might it perceive its options? If the United States had chosen not to respond with nuclear weapons to the original attack, the aggressor may believe that U.S. weapons would be used only when the very survival of the nation is threatened, and thus may believe that it is free to operate with NBC weapons below that threshold—even to the point of inflicting massive destruction on its immediate neighbors. If the United States has responded with nuclear weapons in a limited way, the aggressor may perceive that judicious continued use, calculated not to excite stronger U.S. responses, may be possible. It may believe that calibrating its WMD attacks to the U.S. action by not expanding target sets or levels of destruction might stabilize the conflict while also giving the United States strong incentive to seek a compromise solution. Moreover, however rational its calculations of costs and benefits of various options, the leader may face keen pressures from within his or her regime or elsewhere to strike a decisive, last-ditch blow at the United States, calculating that only the collapse of U.S. will might allow the regime to survive.

The risks for the aggressor are obvious. Continuation of an NBC conflict with the United States would probably remove any lingering opposition in Washington and elsewhere to a strategy aimed at unconditional surrender by the aggressor and removal of the regime in power (if these were not already coalition war aims).[20] An aggressor's continued use of NBC weapons after a U.S. nuclear reply would certainly discredit those within the U.S. decision-making process who had argued for restraint in the U.S. reply, in the hope that an aggressor might reciprocate.

---

19. "Report of the Secretary-General on the status of the implementation of the Special Commission's plan for the ongoing monitoring and verification of Iraq's compliance with relevant parts of section C of Security Council resolution 687 (1991)," United Nations Reference S/1995/864, October 11, 1995.

20. For more on this line of argument, see George Quester's chapter in this volume.

Remaining constraints on U.S. actions would rapidly erode, leading to U.S. efforts to promptly and decisively end the war. At this point in a conflict, the regime's control of its military and society could begin to break down. A leadership core committed to a war of national suicide may find that the military prefers another course, perhaps to fight another day. It may find that the public turns against it, especially elites that heretofore had supported the regime out of a commitment to nationalism and personal status. Loss of control by the aggressive regime would likely create opportunities for negotiated settlements. The fear of a loss of control might well be an important incentive for the regime not to choose to prolong an NBC exchange with the United States. Moreover, even if a state somehow succeeded in inflicting defeat on the United States within the theater, it would have to contemplate the possibility that other states, including nuclear-armed ones, would not stand idly by, given their own stake in an international system in which the United States is the locus of initiative and the provider of security guarantees.

The Military Outcomes of NBC Wars

What does this analytical framework contribute to an understanding of NBC war termination issues? At the very least, it reveals a greater range of outcomes than one-side-takes-all. It is possible to envisage at least six different outcomes of any regional war in which NBC weapons have been used.

First, the aggressor might realize all goals. This might happen if the United States refrains from using its nuclear weapons after an aggressor uses NBC weapons to deny the United States or a U.S.-led coalition military victory, or uses those weapons to attack cities of coalition members, including perhaps the United States, successfully breaking the will of the coalition to proceed.

Second, the aggressor might achieve some goals and retain some spoils of war, but not its full original ambition. This could happen if the aggressor's use of NBC weapons has limited effects—whether on the battlefield, to stalemate or defeat an intervention by U.S. forces or by a U.S.-led coalition, or politically, in hobbling or crushing that coalition. Nuclear use by the United States is unlikely to be a part of this outcome.

Third, the status quo ante might be restored. WMD use by one or both sides produces stalemate on the battlefield and politically. The intervening powers resort to long-term political strategy rather than military solutions to contain further aggression and punish past behavior.

Thus, the aggression is overturned but the aggressor pays no substantial price.

Fourth, the aggression could be overturned with the aggressor paying a substantial price, perhaps the loss of the invasion force, but the regime remains in power. This might happen if the aggressor's use of NBC weapons fails to deter the formation of a coalition or defeat its efforts to reverse the original aggression. In this scenario, the United States would make limited or no use of its nuclear weapons.

Fifth, the aggression might be overturned and the aggressor regime removed from power in a war that the United States opts to pursue by conventional means alone, and in which its forces and allies accordingly suffer substantial WMD-inflicted casualties.

Finally, the aggressor might be defeated and removed from power in a war where both sides use weapons of mass destruction, perhaps substantially.

Which outcomes are most likely? Which least likely? Why? At this point, it is useful to return to the schools of thought sketched at the beginning of this chapter. The view that aggressors will win because their NBC assets are a trump card embodies an important truth. There would be political pressures to end the war before it escalates, or before a second exchange of weapons of mass destruction, and thus to find points of compromise. The undisputed U.S. capacity to inflict decisive military defeat on a regional aggressor combined with the public rage and urge to inflict punishment on it will be juxtaposed with the need to keep a war from getting out of hand. The high human costs associated with all outcomes, even outright U.S. victory, and the uncertain politics of nuclear use, suggest that the United States will be constrained in defining its responses to NBC use and will be pressured to accept outcomes that leave some major issues poorly resolved.

But this school overlooks the many U.S. trump cards, and the fact that it would be likely to enjoy more political and military latitude in meeting the dilemmas of conflict than an aggressor. An aggressor's moment of maximum influence will probably be its first hints at the possible use of NBC weapons, when the fear can be reaped for political concessions. But once the aggressor engages the United States in direct military confrontation with weapons of mass destruction, its war aims are likely to shift from gaining something to survival. The weapons will be risky to use because of the response they may generate. A coup may remove the regime that initiated the aggression. Or that regime may struggle to maintain a domestic power base while abandoning the interests of territory or prestige that seem to be leading the state to a war of national suicide. An uncertain period could come if a leader "turns toward peace"—when it has eschewed victory but has not accepted defeat, and decides on a strategy that maximizes its chances of surviving to fight another day.[21]

Thus the school of thought that the United States will win no matter what also embodies an important truth. The United States would enjoy many advantages of political position and military flexibility and reach over a regional aggressor equipped with a small arsenal of NBC weapons. It did not enjoy such advantages when it faced the Soviet Union, which was armed with what were seen as superior conventional forces and a capacity to inflict assured nuclear destruction on the United States. Nuclear choices by the United States can shape war outcomes in ways that a regional aggressor's cannot.

But if regional NBC war seems somehow easier than a MAD war, it is likely to prove neither easy nor cost-free. Aggressors' options would be far more limited than those of the former Soviet Union, but they would have options nonetheless: to use weapons of mass destruction, perhaps to continue to do so as a conflict evolves, and to target civilians in or beyond the zone of combat. The U.S. ability to prevail at any level of confrontation chosen by the aggressor does not mean that prevailing will be easy, as there will be certain inescapable costs, human and other. Furthermore, the U.S. ability to control an unfolding conflict depends on its military preparations to minimize the leverage of potential opponents or political work to create durable coalitions, actions that it may prove unable to execute satisfactorily.

Basic asymmetries in power, vulnerability, and thus leverage point to the conclusion that how NBC wars will end depends fundamentally on how the United States wants them to end. What conditions might the United States find necessary in defining acceptable war outcomes? At a minimum, the United States would want to extract something from the war termination process that it can call a "win." Stalemates would likely be unacceptable to senior policymakers, who would be concerned that the aggressor and its use of NBC weapons be punished so as not to encourage future aggression, particularly aggression backed by weapons of mass destruction. If the aggressor's use of NBC weapons is seen as more incidental than egregious, the United States might settle for an outcome that reliably eliminates the weapons potential of the state. But U.S. demands would escalate as the damage increases, and the United

---

21. The term is from Fox, "Causes of Peace and Conditions of War," p. 6. See also Iklé, *Every War Must End*, especially chaps. 4 and 5.

States and especially coalition partners neighboring the aggressor would probably also push for the removal of the regime and the reform of the state.

Control of the regime, disarmament of the military, and reform of the state are not modest goals. They are obviously more than battlefield victory: they constitute unconditional surrender. Such surrender is of course well known in history but, in the nuclear era, pushing for unconditional surrender of an NBC-armed opponent has so far been seen as impossible. During the Cold War, the realities of mutual assured destruction and the suicidal nature of waging total war ruled out any possibility of seeking unconditional surrender by the Soviets in any possible nuclear war. But as George Quester argues in this volume, in wars where assured destruction of the United States is not a possibility, it is possible to contemplate unconditional surrender by an enemy equipped with weapons of mass destruction. The United States and the international community should be able to pursue the elimination of regimes whose behavior transgresses global norms and whose continued existence threatens both its neighbors and those norms. They have done so previously in defeating fascism.

To be sure, a skillful aggressor would attempt to steer conflicts in ways that minimize U.S. advantages and exploit U.S. vulnerabilities. But if the foregoing review of likely war outcomes is on the mark, the aggressor seems relatively unlikely to be able to force outcomes on the United States that are distinctly contrary to core U.S. national interests. Outcomes involving some compromise by the United States appear possible, but those inflicting major defeat on the United States are remote possibilities.

Winning the War versus Winning the Peace

The argument to this point implies that successful war termination by the United States requires the exploitation of U.S. advantages in power and leverage. But this is only a part of the puzzle. Fred Iklé again offers a useful perspective, noting

the intellectual difficulty of connecting military plans with their ultimate purpose. Battles and campaigns are amenable to analysis as rather self-contained contests of military power and, to some extent, are predictable on the strength of rigorous calculation. By contrast, the final outcome of wars depends on a much wider range of factors, many of them highly elusive . . . . In part, governments tend to lose sight of the ending of wars and the nation's interests that lie beyond it, precisely because fighting a war is an effort of such vast magnitude. Thus it can happen that military men, while skillfully planning their intricate operations and coordinating complicated maneuvers, remain curiously blind in failing to perceive that it is the outcome of the war, not the outcome of the campaigns within it, that determines how well their plans serve the nation's interests.[22]

There are important differences between defeating enemies and winning wars. The latter requires not just battlefield victory but a winning of the peace that follows. Especially in wars involving weapons of mass destruction, this seems inescapable. As Herman Kahn noted decades ago, even in a war forced upon it by a malevolent and aggressive regime, the United States will be held to a higher political standard than the aggressor.[23] After any war involving weapons of mass destruction, the United States would likely find itself subjected to a type of scrutiny the aggressor would not. This is the price of its status as the more powerful actor with the stronger obligation—and historical claim—to acting in defense of common interests and to using force only by and for agreed principles. As Kahn argued, if the aggression is not punished, if the level of destruction is seen as disproportionate to the stake, if the coalition disintegrates, or if the United States is seen to have acted rashly and in anger, the United States will be blamed—whatever happens to the aggressor. Would U.S. actions be seen as both wise and just by the international community—and by the U.S. public? If the United States fails on either score, it could lose the peace that follows the war.

Kahn could not have anticipated the particular historical epoch in which we now consider this argument. Writing in the 1950s, he was describing a United States leading a new alliance of nations against an aggressive and nuclearizing Soviet Union. But bipolarity has given way to unipolarity—to the moment in history following the disintegration of the Soviet Union when the United States finds itself cast as the world's premier military power, as the only power capable of mobilizing major international institutions such as the United Nations for peace and security purposes, yet somehow less than the superpower it once was with the emergence of other centers and sources of power on the world scene.[24] Today, some fear a United States whose power is unfettered by

22. Iklé, *Every War Must End*, pp. 1–2.

23. Fox, "The Causes of Peace and Conditions of War," p. 10.

24. Charles Krauthammer has written of the United States in its unipolar moment, in which its power is unchallenged but its world role is uncertain. Krauthammer, "The Unipolar Moment," *Foreign Affairs*, Vol. 70, No. 1 (1991), pp. 23–33.

that of a rival peer.[25] Others fear a United States that will give in to its historic temptations of isolationism and withdraw from the world stage, to reappear impulsively and unreliably, if at all. Still others look to the United States as the only country capable of leading the international community and sustaining existing world order institutions.

Thus to win the peace after a major theater NBC war, Washington must ensure that the record of that conflict sends the proper signals about the United States. Put succinctly, it will have to make choices and shape outcomes in ways that avoid the casting of the United States as a nuclear bully or as a cowardly wimp—and that succeed in casting it as a responsible steward of a just world order.

The United States would be cast as a nuclear bully if it were seen to have stumbled unwittingly into nuclear confrontation, allowed itself to be overcome by rage and fear, and then to have used its weapons impulsively and in ways that look heavy-handed or imperious, perhaps imposing destruction later seen as disproportionate to the issues at stake. The "unipolar moment" heightens the risk that U.S. nuclear use in a regional war might be interpreted this way, whatever the aggressiveness and malevolence of the offending state. Those who believe that the United States will grow increasingly arrogant in its use of power and capricious in its world role would interpret any U.S. nuclear use that seemed unduly harsh or overbearing as confirming their expectation. A reputation as a nuclear bully would certainly erode the willingness of other states to side with the United States on international political issues or security concerns. The United States would have to expect a diminution of its standing and a relative decline of its power as other states, whether friendly or not, seek to arm themselves with their own strategic weapons.

Legitimacy won or lost abroad would likely have an impact at home as well, given the U.S. public's sensitivity to how others judge national behavior. However just or necessary U.S. nuclear use might be seen in the White House, the event would likely unleash fluid, chaotic, and contradictory emotions at home—from righteousness to shame, from relief to guilt. The result could be highly unsettling within the U.S. body politic (and damaging to the president's political prospects). The process of coming to terms with U.S. actions could well occupy the nation for years to come, as it has after the Vietnam war.

If the United States comes to be seen abroad and domestically as a nuclear bully, and if U.S. power is delegitimized, the results may prove far-reaching. Herman Kahn anticipated a deep public revulsion toward

---

25. For a discussion of this perspective, see Shahram Chubin, "The South and the New World Order," *Washington Quarterly*, Vol. 16, No. 4 (Autumn 1993), pp. 87–108.

old ways of doing business in the wake of superpower thermonuclear exchange and a surge of opinion that such wars should "never again" be allowed to happen; he believed this would lead to strong pressures for world government.[26] A limited nuclear war might produce a similar revulsion but different political impulses. Abroad, U.S. leadership might meet growing resistance, as all things U.S., from the military to business and cultural institutions, fall into disfavor. At home, the impulse to isolationism might rebound. At the very least, if the United States were to come to be seen as a bully, strong pressures would likely build to counterbalance or roll back U.S. military advantages.

It would be ironic indeed if in prosecuting a war against a "rogue state," the United States were itself to be seen as a belligerent nation, as a state feared because of its nuclear weapons and seemingly driven by a messianic worldview colored by self-delusion and ambition. This outcome seems highly unlikely, but its very possibility is a cautionary reminder that U.S. power and influence rest as much—if not more—on its reputation as a benign power as on its singular military status.

What about the obverse of this problem? How might the United States come to be seen as cowardly wimp? Such a perception might result if the United States were seen to have refrained from using nuclear weapons for fear of a domestic or international political backlash, and therefore compromised a significant national or international interest to appease, in the most venal sense, the ambitions of the aggressor. The possibility that U.S. non-use would be perceived in this way has arguably increased; the passing of the Soviet threat has led some to expect that the United States will again retreat from the world stage. U.S. non-use in a situation where nuclear use was widely seen as necessary and legitimate could fuel this expectation.

The cowardly wimp reputation would also have far-reaching negative consequences among allies, adversaries, and citizens of the United States. Among allies it would raise basic questions about the value of U.S. security guarantees and the reliability of U.S. power in defense of an array of security, political, and economic interests. Among adversaries, the most aggressive would likely be emboldened to test the new limits of U.S. will and endurance. And among the body politic, a political backlash against the prophets of decline would surely result.

Between bully and wimp lies a large middle ground where, it seems likely, U.S. leaders would seek to steer the nation in time of crisis. To demonstrate that the United States remains a nation capable of using its power—and capable of doing so well and for the collective good—the

---

26. Herman Kahn, "Issues of Thermonuclear War Termination," pp. 133–180.

United States is likely to make choices seen and defended in terms of its stewardship of that collective good.

This sense of stewardship might shape U.S. choices in a number of ways. If its use of power is to be seen as responsible, the United States must show flexibility in managing crises by making every effort to avoid especially costly outcomes. But it must also press for a resolution of the root causes underlying the conflict and a long-term removal of the threat if its use of power is also to be seen as purposeful. If it opts to use nuclear weapons, that use must appear as necessary to end the war, proportionate to the issues at stake, and just in terms of lives saved. It must act with the understanding that the record of how the conflict began and how it played out will be scrutinized, even after a victorious war, to assess how well the United States behaved as leader and guarantor and how well it offered its opponents an acceptable way to avert NBC confrontation.

Articulating U.S. choices and building a case for U.S. actions would require that the president help the nation and the world to interpret the crisis and to understand the military, political, and moral context of U.S. choices. Establishing a tenable connection between means and ends, and building some measure of consensus with Capitol Hill, the media, and the U.S. public (and perhaps even within the executive branch) on the chosen course of action would likely be a principal challenge for the president and his or her advisers—especially if that course is nuclear use. Many of the same tasks would be necessary with friends and allies abroad and in major international institutions such as the United Nations and North Atlantic Treaty Organization (NATO). The Cold War debate about nuclear weapons revealed a deep division in U.S. society and internationally about nuclear weapons, with a strong body of opinion that their use would be inconsistent with just war traditions. With so little discussion until now of the role of nuclear weapons in the post–Cold War world, the president would likely face a major challenge in seeking to establish U.S. nuclear retaliation as both necessary and just.[27]

This is a political campaign that cannot be cobbled together at short notice. The United States cannot answer at the last minute—or alone—the questions of power and legitimacy that WMD crises will inevitably pose. Finding the best decisions, actions, and words at the moment of crisis is no substitute for those other initiatives that are always essential to maintaining the U.S. reputation as a reliable and benign power and responsible

---

27. For a discussion of some of these themes, see Brad Roberts, "NBC-Armed Rogues: Is There a Moral Case for Preemption?" in Elliot Abrams, ed., *Close Calls: Intervention, Terrorism, Missile Defense, and 'Just War' Today* (Washington, D.C.: Ethics and Public Policy Center, 1998), pp. 83–108.

steward of common interests. That is, the political requirements of future major theater NBC wars oblige U.S. leaders to attend in the present to the foundations of U.S. power, to U.S. credibility as a great power, to its responsibilities for initiative and leadership, and to the challenge of adapting to a role of first-among-equals even at a time of military supremacy—all tasks that garner slight interest in an inwardly focused nation.

What then is successful war termination? Success means winning the war in a way that answers basic questions about the United States in ways the country would prefer. Victory requires both restraint and the purposeful use of power. It requires recognition *now* that to decide how a war must end is also to decide the nature of the peace that succeeds it, and that winning the peace after a war in which NBC weapons have been used by one or both sides may well prove more difficult than winning the war. Too little thought has been given to the negative consequences for the peace of winning such wars the wrong way.

What are the lessons for the three schools of thought noted above? For those who believe that the aggressor's NBC assets can trump any U.S. action, it is important to recognize the interests created for the United States if it opts to back down in the face of coercive threats. An aggressor's NBC assets will certainly generate pressures on the United States to compromise—but successful war termination requires the purposeful use of U.S. power. This implies that the United States must not allow itself to be deterred or dissuaded by an aggressor's NBC threats and that finding long-term solutions that remove the necessity of coexisting with NBC-armed renegade states will be essential.

For those who believe that "America has all the trump cards," asymmetries in power will certainly imply that wars will end as the United States wants them to. But winning the long-term peace requires that the United States not use overwhelming force.

For those who believe that "it's the president's job" to end such wars, it seems likely that the president would be very sensitive to the political result of the war (and to his or her own place in history). The president is very likely to see the imperative of steering the nation between the poles of bully and wimp on a course of stewardship. The president is unlikely to perceive a clear demarcation between political and military interests and thus is likely to require that military advisers overcome the "intellectual difficulty of connecting military plans with their ultimate purpose" observed by Iklé. In particular, the military should be prepared to endow the president with the options to do the right thing in a regional crisis, and to ensure its own ability to operate within likely political choices and limits set by the White House.

No tool of social sciences enables us to predict the future. We cannot know today how a future regional NBC war will end. But we can know how it might end. We can certainly know how it *should* end. It might end with clear-cut victory for one or the other side, but it is more likely to end after a large cost has been paid by one or both sides to protect basic interests. It should end in a way that secures the peace that follows by teaching the "right" lessons about the utility of NBC weapons and the viability of a U.S.-led international security order—among many other things. From the U.S. point of view, such wars must end in ways that protect key regional interests while also sending the right political signals about the United States as a reliable security guarantor and benign power.

Policy Implications

The outcomes of wars are not dictated by certain inevitable forces of technology, history, or personality. This line of argument puts a large burden on policymakers in Washington and allied capitals to shift the balance of capabilities and perceptions so that when and if war comes they will be able to do what seems right and necessary at reasonable cost. One way to approach the policy problem is to focus on the seven factors enumerated above, which are likely to shape the U.S. choice about how to respond to an aggressor's use of NBC weapons. On each of the seven, policies available today can improve the U.S. position in a future conflict.

The international context can be shaped ahead of time by working with regional allies on political-military strategies to meet the challenges of regional aggressors; with the other members of the UN Security Council to bolster the council's role in replying to NBC aggression; and with other prospective members of international coalitions that might be brought together in response to aggression.[28]

Congressional perspectives can be shaped through an executive-led dialogue with key committees and chairpersons on issues of regional and international security, proliferation concerns, and U.S. foreign and defense policy. It could also be productive to reach out to constituencies with which the Congress interacts on such matters, such as the think tank and academic communities.

The public mood will be shaped largely by events of the moment, but setting out some benchmarks in U.S. declaratory policy and articulating U.S. interests and policies to constituencies in academe and the media could pay dividends in time of crisis.

---

28. Roberts and Utgoff, "Coalitions Against NBC-Armed Aggressors."

The U.S. strategic stake can be kept well focused and well understood internationally by a foreign policy that clearly defines U.S. national interests and foreign policy priorities, and engages both allies and potential adversaries in a dialogue on U.S. perspectives.

Limiting the level and character of destruction of an initial NBC attack requires various approaches. One is working with the probable regional targets of an aggressor's NBC weapons to put in place both active and passive defenses. Another is undertaking international threat reduction measures, including the effective implementation of arms control and nonproliferation mechanisms so that they help constrain the future evolution of the regional NBC threat.

Limiting the capacity for continued destruction is also a task with many parts. One is the development of a better understanding of the strategic personalities of aggressive regimes, so that their ambitions, values, and decision mechanisms can be understood and engaged in time of crisis. Caroline Ziemke's contribution to this volume offers one such effort. This is an intelligence problem as much as a problem of military planning. Another part is the fielding of active and passive defenses by the U.S. military, for the protection of its own forces and as a supplement to the efforts of regional allies.

Preparing the military voice to participate in dialogue at the White House in times of crisis requires careful thought about how WMD wars might unfold, the likely perception of interests both in Washington and in the aggressor's capital, and a readiness to articulate how operational plans can be shaped to help the president achieve acceptable political outcomes.

This list of policies is entirely illustrative; it is not an exhaustive review of the tools available to U.S. policymakers to shape war outcomes. Rather, it illustrates how much the United States can do ahead of time to shape the context in which an NBC crisis will play out, and thus to promote outcomes conducive to U.S. interests, and that it would be difficult or unhelpful to try to do some or all of these things at the very last minute.

Many of the tasks fall under the rubric of the defense counterproliferation initiative.[29] The degree of vulnerability of U.S. military forces to an aggressor's NBC attacks would be a critical determinant of the risk

---

29. This initiative was announced by Secretary of Defense Les Aspin in a speech to the National Academy of Sciences on December 7, 1993. Its purposes are to prevent the proliferation of weapons of mass destruction and roll it back where possible, to deter the use of such weapons, and to change how the United States fights wars in

calculus on both sides. A less vulnerable U.S. force would open up more options for the United States, while a more vulnerable one opens up more options for the aggressor. Indeed, the United States could be compelled to act as a wimp if gaps in U.S. capabilities help the aggressor to fight a war largely on its terms and to its strengths. Similarly, it could be compelled to act as a bully if it is left with no more effective option than strategic nuclear attack to meet the aggressor's use of NBC weapons. Moreover, counterproliferation initiatives jointly pursued with regional allies are the basis of joint action and help to build the political confidence in each other that would be necessary to weather crises. If the United States is capable of sustaining military operations on a battlefield contaminated with chemical and biological weapons, of protecting rear areas and population centers from NBC attack, and of conducting flexible nuclear operations, U.S. action is not foreclosed, and the aggressor may well conclude that an NBC war against the United States or a regional U.S. ally is unwinnable—and thus should not be risked. Thus a successful counterproliferation initiative can change the relative prices of alternative outcomes in ways that make it easier for the United States to do what is right and necessary, and harder for the aggressor to begin an NBC war.

Two more specific policy-related questions also follow from this analysis. What choices in war best secure the U.S. interest in preventing the further proliferation of nuclear weapons? And what does this analysis imply about the problem of deterrence?

What is best for the future of the nuclear nonproliferation effort is not obvious. Desert Storm has left a strong conviction that the cause of nuclear nonproliferation is best served by prosecuting wars against NBC-armed adversaries with only conventional weapons. After an aggressor uses weapons of mass destruction, however, convictions may change. This seems especially likely to be so if nuclear retaliation is the only means to inflict commensurate punishment on that aggressor or to deny its war aims at costs in human terms acceptable to the victims of aggression and to those called upon to reverse it. Moreover, whether or not nuclear weapons are used by the United States seems likely to prove a less significant determinant of the nuclear future than whether other states come to view the United States as a bully, wimp, or steward.

How should the United States seek to deter an NBC-armed aggressor? This chapter has illustrated a variety of ways that deterrence might not work as hoped, as an aggressor's risk calculus downplays or disre-

---

the presence of NBC threats so that it can achieve its war aims despite that presence. See *Proliferation: Threat and Response* (Washington, D.C.: Office of the Secretary of Defense, November 1997).

gards U.S. capabilities or declaratory policies. It demonstrates that there is benefit to be reaped in "rising above deterrence" to assemble the capability to successfully prosecute wars against NBC-armed regional adversaries without merely relying on the dissuasive effects of U.S. conventional and nuclear superiority.[30] But it also points to some specific lessons for specific situations.

In deterring an aggressor's first use of NBC weapons, perceptions of U.S. will are likely to affect the aggressor's choice more than its perceptions of U.S. capabilities. The United States could strengthen deterrence of NBC first use by an aggressor by working to dispel the notion that retreats from conflict, as in Somalia and Lebanon, imply that the United States will back down in a conflict over a major interest. Developing a better understanding of the U.S. public's sensitivity to casualties could also contribute to deterrence. More generally, an aggressor's expectations about how U.S. leaders will behave in crisis is shaped fundamentally by its reading of U.S. culture and politics. An aggressor that perceives a complacent and vulnerable United States may stimulate a crisis and then initiate NBC use in the hope of reaping local gains. One that perceives a nervous but unprepared United States may opt to strike before U.S. advantages are secured. One that sees the U.S. president as timid or foolish, or the U.S. body politic as divided and manipulable, may seek confrontation in the hope of dealing a sharp blow to U.S. prestige and influence. But a potential aggressor that has been led through how the United States would assess its alternative courses of action in an unfolding crisis may conclude that any route that might lead to confrontation with the United States is suicidal. Thus, deterrence of first-use would be strengthened by clarification of the interests that would be created or magnified for the United States by any act of regional aggression backed by NBC weapons.

Deterring an aggressor's second use of WMD in reply to U.S. retaliation entails a different set of problems. The task of coercing leaders who do not hold their populations dear poses difficult nuclear use questions for the United States. Countervalue strategies seem to offer less promise than counterforce ones—which is to say that the United States seems likely to use nuclear weapons not to coerce but to defeat. The United States may find this stage of conflict ripe for trying to separate a warmaking leader from his or her military and societal sources of support, as noted above. Deterrence will be well served by conveying the capacity of the United States (and its coalition partners) to survive the worst blows that the aggressor can contemplate delivering and the reality that every

---

30. Utgoff and Wallis, "Rising Above Deterrence."

step up the escalation ladder gives the United States more room to maneuver while worsening the aggressor's own situation.

Conclusions

The variety of policy directions suggested above demonstrates that the United States can indeed take actions ahead of time that will influence how a war fought with NBC weapons will end. But these actions require a strategic view of the problem well beyond the current conventional wisdoms that clutter the debate. No longer a problem that is "truly unthinkable" in a war that "cannot be won and therefore must not be fought," such wars require serious analytical scrutiny in an effort to prevent them or minimize their costs. The United States should plan for the termination phase of such wars by consulting with allies and prospective coalition partners on the kinds of interests likely to shape war-time coalition dynamics, and working backwards from that to develop the right military capabilities. It should attempt to shape the termination phase of such wars by limiting the vulnerabilities that an aggressor might target with NBC weapons, by dispelling myths about the strategic personality of the United States, and by putting together a posture of military force and declaratory policy suited to the deterrence requirements of specific operational theaters. Our preference to focus on how to secure the nuclear peace does not relieve us of the obligation to think through these choices about nuclear war—indeed, the two tasks are integral.

What does this analysis of the war termination problem imply for the larger questions raised in this volume? First, it is another reminder of the way in which wishful thinking, conventional wisdom, and political fashion narrow the necessary debate. The wishful thinking is that major war, especially war that threatens the United States with mass casualties, is an anachronism. The conventional wisdom is that nuclear wars will not be fought because they cannot be won. The political fashion dictates that problems of nuclear war are somehow beyond the bounds of polite and proper discourse. Were it only so. Wars employing nuclear, biological, or chemical weapons and involving U.S. forces are more likely today than in decades past, given the growing prevalence of asymmetric counters to superior U.S. conventional military power in precisely those regions where the United States offers security guarantees. These issues belong on the policy agenda—however unpopular or seemingly obscure.

Second, this analysis reinforces a theme in many of the chapters in this volume—that the first major regional war with NBC weapons, if it occurs, will be a defining event. At the very least, ideas based on pre-war perceptions, especially those carried over from the Cold War, seem likely to be driven aside by new concerns. Perceptions of the utility of weapons of mass destruction, of the nature of the retaliatory problem, and of the international role of the United States would be catalyzed by events. Indeed, any confrontation seems likely to be in large part a political effort by the combatants to shape those perceptions. Mishandled, it would cast the United States as either bully or wimp, thus weakening the foundations of world order, by teaching states that they must organize either against or without the United States. Handled well, it would solidify U.S. leadership, confirm its stewardship of common interests, and deter future regional challengers.

Chapter 10

The Coming Crisis: Nuclear Proliferation, U.S. Interests, and World Order—A Combined Perspective

Victor A. Utgoff

This collection of essays has addressed a fundamental question: What changes might the proliferation of nuclear, biological, and chemical (NBC) weapons create in the long run for international relations? While only time will answer this question completely, the importance of anticipating the possibilities grows with each new NBC-armed state.

Our main concern with this question is that proliferation will eventually lead to a crisis in which an NBC-armed state tries to force the United States to sacrifice a vital interest. As Rosen, Posen, Quester, and Roberts all argue, how the United States and its allies act in the face of such a challenge will say a great deal about the likely nature of the subsequent world order. Wisdom and resolve could preserve the prospects for a reasonably safe and improving world order; their absence could turn the world toward further NBC-backed aggression, accelerated proliferation, and the breakdown of important forms of international cooperation. This book has sought to present insights that can help to foster the needed wisdom and resolve.

Addressing the question of how NBC weapons may affect the world order requires exploration of an enormous expanse of intellectual terrain, much of which remains terra incognita. As one would expect from seasoned explorers, the trip reports presented by the authors in this volume are filled with descriptions of important and interesting features of this terrain. Equally interesting are the very different perspectives through which they viewed the terrain. Most explored primarily with the eyes of realists, concerning themselves with relative balances of power among states involved with NBC weapons. One viewed the terrain in terms of strategic personality, a nation's historically derived sense of how it should interact with other nations. Others saw the terrain through mixed perspectives including those of realism, international norms, bureaucratic interests, and the attractions of international prestige.

What can be learned from these trip reports? The most important message is that the conventional wisdom—which posits that the continued proliferation of weapons of mass destruction will cause the United States to become a far more reluctant sheriff in the international arena than it has been in the past—seems wrong.[1] While the essays do provide some arguments that are consistent with this view, the larger weight of their arguments points in the opposite direction. A second important message is that most of the unconventional wisdom presented by the authors arises from perspectives other than realism. This raises an important question: Should realism continue to play so strong a role in assessments of likely U.S. international behavior as it has in the past?

I believe it should not. Throughout its history, the United States has disdained Europe's historical emphasis on balance of power international politics. While U.S. international behavior during the Cold War was strongly conditioned by concerns about the enormous military power of the Soviet Union, the collapse of that empire has largely freed the United States to revert to type. The United States is the world's strongest power. It has far more capability than ever before to exert its will around the globe, and thus to preserve and improve the aspects of the global order that it believes are right. Accordingly, projections of future U.S. behavior in the international arena should now focus more attention on what this nation sees as right behavior in any particular circumstance.[2]

How can we understand what kind of behavior the United States is disposed to see as right?[3] One way is to adopt the approach taken by Caroline Ziemke in her essay on Iran and construct a picture of the U.S. collective view or "strategic personality." Such a construct must necessarily include a concept of the nation's most important values and how those values enter into its perceptions and decision-making.

This chapter summarizes the main features of the strategic personality type the United States shares with several other nations. It describes

---

1. As will be argued below, the United States does not seem that reluctant a sheriff. See Richard N. Haass, *The Reluctant Sheriff: The United States after the Cold War* (Council on Foreign Relations, 1997).

2. Kissinger argues that without the threat of the Soviet Union to leaven its Wilsonian idealism, the United States needs to pay more conscious attention to balance of power considerations in framing its post–Cold War foreign policy. This chapter argues in favor of understanding and being more conscious of the role of the U.S. value-oriented strategic personality in motivating key foreign policy choices. Henry Kissinger, *Diplomacy* (New York: Simon & Schuster, 1994).

3. Unfortunately, neither the theory nor the practice of political policy analysis has yet developed rigorous tools for combining realism with other perspectives. See Gregory A. Raymond, "Problems and Prospects in the Study of International Norms," *Merschon International Studies Review*, No. 41 (1997), pp. 205–245.

how that personality type is manifested by the United States, emphasizing characteristics that will influence its behavior in the face of challenges involving NBC weapons. It then explores how this strategic personality seems likely to reinforce or counterbalance other more commonly considered factors that would influence the outcome of a confrontation between the United States and an NBC-armed regional aggressor. And, finally, it draws some larger policy implications from this analysis.

Characterizing the U.S. Strategic Personality

Caroline Ziemke argues that a state's strategic personality describes its distinctive ways of thinking, and how its ways of thinking influence its international conduct.[4] She argues that the strategic personality of a state reflects its entire history of adapting to its physical surroundings, maintaining the internal stability of its society, and meeting the challenges that come from interactions with other states. Thus, the strategic personalities of long-established states are very stable.

Ziemke divides strategic personalities into types based on three characteristics. The first characteristic is how a state views interactions with others. The second is what information it selects and remembers from the infinity of events that constantly inundate it. The third is the basis upon which it selects courses of action. Ziemke shows that the U.S. strategic personality type is "extroverted, intuitive, and feeling."[5]

Extroverted states extol individualism and competition as the engine for domestic progress. Similarly, they see interactions of their state with others as desirable for themselves and for the international system as a whole. Extroverted states view history as an irreversible march toward some common progressive end goal for all humankind. They believe that they must move forward, grow, and change, or that they will stagnate and be overtaken by the forces of progress in history.

In their interactions with other states, extroverted states commonly challenge, push, and prod until they encounter resistance or get pushed back. That is, they have a weak sense of boundaries, which is perhaps the most consistent source of international conflict for such states. The social structures of extroverted states call on individuals to adapt their needs and ambitions to those of the others with whom they interact. This

---

4. Caroline F. Ziemke, *The Strategic Personality Types: Using National Myth to Project State Conduct*, forthcoming.

5. Ziemke identifies Portugal and the Netherlands as having the same strategic personality type as the United States. Ziemke, *Strategic Personality Types*. Clearly any specific personality type can manifest itself in a wide variety of ways.

internal social norm translates into the expectation that the international system also should employ negotiation and compromise rather than violence to resolve interstate differences. Over the course of their histories, the extroverted states have arrived at a consensus that war is not the way to settle their differences. Finally, extroverted states see themselves as relatively adaptable and can fail to appreciate that not all states are. They care very much about how the rest of the international system sees them.

Intuitive states tend to have national myths defining their origins and ultimate goals in terms of a particular idea or doctrine. They usually see themselves as working toward some future golden age and their national myths often contain a promised land, paradise, or utopia. Accordingly, they pay most attention to the patterns and meanings in unfolding events that seem to provide insights into whether the future is evolving as desired. These states define evil as any force that stands in the way of achieving their ideals. For intuitive states, patriotic rituals are important for reinforcing the guiding ideal. Residence and citizenship requirements are usually defined in terms of acceptance of the guiding ideal. Physical boundaries are less important than ideological ones.

Because they seem to have an instinctive sense of where events are headed, extroverted intuitive states are natural leaders. They are particularly effective in promoting new approaches to addressing problems, inspiring enthusiasm, and drawing other states to follow their vision, even if their followers do not completely grasp or accept it.

Feeling states judge alternative conclusions or courses of action in terms of their consistency with a hierarchy of national values. While logic enters their assessments, they are suspicious of it as a guide to conduct. Thus, where logic and values conflict, feeling states are more likely to act in accordance with their values.[6]

When extroverted feeling states are forced by circumstance to follow the dictates of logic or realism rather than their values, they feel uncomfortable, and can be expected to change their positions if an opening presents itself. With respect to the international system, the extroverted feeling state is guided more by values defining how the system should work than by any logical picture of how the system does work. While the value structures of extroverted feeling states are rooted in both their

---

6. Huntington notes that "throughout the history of the United States a broad consensus has existed among the American people in support of liberal, democratic, individualistic and egalitarian values." He also states that "being human, Americans have never been able to live up to their ideals; being Americans, they have been unable to abandon them." In Samuel P. Huntington, "American Ideals versus American Institutions," *Political Science Quarterly*, Vol. 97, No. 1 (Spring 1982).

basic culture and history, they are constantly adapted to their experience in interacting with others and the implications of that experience for their values. This constant evolution in values and especially their context-dependent application can frequently make extroverted feeling states seem fickle and sometimes hypocritical.

For feeling states, righteousness comes from conducting one's affairs virtuously and from good works. The legal codes in feeling states are generally designed to promote and enforce a particular hierarchy of values, and penal codes often go beyond mere justice and seek rehabilitation or retribution. Government in the feeling states is often primarily a caretaker of the cherished values and arbiter of differences. For the feeling state, being admired and recognized internationally for its virtue is important to its sense of national worth and strength.

U.S. MANIFESTATION OF ITS STRATEGIC PERSONALITY TYPE

The United States manifests its extroverted, intuitive, feeling personality type in specific ways that say a great deal about how it would be disposed to act toward aggression backed with NBC weapons.[7]

First, the fundamental shared ideal upon which the United States as an intuitive feeling state is based is liberty for the individual. This preeminent value can explain most of the significant political-economic features of the United States. It is the basis of U.S. law. It explains the U.S. emphasis on negotiation and compromise as the only legitimate means for resolving disputes. Its best protection is democratic government. It leads directly to a free market economy and rigorous protection of property rights.

The key theme of U.S. domestic history has been the pursuit of a more perfect liberty. Most of the larger initial flaws in how liberty for the individual was defined, practiced, and protected have been corrected. Slavery ended in the Civil War, initial restrictions on suffrage have been largely eliminated, and the threats to liberty that came with unregulated capitalism have been contained.

Most important, individual liberty is seen as the fundamental key to the enormous success of the United States as a society. Liberty opens great opportunities to create wealth, institutions, and other values, and imposes primary responsibility on the individual for determining his or her

7. This section was informed by: Walter A. McDougall, *Promised Land, Crusader State: The American Encounter with the World Since 1997* (Boston: Houghton Mifflin, 1997); Alan Brinkley, *The Unfinished Nation: A Concise History of the American People*, 2nd ed. (New York: Knopf, 1997); George Wilson Pierson, *Tocqueville in America* (Baltimore, Md.: Johns Hopkins University Press, 1938); and Richard Slotkin, *Regeneration through Violence: The Mythology of the American Frontier, 1600–1860* (New York: Harper, 1972).

own fate and meaning. This combination of opportunity and responsibility is seen as the internal engine for continuous U.S. innovation and growth. Not surprisingly, the United States sees itself in the lead on humankind's long march toward the perfect society.

Religious liberty for the individual is a particularly important component of the U.S. strategic personality. It places greater responsibility on the individual to decide the larger questions of right and wrong than felt by individuals within states with an official religion. While a majority of U.S. citizens may accept the teachings of the New Testament as an ideal, they seem most often to practice the eye-for-an-eye philosophy of the Old. Indeed, most historical U.S. heroes did not turn the other cheek when insulted or assaulted, but gave back as good as they got. Justice and the need for revenge usually require commensurate punishment in U.S. society.

As an extroverted state, the United States is very active in the international arena. It sees interaction and competition among states as important for healthy adaptation and growth for itself and for others. More generally, the United States has strong views about how the international system of states should work—a guiding vision of an international "City on the Hill." In particular, as an extroverted state, the United States wants the world system to mirror its own main characteristics.[8] For example, the world economic system should have open global markets. The rule of law should be the basis for global political stability rather than balances of power among nations. Negotiation and compromise should be the means by which states settle their differences. And the world should be composed of democracies that cherish liberty and human rights.[9]

For many decades the United States has been active in creating and participating in numerous institutions to support such a world. Providing for international security came first, with the United Nations, NATO, the Conference on Security and Cooperation in Europe, and a variety of other types of security regimes. The search for global economic security has led to institutions such as the World Trade Organization, the World Bank, the

---

8. Ruggie argues that "there is a certain congruence between the vision of the world order invoked by American leaders when 'founding' a new international order has been at stake, and the principles of domestic order at play in America's understanding of its own founding, in its own sense of political community." John Gerard Ruggie, "The Past as Prologue? Interests, Identity, and American Foreign Policy," *International Security*, Vol. 21, No. 4 (Spring 1992). See also David C. Hendrickson, "In Our Own Image: The Sources of American Conduct in World Affairs," *The National Interest* (Winter 1997/98).

9. Tony Smith, *America's Mission: The United States and the Worldwide Struggle for Democracy in the Twentieth Century* (Princeton, N.J.: Princeton University Press, 1994).

International Monetary Fund, and the Group of Eight. And, reflecting the extrovert's minimal regard for national and cultural boundaries, the United States has pursued an international campaign for human rights that has made many nations extremely uncomfortable.

Perhaps most important, the United States has been exporting and defending democracy abroad for the past hundred years.[10] For example, when the United States became a colonial power with the Spanish-American War of 1898, President William McKinley set as U.S. goals to "establish liberty, justice and good government in our new possessions." And, after proposing a postwar League of Nations to maintain the peace, President Woodrow Wilson took the United States into World War I to "make the world safe for democracy." President Franklin Roosevelt took the United States into World War II and led the effort to defeat and then make democracies of the totalitarian regimes that began it. And for more than forty years after World War II, the United States led an international campaign to contain the global challenge to democracy mounted by the Soviet Union. That this campaign ended without another global war, but with freedom and budding democracy for many of the states that had been part of the Soviet Union's empire and for Russia itself, is an enormous source of satisfaction for the United States and its allies.

Since the end of the Cold War, the defense and enlargement of the community of democracies has been a stated goal of the United States. The United States has carried out at least three military interventions abroad to protect, restore, or establish democracy in the post–Cold War era.[11] And the Clinton administration supports the expansion of NATO to assure new allies that it will protect their newly won freedom and democracy.

Looking back, the United States takes pride in the results of its campaign to transform the world into a closer image of itself. To be sure, it has passed up a variety of possible opportunities to protect and expand liberty and democracy abroad. Some efforts it did undertake failed miserably. Still, U.S. support for liberty and democracy, most often simply by providing an example of its potential advantages, has had a remarkably positive effect on the world. Most of the nations, with a majority of the global population, are democratic today, where only a few were a hundred years ago. And the world seems a closer image of the United States

---

10. Quotes from McDougall, *Promised Land, Crusader State.*

11. The three interventions were in Somalia, Bosnia, and Haiti. The 1990–91 intervention against Iraq might be considered a fourth such intervention. Although neither Kuwait nor Saudi Arabia is a democracy, the purpose of the intervention was also to protect the larger democratic community. Many other lesser interventions have taken place in addition to these relatively high-profile interventions.

in other important ways. Free markets have spread widely. Much of the world has adopted a technical culture similar to that seen in the United States.

In sum, the preeminent value for the United States in its international role has been the extension to the rest of the world of its own system of liberty, democracy, free markets, and the rule of law. The United States has campaigned for these values for more than a century at considerable cost in treasure and blood, and the campaign appears to have been very successful.

RELIABILITY OF THE U.S. STRATEGIC PERSONALITY AS A PREDICTOR OF U.S. BEHAVIOR PREFERENCES

If the U.S. strategic personality is to be useful as a predictor of its preferred behavior, at least two things must be true. First, the mechanisms by which it makes itself felt must be reasonably reliable. Second, it must be reasonably stable.[12]

ON RELIABLE MECHANISMS The U.S. strategic personality makes its influence felt on the nation's decisions through communication, debate, and leadership. The communications media, which constantly searches for issues that will capture the public's attention, is drawn to issues that trigger fundamental U.S. values. In turn, debate on such issues inevitably surfaces a wide range of suggestions for how to resolve the issue. Those that resonate best with U.S. values are the most likely to be selected.

Decisions that the nation's leaders make in secret do not benefit from wide public debate, but a wise leadership will do its best to act as it understands the larger body politic would want. Further, whatever interpretations and actions are decided upon are likely to become public before long. This also argues for decisions that will appeal to U.S. values.

---

12. How rapidly the U.S. strategic personality may be changing is disputed. Arthur Schlesinger, Jr., argues that "in its essential, the national character will be recognizable much the same as it has been for a couple of centuries. People seeking clues to the American Mystery will still read, and quote, Tocqueville." In Arthur Schlesinger, Jr., "Has Democracy A Future?" *Foreign Affairs* (September–October 1997). Samuel Huntington has also argued that "despite their seventeenth- and eighteenth-century origins, American values and ideals have demonstrated tremendous persistence and resiliency in the twentieth century. Defined vaguely and abstractly, these ideals have been relatively easily adapted to the needs of successive generations." Samuel Huntington, "American Ideals versus American Institutions." William H. McNeill provides an interesting essay on the state of the U.S. public myth (a concept closely related to strategic personality) in "The Care and Repair of Public Myth," in William H. McNeill, *Mythistory and Other Essays* (Chicago: University of Chicago Press, 1986).

The president's leadership also tends to reinforce U.S. behavior that is consistent with its strategic personality. Especially when risking war, a president wants the broadest and deepest support possible, and the most effective way to win it is to appeal to interpretations and values that are consistent with the U.S. strategic personality. Presidents are also very conscious of how they will be compared to their predecessors in the future—and the greatest presidents have articulated appealing values for the nation, led the country in heroic actions to further those values, and thus helped to strengthen and shape the U.S. strategic personality.

Of course, a president who has great public credibility can occasionally deviate from the interpretations and actions called for by the U.S. strategic personality. However, doing so risks the quick evaporation of public support if the chosen action does not go well. In contrast, actions that are consistent with U.S. values may be supported long after they have been shown to be unwise.[13] In sum, then, it seems likely that the mechanisms that conform U.S. behavior to its strategic personality are fairly reliable.

ON THE STABILITY OF THE U.S. STRATEGIC PERSONALITY. The most fundamental characteristics that make the United States an "extroverted-intuitive-feeling" state, and the more basic ways in which the United States manifests its personality type, change very slowly.[14] The U.S. personality is the product of over 200 years as an independent nation, and nearly 400 years' experience with the special challenges and opportunities of life in the New World. Fundamental changes to the personality of such a well-established state usually require comparable lengths of time to evolve. Less fundamental changes can happen more quickly. However, there are good reasons to believe that any such changes to the U.S. strategic personality are evolving gracefully over a matter of many decades. And they may be offset in part by other trends.

First, the outside world does not seem to be the source of large pressures for change in the U.S. strategic personality. The United States has been and remains relatively successful in the international arena, and many important states seem to depend upon U.S. global involvement. Some observers worry that the United States will ruin its wide acceptance

13. As examples, President Dwight D. Eisenhower's credibility as a military leader was very important in bringing the United States to understand and accept the notion of limited war in the aftermath of World War II. The credibility of Presidents Kennedy, Johnson, and Nixon worked to extend the U.S. involvement in Vietnam.

14. Ziemke, *Strategic Personality Types*.

and influence by abusing its current preeminence of power. This seems unlikely; reflecting its basic extroverted nature, the United States seems too sensitive to world opinion to let itself become a pariah, and it attaches fundamental value to the concept of democracy among the states.

While internal change could also lead to a shift in the U.S. strategic personality, this does not seem a significant worry either. The values that are the basis for the effective functioning of U.S. society are remarkably few and simple, as indicated above. Most of the body politic takes them for granted, and most groups that do not generally seem to be seeking to have those values applied more beneficially to them.

There is some worry that families, schools, churches, service organizations, job environments, and the media generally are not inculcating traditional U.S. values effectively. Troubling signs include a reduced sense of responsibility, as reflected in more out-of-wedlock births, and a reduced respect for solving problems without violence, as reflected in increased violence in public schools. At the same time, however, there are good reasons to believe that the core U.S. values continue to be accepted by the great majority of the public. Vastly improved communications seem on balance to be a positive force for building and reinforcing common values.[15] The public applauds traditional heroic behavior that seems no less frequent in today's news than in decades past. The continuing demand and respect for traditional heroes seems evident in positive public reactions to well-attended motion pictures featuring the likes of Bruce Willis, Tom Hanks, Sigourney Weaver, Jodie Foster and others playing roles in which they perform impressive feats and make great sacrifices to protect the lives, freedom, dignity, and general welfare of their fellow citizens.

Some observers argue that the United States is becoming more divided along ethnic lines, particularly as a result of high rates of immigration from south of the border. While this immigration is having some effect on U.S. culture, it seems doubtful that it will lead to substantial changes in the larger U.S. strategic personality. The great majority of the immigrants are attracted to the United States by features that are fundamental to this personality. Further, leaving home requires an intuitive

---

15. The role of communications in encouraging a sense of nationhood among large numbers of people who know only a minute fraction of their fellow citizens is discussed in Anderson Benedict, *Imagined Communities: Reflections on the Origin and Spread of Nationalism* (London: Verso, 1983). In particular, Benedict analyses the connections between the appearance of inexpensive newspapers and modern nations. A pessimistic assessment of the current state of some of the social mechanisms that nourish U.S. democracy is provided by Robert D. Putnam, "Social Capital and Democracy," *Braudel Papers*, Fernand Braudel Institute of World Economics, No. 9 (1995).

person's imagination and willingness to take risks, which biases immigration toward those who can readily fit into U.S. society. And, as with all previous waves of immigrants, subsequent generations fit in even better.

Francis Fukuyama has recently suggested that the increasing number of women in influential positions in the United States may be leading toward less aggressive U.S. international behavior.[16] He notes that this could put the United States at a disadvantage in dealing with upstart states that are led by more aggressive men. However, such change is occurring very slowly, particularly in the U.S. national security apparatus. Further, women who win senior positions do not seem significantly less competitive than their male counterparts, or likely to misunderstand the imperatives of dealing effectively with serious external aggression.

Edward Luttwak has argued that the U.S. willingness to go to war has been reduced because U.S. families usually have far fewer children than they did decades ago, and parents with few offspring are less willing to risk their loss to war.[17] On the other hand, the prospects of dying in battle are far smaller than they were in past generations as a result of increased investments in protecting the combatants and the improved speed and quality of medical care. Perhaps equally important, volunteers are more justifiably sent into combat than draftees were.

Finally, yet another stabilizing factor is at work. The United States considers itself a highly adaptive problem-solving society. Consistent with its intuitive feeling character, whenever the U.S. public becomes seriously concerned that its society is losing its competitive edge or other important qualities, political support develops for corrective measures that often have useful effect.

In summary, then, it seems likely that the U.S. strategic personality as we have known it for many decades will remain a reasonably reliable predictor of U.S. behavioral preferences.

Potential Influence of the United States' Strategic Personality on a Crisis with an NBC-Armed Regional Aggressor

Ziemke's approach to typing the personalities of states shows great promise for projecting the more likely behavior of states. While her approach is new, this characterization of the U.S. strategic personality type and its

---

16. Francis Fukuyama, "What if Women Ran the World?" *Foreign Affairs* (September/October 1998).

17. Edward N. Luttwak, "Where Are the Great Powers? At Home with the Kids," *Foreign Affairs* (July/August 1994).

manifestation spelled out above ring true. It suggests that the United States has a strong attachment to its vision of how the international system should work, and to the states that have come to accept that vision. This strong attachment seems likely to weigh heavily in U.S. decision-making on how to deal with aggression by regional states armed with weapons of mass destruction.

I examine this hypothesis by considering four questions that seem particularly important in characterizing likely U.S. behavior in this kind of situation: First, would the United States likely prove resolute in facing future aggression by a regional challenger armed with NBC weapons? Second, would other regional nations prove willing to stand with the United States against such aggression? This question also depends upon the U.S. strategic personality and how potential allies perceive it. Third, how might the United States retaliate for an aggressor's devastating use of NBC weapons? And, finally, would the use of such weapons lead the United States and its allies to commit to the total defeat of such an aggressor? In each case, I will juxtapose the answers that are suggested by the U.S. strategic personality with those suggested by other perspectives represented in this volume, especially realism.

U.S. RESOLVE

Would the United States likely prove resolute in facing aggression by a regional challenger armed with NBC weapons? The U.S. strategic personality suggests that it would. Consider how the United States might feel about failing to come to the aid of a state, especially an ally or a democracy that is threatened by NBC-backed aggression. Tolerating such aggression would be seen as a first step toward abandoning the largely democratic world order that the United States sees itself as having been instrumental in creating. Not intervening would also be seen as deeply compromising the U.S. role as the preeminent world leader, and specifically its role as the defender of the democratic faith. Such a failure would be felt as a source of shame for the United States.[18] The United States has generally played the role of sheriff in "democracy town," and is stronger than anyone else. Thus, other states and the United States itself would expect it to form a posse to go after the villain, if not to face the challenger single-handed.

Fears of being thought a coward by itself or by other states would drive the United States toward intervention as well. A United States that

18. For an interesting discussion of the powerful role of honor in motivating national behavior, see Elliot Abrams, ed., *Honor Among Nations: Intangible Interests and Foreign Policy* (Washington, D.C.: Ethics and Public Policy Center, 1998).

has the capacity to extinguish any state would feel that it was cowardly to back down from confronting an evil that at worst can impose great pain on the United States, but not destroy it. Paradoxically, U.S. fears of being attacked with nuclear weapons could also prove a source of courage: the United States could readily believe that the offender would never escalate to NBC warfare, as it must be even more fearful of U.S. nuclear capabilities to annihilate it in retaliation.

In sum, the U.S. strategic personality suggests that it would feel strongly that standing up even to NBC-backed aggression is the right thing to do. In fact, one could argue that, all else being equal, the United States could find it even more difficult to tolerate aggression when it is backed with NBC weapons than when it is not.[19]

Rosen and Posen address this question directly from their realist perspectives. Rosen argues that the risks to the United States of confronting a regional power armed with nuclear weapons are very high and that the end of the superpower competition with the Soviet Union has dramatically reduced U.S. incentives to accept such risks. He also states that the United States would see no good options for prosecuting such a confrontation, and that the U.S. military leadership would argue to the president that the risks of intervening against such an aggressor are unacceptably high. However, he also observes that protecting other nations overseas has been the stated purpose of U.S. armed forces. While military leaders may see high risks in confronting an NBC-armed aggressor, they could also worry that reluctance to do so could raise serious questions about why the United States should maintain expensive intervention forces. In other words, the United States could have significant bureaucratic interests in not ducking this kind of challenge.

Posen takes a very different position on the question of U.S. resolve. He maintains that the United States would judge as unacceptable the long-run consequences of not standing up to an NBC-armed challenger and thus demonstrating that nuclear weapons can enable aggression. Posen also remarks that if the United States wished to avoid the painful dilemma of either confronting a nuclear challenger or of accepting the painful consequences of not doing so, it would require diplomatic strategies that make inaction appear differently. Developing and implementing such strategies would be no small feat given the obligation the United States feels to protect at least the democracies, and the honor that it sees in this role.

---

19. See Lawrence Freedman, "Great Powers, Vital Interests and Nuclear Weapons," *Survival* (Winter 1994–95), pp. 35–52.

Posen's position that the United States would likely see that it must be resolute seems the more persuasive of these two realpolitik perspectives; when we consider the behavior suggested by the U.S. strategic personality, resolute behavior seems far more likely than not.

REGIONAL ALLIES

Would other regional states prove willing to stand with the United States against NBC-backed aggression? The U.S. strategic personality influences the answer to this question in several ways. First, Walt argues that regional states that do not have their own nuclear weapons and are threatened by those who do would almost certainly be willing to ally with the United States, provided several conditions are met. The most important of these conditions are, first, that the United States does not propose to initiate attacks on the threatening state. Second, they would need to see that the United States can be counted on as a reliable and capable protector.

Threatened regional states considering alliance with the United States should derive comfort from a good appreciation of its strategic personality. First, the United States places a very high premium on not resorting to violence against any state that has not yet broken the peace. Second, the United States attaches a very high value to honoring commitments; it sees abandoning an ally to aggression as shameful. Third, it cares deeply about its reputation in the international community.

The second way in which the U.S. strategic personality should affect the willingness of other states to join a U.S.-led coalition arises from the substantial mismatch between the strategic personality of the United States and those of many of the regional states that might seek its help. The great majority of the states outside the Western Hemisphere and Europe are introverted, and most of them are sensing, feeling states. Their highest values are the safety and internal stability of their societies. They see the Western emphasis on freedom of the individual and on progress through competition as serious threats to that stability. They have historically dealt with corrupting influences from the outside by withstanding, excluding, and eventually expelling foreigners.[20]

Such states can be expected to be quite reluctant to allow the United States to station its forces among their populations. Thus, they would likely delay any such action until the threat of aggression became unmistakable and imminent. And when U.S. troops are on their soil, these states can be expected to keep them at arm's length. Some of the more common mechanisms are isolation, efforts to suppress displays of behavior that

---

20. Ziemke, *Strategic Personality Types.*

are normal in the United States but offensive to the host society, and active reinforcement of the host population's belief in the superiority of its own culture. After the threat has passed, these countries can be expected to control contact with their populations even more stringently and to encourage the United States to go home. However understandable such actions are, they can suggest a lack of appreciation for its culture that the United States can find grating, particularly given the importance the United States attaches to being appreciated and respected. Saudi Arabia's treatment of U.S. forces deployed on its territory during and after the Persian Gulf War of 1990–91 is a good example.

Thus the need to form alliances can create conflicts among U.S. values. Clearly, the United States wants to slow NBC proliferation that could lead to a very threatening and violent world. This suggests that alliance with the United States should be made as easy as possible for countries that would accept U.S. protection in lieu of their own NBC forces. But the price of alliance can be acceptance of the host country's resistance to U.S. culture, which limits U.S. opportunities to win converts to its creed.

A second conflict of values would arise when the United States considers how to assemble a coalition to stand up to NBC-backed regional aggression. Consistent with its democratic ideals, the United States attaches a high value to legitimizing important international actions by seeking the support of as many members of the international community as possible. At the same time, it would be deeply upset if their involvement in a U.S.-led action were to lead to massive destruction for those supporters. Balancing these conflicting values may require the United States to minimize the number of active coalition partners it would seek in such a confrontation and to settle for less direct or delayed political support from other sympathetic states.

Taken together, these observations suggest that the nature of the U.S. strategic personality should encourage alliance with the United States, but that the United States should not expect a close embrace that would indicate gratitude from its regional allies. The United States may want to limit the numbers of allies it accepts into the coalition to those that have essential contributions to make or can protect themselves reasonably well from NBC attacks.

The realpolitik arguments some of the authors address to this question are contradictory. Several suggest that continued nuclear proliferation could make the United States seem a less reliable protector than during the Cold War. Rosen further suggests that most of the world would no longer see its protection as a vital U.S. interest, at least in realpolitik terms. Thus, they would see little prospect that the United

States would take the kinds of measures that might make an offer of protection credible, such as stationing U.S. forces on their territory during or even before a crisis.

As noted above, Walt's arguments point in the opposite direction. Walt also notes Posen's key point—the United States would not want nuclear weapons to be demonstrated as capable of enabling successful conventional aggression. He points out that this obvious U.S. realist interest should help to convince potential allies of the U.S. commitment to stand by them, and argues that great power allies should share this basic concern. Finally, Walt points out that many states have been willing to stand up to nuclear-armed powers in the past, and have even attacked them.

Walt's realpolitik arguments seem very persuasive. Non-NBC-armed regional states facing an NBC-armed aggressor would be at a tremendous disadvantage in terms of the local balance of power, and realpolitik arguments would weigh heavily in their decisions. Alliance with a willing and reliable state that is strong enough to overwhelm the regional aggressor but would not start a war itself is an obvious choice under the circumstances. And the U.S. strategic personality suggests that it will continue to meet these conditions very well.

U.S. RETALIATION

Would the United States retaliate in kind if an opponent's use of NBC weapons were to cause great damage to U.S. forces or to U.S. allies? The U.S. strategic personality suggests in three ways that the United States would be disposed to strike back with nuclear weapons if an opponent were to make highly destructive use of NBC weapons. First, reflecting its intuitive nature, the United States seems likely to conceive of its role in confronting NBC-backed aggression as one of good against evil: once again it would be required to defend the hard-won realm of democracy against totalitarian evil. This characterization could be accentuated to the extent that the opposing leadership had been demonized and to the extent that the use of NBC weapons had become even more stigmatized than it is today. Highly destructive use of NBC weapons in such a situation would seem to provide clear illumination of an evil state that richly deserves nuclear retaliation, at least against its ruling elite.

Second, the U.S. strategic personality includes a strong sense of justice and the need for revenge that would motivate commensurate punishment. Consistent with the U.S. conception that criminal offenders should make amends to society by accepting punishment, an offending state will be seen to deserve nuclear retaliation for the sin of first use, and having suffered it, may find quicker forgiveness from the United States and other like-minded societies.

Third, the United States seems likely to have made at least implicit promises to its regional allies that it would retaliate in kind for any devastating NBC attacks made against them. In the event of such attacks, the allies could call on the United States to honor such promises. As a feeling nation, the United States would want to choose a course of action that is consistent with the values it cherishes, which include honor and reliability. And, as an extroverted feeling state, the United States values highly its reputation with its allies and the larger international community. Nuclear retaliation seems more consistent with such values than any other course of action.

In contrast, the conventional wisdom regarding U.S. willingness to retaliate with nuclear weapons appears to point in the opposite direction. Rosen notes an extensive series of RAND Corporation war games played with a cross-section of experts and officials in the United States.[21] The war games posed situations in which a regional aggressor had employed nuclear weapons to oppose a U.S.-led intervention.

RAND's analyses of the play of these war games suggests an overwhelming reluctance by the U.S. policy community to recommend retaliation with nuclear weapons, provided that great damage had not been done to the United States itself. The players also exhibited a remarkable lack of consensus that it was in the U.S. interest to punish nuclear proliferation, or even the use of nuclear weapons. Moreover, faced with the prospect of a regional confrontation that could lead to the use of nuclear weapons, the players showed a marked reluctance even to recommend that U.S. forces become engaged.

Rosen suggests that these games may point to a significant shift in the thinking of U.S. policymakers. He also suggests that they may reflect a disinclination in the U.S. policy community to come to grips with the difficulties of handling hostile regional powers armed with nuclear weapons. Finally, he notes that an unwillingness to address these problems is also evident in the scant attention given to them by published U.S. military doctrine.

Of course, these war games, which probably constitute the broadest and most systematic U.S. post–Cold War effort to address the problems of intervening against nuclear-armed regional aggressors, have sampled the opinions of only a tiny fraction of the community with an interest

21. Marc Dean Millot, Roger Molander, and Peter Wilson, *"The Day After . . ." Study: Nuclear Proliferation in the Post–Cold War World* (Santa Monica, Calif.: RAND Corporation, 1993).

and expertise in such matters. Moreover, very few of the players have held such senior positions as to be actually responsible for the safety of the nation. Still, arguments that nuclear weapons are becoming unusable even under the most extreme circumstances are becoming increasingly common. Perhaps the most vocal supporter of this position is General Lee Butler, an excommander of the U.S. Strategic Air Command and its successor, the Strategic Command.[22]

Rosen himself sees "American blood boiling for retaliation in the first few days" if a regional opponent were to initiate the use of nuclear weapons to cause great damage. But he argues that other considerations would come into play afterwards, such as concerns for the morality of U.S. actions, or the potential effects of future relations with allies.

Quester and Roberts make a variety of points suggesting U.S. actions would be strongly influenced by its values, though they do not employ the strategic personality idea. Quester suggests that the U.S. concepts of justice and revenge embedded in its criminal justice system would point toward nuclear retaliation. He notes that the U.S. "instinct for revenge" can heat quickly and take considerable time to cool. He cites the substantial support shown by U.S. citizens in the months after Japan's surrender in World War II for the proposition that more atom bombs should have been dropped. He also argues that had the United States been destroyed as a result of a nuclear attack by the Soviet Union, massive nuclear retaliation would have been virtually certain just for revenge. Finally, Quester argues that the primary effect of an aggressor's use of nuclear weapons against the United States or its allies would likely be to convince them that they could not live with such a state, a view that would also seem to reduce U.S. inhibitions against nuclear retaliation.

Roberts sees U.S. nuclear retaliation against a regional opponent who causes great damage with nuclear weapons as a real possibility, primarily for reasons that seem to be a matter of the U.S. personality. He argues that U.S. nuclear retaliation could be motivated by deep anger, demands that might come from allies and coalition partners, and the prospect that nuclear retaliation might suppress the opponent's further use of its weapons. It could also be motivated by a desire to restore a broken taboo against the use of nuclear weapons, or to expunge an opponent whose use of such weapons had marked it as the epitome of evil. Roberts doubts that a U.S. president would be willing to refrain from nuclear retaliation if it is seen as likely to reduce significantly the number of U.S. lives lost

---

22. See General Lee Butler, *The Risks of Deterrence: From Super Powers to Rogue Leaders, An Address to the National Press Club,* Washington, D.C. (February 2, 1998).

in meeting necessary military goals. He argues that the president might see not retaliating as politically impossible.

Roberts also argues that the United States should decide on the nature of its retaliation in light of how it wants to judge its own actions, and have them judged by others, in the aftermath of the conflict. Thus, the United States should not succumb to the opposite temptations of apocalyptic revenge, or of quitting the fight, but rather should aim toward a middle course of being a responsible steward of its own and the world's long-term security interests.

Walt recommends that the United States make clear to any regional aggressor that it would suffer overwhelming retaliation if it were to make highly destructive use of nuclear weapons. He also argues that in the aftermath of such use, any revulsion at the idea of nuclear retaliation seems likely to disappear.

In summary, then, while no one can know for certain, the United States seems far more likely than not to retaliate in kind if an opponent were to use NBC weapons to do great damage to the U.S. or its allies. Many strong incentives to retaliate have to do with U.S. values as reflected in its strategic personality. Indeed, staying cool enough to keep the response measured in any sense would be a challenge for U.S. leaders, especially given the likely perception of the perpetrator of NBC use as the embodiment of evil.

UNCONDITIONAL SURRENDER

Would the United States and its allies commit to the total defeat of a regional aggressor that had used NBC weapons to cause great damage? The U.S. strategic personality would strongly influence the answer to this question, which is closely related to the last one. Again, the opponent's use of NBC weapons in support of aggression would be a most dramatic insult to the democratic world order, which the United States sees as an extension of itself. By threatening the fundamental U.S. concept of how it is doing good for the world, the opponent would be casting itself as evil in the eyes of the United States; as noted above, that image would be intensified enormously by an aggressor's devastating use of NBC weapons.

Thus, the United States would see making peace as an accommodation with the devil. And peace could leave the United States and its allies feeling that justice for so large a crime had not been done. Moreover, for states with extroverted strategic personalities such as the United States, human progress is supposed to be linear and irreversible. Accordingly, problems are supposed to be solved once and for all, rather than through the acceptance of partial solutions with their likelihood of repeated trouble and continued costs for the indefinite future. These arguments suggest that the United States would want to pursue the opponent's total defeat.

The United States' regional allies could feel very differently. They would bear the greatest risks of further NBC attacks by the aggressor, and the probability of such attacks would likely rise as the opponent sees its defeat approaching. For many possible regional allies, even a few NBC attacks against key undefended targets could end the state's existence. And the introverted nature of most of the states that are neither European nor descendants of European states does not point toward their favoring the total defeat of the opponent. In contrast to the extroverted United States, introverted states tend to see life as cyclical with few if any final solutions short of paradise.[23]

Quester's answer to the question of total defeat relies more on U.S. values than on realist considerations. He argues that so long as the existence of the United States is not at risk, it would want to defeat, disarm, and reform aggressors that initiate highly destructive use of nuclear weapons. Quester posits that such crimes would convince the allies that they cannot live with such aggressors. In supporting his position, Quester points to the analogous policies of the U.S. criminal justice system for dealing with armed robbers. He notes the unconditional surrender strategy that was followed in World War II, even though the Allies knew that they could avoid the near-term loss of millions more lives by accepting the continued existence of Nazi Germany and Imperial Japan.

Roberts and Walt take similar positions. Roberts argues that the opponent should suffer a major loss for its aggression, and that if it had caused great damage with its nuclear weapons, it should be treated as Quester suggests: with defeat, disarmament, and reform. Walt states that while the United States should let such opponents know that it will not try to overthrow them so long as they refrain from using nuclear weapons, it should assure opposing leaders that it will oust them if they use nuclear weapons to cause great destruction.

Other largely realpolitik considerations could help to motivate a decision to seek the total defeat of the opponent. U.S. and allied leaders could appreciate the large costs, risks, and political complexities of reassembling the forces needed to defeat the offending state if it were to engage in aggression again. And the prospects of further aggression would seem high if the same offending government remains in power and nurses the resentment generated by its original failure. Further,

---

23. Ziemke, *Strategic Personality Types*.

choosing not to seek total defeat of the aggressor for fear of spurring additional use of NBC weapons would send a general message that such weapons can provide a safety net that at least limits the risk of offensive military campaigns.

Still, all of these points do not produce an obvious answer to the last question. While the United States would seem inclined to seek total defeat for an aggressor that makes highly destructive attacks with NBC weapons, its regional allies may prefer lesser goals. Anticipating this possible divergence of views, the United States might want to accept the minimum number of regional allies. On the other hand, regional allies might see that gaining influence over the U.S. strategy, and this decision in particular, would be an important reason to try to join the U.S. coalition.

Observations and Conclusions

So, will the world order be fundamentally changed if the proliferation of NBC weapons continues? Yes, but probably not as most people seem to believe. The conventional wisdom says that obtaining NBC weapons would enable the smaller powers to veto interventions by the United States and the other major powers. The United States and the other major powers would be forced to compromise or abandon their regional interests, and aggressive regional states armed with NBC weapons would be free to dominate if not conquer their less well-armed neighbors.

This book tells a different story. That story suggests that the United States is very likely to understand and reject the outcomes projected by the conventional wisdom. It will be motivated by a realist's perception of the dangerous consequences of allowing conventional aggression backed by NBC weapons to go unopposed. And it will be motivated to reject them because to do otherwise would mean abandoning its long-standing role as promoter and defender of the democratic faith. Retreating could also require the United States to repudiate other more specific values such as its sense of its honor, bravery, and reliability. These are deep-seated values. Together, the essays in this volume, including this one, suggest that the United States is not nearly so reluctant to play its role as global sheriff as its occasional complaints about the burdens of the role might indicate.

This unconventional picture of the consequences of NBC proliferation and aggression backed by such weapons has some important implications. First, the United States needs to develop a deeper and richer understanding of the consequences of continued NBC proliferation, one that combines and balances all the relevant perspectives. The strategic personality concept can provide a better understanding of how U.S.

values bear on its likely behavior in an increasingly proliferated world, how other involved nations may perceive that world, and how they might be inclined to act in it. It should also facilitate an improved understanding of how the states involved in this problem perceive each other's words and actions.

Second, major differences in how the United States, potential challengers, and other involved states see the stakes at risk in a future confrontation where NBC use is a possibility are a recipe for catastrophe. All international actors must be made aware of the depth of the U.S. commitment to the evolving democratic world order, and that it is likely to see such a confrontation as a mortal threat to that order. Thus, the United States should not only develop the clearest understanding of the stakes it would see at risk, but communicate this understanding to all who might become involved.

Third, to the extent that the U.S. stakes in opposing NBC-backed aggression are as high and its bias toward defending them as strong as argued here, the United States should abandon any delusions that it can simply decide not to become engaged in such confrontations. Instead, it needs to be well prepared to defend against all forms of aggression, including those where the aggressor is armed with NBC weapons. Not surprisingly, the U.S. government has made a good start in this direction.[24] Still, it has a long way to go to meet the minimum goal of protecting its intervention forces well enough to avoid their defeat should a well-armed regional aggressor make substantial use of NBC weapons. And other states that the United States might need as allies will need to be protected as well. This goal is feasible. Defending against the modest nuclear arsenals that a new proliferator might plausibly afford is far easier than defending against a nuclear superpower's arsenal. And defending against attacks with chemical and biological weapons is far easier than defending against a modest nuclear arsenal.[25]

Protecting U.S. and allied intervention forces from defeat with NBC weapons has a variety of advantages. It could discourage an opponent from attempting aggression and could undermine some of the incentives to obtain NBC weapons in the first place. Especially important, strong

24. See *Proliferation: Threat and Response*, Office of the Secretary of Defense (Washington, D.C.: U.S. Department of Defense, November 1997), and *Report on Activities and Programs for Countering Proliferation and NBC Terrorism*, Counterproliferation Program Review Committee (Washington, D.C.: U.S. Department of Defense, May 1998).

25. See Joshua Lederberg, ed., *Biological Weapons: Limiting the Threat*, BCSIA Studies in International Security (Cambridge, Mass.: MIT Press, 1999); and Victor A. Utgoff, "Nuclear Weapons and the Deterrence of Biological and Chemical Warfare," Henry L. Stimson Center Occasional Paper No. 36 (October 1997).

defenses could limit the damage done by an opponent's NBC weapons to a small fraction of the damage that might otherwise be possible. As a result, the pressures for nuclear retaliation would be lower, and the amount of retaliatory damage that might seem necessary would be reduced. Further, by reducing the potential risks of further NBC use, strong defenses could open the door to imposing total defeat on a challenger that had initiated the use of such weapons.

Fourth, the United States should rededicate itself to preventing and rolling back proliferation in any way it can. Past nonproliferation efforts have led to many important successes. The current efforts to get India and Pakistan to halt further development of their nuclear capabilities are important. And there will be future successes. Still, nonproliferation has too often been sacrificed to other interests. Such sacrifices could cost the world dearly if they increase the potential for the kinds of dangerous confrontations that are our concern here.

Finally, while the United States should continue to work toward reducing its dependence on nuclear weapons, it must be careful not to get the cart before the horse. If the United States is as likely to retaliate with nuclear weapons for an aggressor's highly destructive use of NBC weapons as suggested here, it should be very leery about suggesting otherwise. And unless and until the world finds a safe way to eliminate all NBC weapons, the United States should never let doubts develop about its capability to impose nuclear punishment when it is justified. In the end, while proliferation of NBC weapons remains a threat to peace, the threat of retaliation in kind remains an essential underpinning for the stability of the world order as we know it.

Contributors

Richard K. Betts is the Leo A. Shifrin Professor and Director of the Institute of War and Peace Studies at Columbia University, and Director of National Security Studies at the Council on Foreign Relations. He was formerly a Senior Fellow at the Brookings Institution and has taught at the Johns Hopkins Nitze School of Advanced International Studies and Harvard University. Among his books are *Military Readiness* (Brookings, 1995); *Nuclear Blackmail and Nuclear Balance* (Brookings, 1987); *Surprise Attack: Lessons for Defense Planning* (Brookings, 1982); and *Soldiers, Statesmen, and Cold War Crises* (Harvard University Press, 1977). In January 1999, Senate Minority Leader Tom Daschle appointed Betts a member of the National Commission on Terrorism.

Barry R. Posen is Professor of Political Science at the Massachusetts Institute of Technology. He has written two books, *Inadvertent Escalation* (Cornell University Press, 1991), and *The Sources of Military Doctrine* (Cornell University Press, 1984). Other publications include "Nationalism, the Mass Army and Military Power," in *International Security*, Vol. 18, No. 2 (Fall 1993); "Competing U.S. Grand Strategies," with Andrew L. Ross, in *Eagle Adrift: American Foreign Policy at the End of the Century*, edited by Robert J. Lieber (HarperCollins, 1996); and "Military Responses to Refugee Disasters," in *International Security*, Vol. 21, No. 1 (Summer 1996). Dr. Posen's current activities include research on innovation in the U.S. Army from 1970–80. He is affiliated with the MIT Security Studies Program.

George H. Quester is a Professor of Government and Politics at the University of Maryland, where he teaches courses on international relations, American foreign policy, and international military security. He has taught previously at Cornell and Harvard Universities, at UCLA, and in the Department of Military Strategy at the National War College. From 1991 to 1993, he served as the Olin Visiting Professor at the U.S. Naval Academy. Dr. Quester is the author of a number of books and articles on international security issues, and on broader questions of international relations, and has served as a consultant for the Institute for Defense Analyses (IDA).

Brad Roberts is a member of the research staff at IDA, where he contributes to studies for the Office of the Secretary of Defense and the joint military staff. His areas of expertise are counterproliferation, nonproliferation, counterterrorism, and NBC weapons. He is also an adjunct professor in the Elliott School of International Studies at George Washington University, chairman of the research advisory council of the Chemical and Biological Arms Control Institute, a member of the executive committee of the Council for Security Cooperation in the Asia Pacific, and a consultant to Los Alamos National Laboratory.

Stephen Peter Rosen is the Kaneb Professor of National Security and Military Affairs and Associate Director of the Olin Institute for Strategic Studies at Harvard University. Previously, he was the civilian assistant to the Director, Net Assessment in the Office of the Secretary of Defense; the Director of Political-Military Affairs on the staff of the National Security Council; and a professor in the Strategy Department at the Naval War College. He was a consultant to the President's Commission on Integrated Long Term Strategy, and to the Gulf War Air Power Survey sponsored by the Secretary of the Air Force. He is the author of *Societies and Military Power: India and its Armies* (Cornell University Press, 1996) and *Winning the Next War: Innovation and the Modern Military* (Cornell University Press, 1991). He is currently working on a project on strategy and the biology of cognition.

Scott D. Sagan is Associate Professor of Political Science and Co-Director of Stanford's Center for International Security and Cooperation. Dr. Sagan is the co-author (with Kenneth N. Waltz) of *The Spread of Nuclear Weapons: A Debate* (W.W. Norton, 1995) and the author of *The Limits of Safety: Organizations, Accidents, and Nuclear Weapons* (Princeton University Press, 1993) and *Moving Targets: Nuclear Strategy and National Security* (Princeton University Press, 1989).

Victor A. Utgoff is Deputy Director of the Strategy, Forces and Resources Division of IDA. During 1998–99 he established the Advanced Systems and Concepts Office for the Defense Threat Reduction Agency. He is the author of *The Challenge of Chemical Weapons: An American Perspective* (MacMillan, 1990), co-author (with Barry M. Blechman) of *Fiscal and Economic Implications of Strategic Defenses* (Westview, 1986), and has published widely on issues posed by the proliferation of nuclear, biological and chemical weapons. In 1999, he received the Andrew J. Goodpaster Award for Excellence in Research.

Stephen M. Walt is Kirkpatrick Professor of International Affairs at the John F. Kennedy School of Government at Harvard University. Previously, he was

Professor of Political Science and Master of the Social Science Collegiate Division at the University of Chicago. He is the author *of The Origins of Alliances* (Cornell University Press, 1987), which received the Edgar S. Furniss National Security Book Award, and numerous articles on international relations and security studies.

General Larry D. Welch, USAF (Ret.) is the President and Chief Executive Officer of IDA. He took this position after a long military career that included assignments as Commander in Chief of the Strategic Air Command and Chief of Staff of the U.S. Air Force. He participates in many senior defense advisory activities including the Commission to Assess the Ballistic Missile Threat to the United States, the Commission on Maintaining U.S. Nuclear Weapons Expertise, and the U.S. Strategic Command Advisory Group. He is chairman of the President's Security Policy Advisory Board, and the Department of Defense Threat Reduction Advisory Committee. He chaired the 1998 Defense Science Board Task Force on Nuclear Deterrence.

Caroline F. Ziemke is a Research Staff Member at IDA. She has a doctorate in Military History and Strategic Studies from Ohio State University. She is working on a book on the influence of historical experience on the strategic conduct of states entitled *The Strategic Personality Types: Using National Myth to Project State Conduct.*

Index

M

N

in NPT, 25, 44, 69–70, 166,
232–233
states, nuclear
in NPT, 25–26, 69–70, 232
nuclear weapons doctrine in
new, 127–131
See also China; France; India;
Pakistan; South Africa;
United Kingdom; United
States
strategic personality
potential influence on NBC
aggressor of U.S, 289–299
types based on charac-
teristics, 281
of United States, 280–289, 294
Suez Crisis (1956), 41–42
Sundarji, K., 127–129, 247–248
Sunni Islam, 105
Syria, 102

T

Taiwan, 166
terrorist groups
as exception to utopian realism, 77
potential use of WMDs by,
77–78
Thatcher, Margaret, 172
Tolubko, Volodomyr, 45
Turkey, 182

U

Ukraine
bartering of nuclear weapons for economic aid, 56
nuclear restraint, 24, 44–47,
188

uncertainty
in assessing threat by rogue
state, 219–220
as reason for alliance formation, 208–209
unipolarity supplants bipolarity, 267–268
United Arab Emirates (UAE),
102
United Kingdom
development of nuclear
weapons, 21, 67
United Nations
IAEA inspectors operating in
Iraq under, 134
program to root out Iraq's
NBC programs, 4
United States
ability to form and lead coalitions, 203–204
advantages in dealings with
Iran, 118–119
commitment under Article
VI of the NPT, 37, 48
counterproliferation efforts,
111–112, 118
creation of defenses against
NBC weapons, 5
decision not to intervention
directly in Iraq, 182
development of atomic
bomb, 21
dual containment strategy,
118
efforts to isolate Iran and
Iraq, 206
as hypothetical cowardly
wimp, 269, 274
as hypothetical nuclear bully,
268–269, 274

BCSIA Studies in International Security

Published by The MIT Press

Sean M. Lynn-Jones and Steven E. Miller, series editors
Karen Motley, executive editor
Belfer Center for Science and International Affairs (BCSIA)
John F. Kennedy School of Government, Harvard University

Allison, Graham T., Owen R. Coté, Jr., Richard A. Falkenrath, and Steven E. Miller, *Avoiding Nuclear Anarchy: Containing the Threat of Loose Russian Nuclear Weapons and Fissile Material* (1996)

Allison, Graham T., and Kalypso Nicolaïdis, eds., *The Greek Paradox: Promise vs. Performance* (1996)

Arbatov, Alexei, Abram Chayes, Antonia Handler Chayes, and Lara Olson, eds., *Managing Conflict in the Former Soviet Union: Russian and American Perspectives* (1997)

Bennett, Andrew, *Condemned to Repetition? The Rise, Fall, and Reprise of Soviet-Russian Military Interventionism, 1973–1996* (1999)

Blackwill, Robert D., and Michael Stürmer, eds., *Allies Divided: Transatlantic Policies for the Greater Middle East* (1997)

Brom, Shlomo, and Yiftah Shapir, eds., *The Middle East Military Balance 1999–2000* (1999)

Brown, Michael E., ed., *The International Dimensions of Internal Conflict* (1996)

Brown, Michael E., and Šumit Ganguly, eds., *Government Policies and Ethnic Relations in Asia and the Pacific* (1997)

Elman, Miriam Fendius, ed., *Paths to Peace: Is Democracy the Answer?* (1997)

Falkenrath, Richard A., *Shaping Europe's Military Order: The Origins and Consequences of the CFE Treaty* (1994)

Falkenrath, Richard A., Robert D. Newman, and Bradley A. Thayer, *America's Achilles' Heel: Nuclear, Biological, and Chemical Terrorism and Covert Attack* (1998)

Feldman, Shai, *Nuclear Weapons and Arms Control in the Middle East* (1996)

Forsberg, Randall, ed., *The Arms Production Dilemma: Contraction and Restraint in the World Combat Aircraft Industry* (1994)

Hagerty, Devin T., *The Consequences of Nuclear Proliferation: Lessons from South Asia* (1998)

Heymann, Philip B., *Terrorism and America: A Commonsense Strategy for a Democratic Society* (1998)

Kokoshin, Andrei A., *Soviet Strategic Thought, 1917–91* (1998)

Lederberg, Joshua, *Biological Weapons: Limiting the Threat* (1999)

Shields, John M., and William C. Potter, eds., *Dismantling the Cold War: U.S. and NIS Perspectives on the Nunn-Lugar Cooperative Threat Reduction Program* (1997)

Tucker, Jonathan B., ed., *Toxic Terror: Assessing Terrorist Use of Chemical and Biological Weapons* (2000)

Utgoff, Victor A., ed., *The Coming Crisis: Nuclear Proliferation, U.S. Interests, and World Order* (2000)

The Robert and Renée Belfer Center for Science and International Affairs

Graham T. Allison, Director
John F. Kennedy School of Government
Harvard University
79 JFK Street, Cambridge MA 02138
Tel: (617) 495–1400; Fax: (617) 495–8963
http://www.ksg.harvard.edu/bcsia bcsia_ksg@harvard.edu

The Belfer Center for Science and International Affairs (BCSIA) is the hub of research, teaching and training in international security affairs, environmental and resource issues, science and technology policy, human rights, and conflict studies at Harvard's John F. Kennedy School of Government. The Center's mission is to provide leadership in advancing policy-relevant knowledge about the most important challenges of international security and other critical issues where science, technology and international affairs intersect.

BCSIA's leadership begins with the recognition of science and technology as driving forces transforming international affairs. The Center integrates insights of social scientists, natural scientists, technologists, and practitioners with experience in government, diplomacy, the military, and business to address these challenges. The Center pursues its mission in five complementary research programs:

- The **International Security Program** (ISP) addresses the most pressing threats to U.S. national interests and international security.

- The **Environment and Natural Resources Program** (ENRP) is the locus of Harvard's interdisciplinary research on resource and environmental problems and policy responses.

- The **Science, Technology and Public Policy Program** (STPP) analyzes ways in which science and technology policy influence international security, resources, environment, and development, and such cross-cutting issues as technological innovation and information infrastructure.

- The **Strengthening Democratic Institutions Project** (SDI) catalyzes support for three great transformations in Russia, Ukraine and the other republics of the former Soviet Union—to sustainable democracies, free market economies, and cooperative international relations.

- The **WPF Program on Intrastate Conflict, Conflict Prevention and Conflict Resolution** analyzes the causes of ethnic, religious, and other conflicts, and seeks to identify practical ways to prevent and limit such conflicts.

The heart of the Center is its resident research community of more than 140 scholars: Harvard faculty, analysts, practitioners, and each year a new, interdisciplinary group of research fellows. BCSIA sponsors frequent seminars, workshops and conferences, maintains a substantial specialized library, and publishes books, monographs and discussion papers.

The Center's International Security Program, directed by Steven E. Miller, publishes the BCSIA Studies in International Security, and sponsors and edits the quarterly journal *International Security*.

The Center is supported by an endowment established with funds from Robert and Renée Belfer, the Ford Foundation and Harvard University, by foundation grants, by individual gifts, and by occasional government contracts.